Lecture Notes in Physics

Edited by J. Ehlers, München, K. Hepp, Zürich
R. Kippenhahn, München, H. A. Weidenmüller, Heidelberg
and J. Zittartz, Köln
Managing Editor: W. Beiglböck, Heidelberg

117

Deep-Inelastic and Fusion Reactions with Heavy Ions

Proceedings of the Symposium
Held at the Hahn-Meitner-Institut
für Kernforschung, Berlin
October 23–25, 1979

Edited by W. von Oertzen

Springer-Verlag Berlin Heidelberg GmbH 1980

Editor

Wolfram von Oertzen
Hahn-Meitner-Institut
für Kernforschung Berlin GmbH
Bereich Kern- und Strahlenphysik
Postfach 390128
D-1000 Berlin 39

ISBN 978-3-540-09965-9   ISBN 978-3-540-39177-7 (eBook)
DOI 10.1007/978-3-540-39177-7

Library of Congress Cataloging in Publication Data. Main entry under title: Deep-inelastic
and fusion reactions with heavy ions. (Lecture notes in physics; 117) "Sponsored jointly by
HMI Berlin, Danfysik, Denmark, and Scanditronix AB, Sweden." Bibliography: p. Includes
index. 1. Deep inelastic collisions--Congresses. 2. Nuclear fusion--Congresses. 3. Heavy
ion collisions--Congresses. I. Oertzen, W. von, 1939- II. Hahn-Meitner-Institut für Kernforschung, Berlin. III. Danfysik A/S. IV. Scanditronix AB. V. Series: Lecture notes in
physics (Berlin); 117. QC794.6.C6D43. 539.7'64. 80-12868

Otto Hahn   Lise Meitner, Lindau 1962

FOREWORD

An accelerated beam was obtained in the VICKSI heavy ion accelerator
at HMI Berlin for the first time in 1978. After almost one year of
regular operation, it seemed the appropriate time for an official dedication and the presentation and discussion of our research interests
and some results of nuclear reactions with heavy ions. Therefore we
arranged a symposium for colleagues working in our field of interest.

The dedication ceremonies were held on Monday, October 22. The symposium
itself, with approximately 45 participants from outside of Berlin, was
held October 23 to 25, 1979. The symposium also presented the opportunity of commemorating the 100th anniversary of the birth of Lise Meitner
and Otto Hahn, whose names our Institute carries. Thus the first lecture
given by Prof. J. Huizenga from Rochester was dedicated to the memory
of these two scientists whose pioneering work on nuclear fission established a basis for heavy ion physics.

The symposium was organized by physicists from the HMI physics division:
Fuchs, Gross, Hilscher, Homeyer, Jahnke, Lipperheide, Lindenberger, von
Oertzen, and was sponsored jointly by HMI Berlin, Danfysik, Denmark,
and Scanditronix AB, Sweden.

These proceedings contain the talks presented at the symposium. The
topics of the various sessions were chosen so as to achieve a consistent
and fresh view of the state of the art. We at HMI are indebted to the
participants for their active approach to this symposium, and we hope
that these proceedings will help sustain discussions on the subjects
covered.

For the organizing committee, the editor

Berlin, December 1979

W. von Oertzen

TABLE OF CONTENTS

Session VI. Fusion Reactions with Heavy Ions

PARTICIPANTS FROM HMI

H.G. Bohlen

W. Bohne

U. Brosa

M. Bürgel

M. Clover

C. Egelhaaf

H. Fuchs

P. Fröbrich

A. Gamp

B. Gebauer

K. Grabisch

D. Gross

H. Homeyer

G. Ingold

U. Jahnke

S. Kachholz

P. Kaufmann

C. Kluge

H.J. Krappe

H. Lehr

H. Lettau

H. Lindenberger

R. Lipperheide

K. Möhring

H. Morgenstern

W. von Oertzen

H. Ossenbrink

H. Rossner

H. Siekmann

W. Stöffler

G. Thoma

U. Wille

Symposium on Deep-Inelastic and Fusion Reactions with Heavy Ions
Hahn-Meitner-Institut für Kernforschung Berlin
October 23 - 25, 1979

R. Bass, Frankfurt

G. Baur, Jülich

J.R. Birkelund, Rochester

R.K. Bhowmik, Birmingham

J. Bondorf, Kopenhagen

H.C. Britt, München

Y. Civelekoglu, Heidelberg

C.H. Dasso, Kopenhagen

P. Doll, Darmstadt

T. Døssing, Kopenhagen

J. van Driel, Groningen

H. Friedrich, Münster

J. Galin, Orsay

G. Gaul, Münster

K. Gelbke, East Lansing

G. Graw, München

M.L. Halbert, Oak Ridge

S. Harar, Gif sur Yvette

W. Hering, München

J.R. Huizenga, Rochester

H. Ho, Heidelberg

H. Hofmann, München

P. Kienle, München

W. Kühn, Heidelberg

M. Lefort, Orsay

H. Löhner, Münster

U. Lynen, Heidelberg

J. Maruhn, Frankfurt

G.C. Morrison, Birmingham

U. Mosel, Giessen

C. Nemes, Heidelberg

W. Nörenberg, Darmstadt

A. Olmi, Darmstadt

H. Oeschler, Gif sur Yvette

L. Papineau, Gif sur Yvette

D. Pelte, Heidelberg

F. Pühlhofer, Marburg

K.E. Rehm, München

J.C. Roynette, Orsay

G. Rosner, Heidelberg

G. Schrieder, Darmstadt

R.H. Siemssen, Groningen

N. Stelte, Marburg

D. Trautmann, Basel

J. Wilczynski, Krakow, (Groningen)

J.P. Wurm, Heidelberg

HEAVY-ION REACTIONS: A NEW FRONTIER OF NUCLEAR SCIENCE[*]

JOHN R. HUIZENGA
Departments of Chemistry and Physics
and
Nuclear Structure Research Laboratory
University of Rochester
Rochester, New York 14627 U.S.A.

I. INTRODUCTION

The year 1979 marks the fortieth anniversary of the first announcement of the discovery of a radically new nuclear process whereby a heavy nucleus divides into two parts. Otto Hahn and Lise Meitner, working here in Berlin, played a major role in the sequence of events that led to this major scientific breakthrough. Hahn was born one century ago and Meitner a year earlier in 1878. The present year 1979 also marks the one hundredth anniversary of the birth of Albert Einstein, a man who profoundly influenced the shape of science and the course of history. On April 22, I witnessed the dedication of Robert Berk's impressive memorial statue to Albert Einstein located in the front of the National Academy of Sciences' building on Constitution Avenue in Washington, D.C. I recommend that you see this attractive statue when in Washington.

Hahn and Meitner made Berlin a major center of nuclear science in the thirties. I congratulate those of you at the Hahn-Meitner Institute for following the early Berlin tradition in nuclear science as you dedicate VICKSI and move quickly into the forefront of heavy-ion research, a new frontier of nuclear science.

II. EARLY HISTORY OF FISSION RESEARCH

Those of us who have worked in the field of very heavy-ion reactions during the present decade will recognize some parallels with the excitement in the thirties of those studying the reactions of neutrons with heavy elements. Enrico Fermi, working in Rome, reasoned that neutrons because of their lack of charge, should be effective in penetrating nuclei, especially those of high atomic number which repel charged particles strongly. Fermi[1] realized that neutron bombardment of uranium might produce isotopes of new elements by one or more beta disintegrations. On bombardment of thorium and uranium with neutrons, the Italian group[2] found species with several

---
[*]Meitner-Hahn Memorial Lecture delivered on October 23, 1979 at the Hahn-Meitner Institut für Kernforschung, Berlin.

JOHN R. HUIZENGA

different half-lives. However, despite the fact that so many different radioactive
species were present, Fermi and his collaborators did not expect possible neutron
reactions other than those already established. Failing to identify two of the new
activities (13- and 90-minute half-lives) with uranium or any other of the known
elements immediately below uranium, Fermi incorrectly supposed that these activities
were due to an element of atomic number higher than uranium.

The results of Fermi's group stimulated similar neutron experiments all over
the world. The reported discovery of transuranic elements was of particular interest
to chemists. Noddack[3], for example, criticized Fermi's conclusions on the ground
that his chemical separations were non-specific. She suggested already in 1934 that
the bombarded nuclei might split to form elements of lower atomic number, so that
proof of the discovery of transuranic elements required more elaborate chemical tests
in order to exclude all known elements. If Noddack's early suggestion was more than
mere speculation, it is regrettable that she did not develop the arguments supporting
her suggestions. In retrospect, Noddack's early suggestion of fission seems to have
been offered more by way of pointing out a lack of rigor in the argument for the
existence of transuranic elements than as a serious explanation of the experimental
observations. In any case, her suggestion was largely neglected and seems to have
had little, if any, influence on the subsequent course of events.

Confusion reigned in the field of transuranic elements from 1934 through most
of 1938. Further investigations of the 13- and 90-minute activities had the significant result of interesting Hahn and Meitner in the question. During the period
1935-38 they, along with Strassmann, published a large number of papers dealing with
the activities produced by neutron irradiation of thorium and uranium (for a listing
of these papers, see the review article by L.A. Turner[4]). After an extensive series
of experiments with different times of irradiation, the use of fast and slow neutrons,
and a great variety of chemical tests, Meitner, Hahn and Strassmann[5] concluded that
the neutron irradiation of uranium produced three different active isomers of
uranium, each of which decayed by successive beta-disintegrations. The suggested
production of three isomers of ^{239}U was difficult to understand and led to a wave of
new carrier-type experiments to conclusively prove the chemical identity of the
observed activities. Curie and Savitch[6] showed that a 3.5-hr activity had chemical
properties like lanthanium and were puzzled where a chemical element of such properties could be fitted into the periodic table beyond uranium.

It was left for two chemists, Hahn and Strassmann, working here in Berlin to
identify positively an isotope of barium as one of the products obtained by irradia-
ting uranium with neutrons, and to announce the discovery of nuclear fission to the
world in a paper[7] published in January 1939. As Turner[4] states in his review article,

these unexpected and startling results, which seemed to be incompatible with the then known properties of nuclei, were offered with much reserve. The authors felt that it was possible that some series of unusual accidents might have combined to give mis- leading results. Data of other experimenters very quickly confirmed the correctness of Hahn and Strassmann's conclusions. A second paper[8] by the latter authors showed beyond a doubt that their assignment of the new activities to barium rather than radium was correct. Hahn received the Nobel prize in Chemistry in 1944 for what is no doubt one of the most important scientific discoveries of this century.

Fermi, after the discovery of fission, is recorded by his wife in his biography "Atoms in the Family" as saying: "We did not have enough imagination to think that a different process of disintegration might occur in uranium from that in any other element, and we tried to identify the radioactive products with elements close to uranium in the periodic table of elements. Moreover, we did not know enough chemistry to separate the products of uranium disintegration from one another". As is well known, Fermi was awarded the Nobel prize for misinterpreted results. However, the Nobel Committee need not have experienced any embarrassment at this rare error on their part, since Fermi amply distinguished himself before and after the award and I know of no one more deserving of a Nobel prize.

The discovery that the capture of a thermal or low-energy neutron by a heavy nucleus resulted in the rupture of the nucleus into fragments of intermediate mass raised new theoretical problems. Meitner and Frisch[9] were the first to suggest a theoretical explanation on the basis of a nuclear liquid-drop model. They pointed out that just as a drop of liquid which is set into vibration may split into two drops, so might a nucleus divide into two smaller nuclei. These authors treated the stability of nuclei in terms of cohesive nuclear forces of short range, analogous to a surface tension, and an electrostatic energy of repulsion. They went on to estimate that nuclei with $Z \approx 100$ would immediately break apart. Since uranium had only a slightly smaller charge, they argued that it was plausible that this nucleus would divide into two nuclei upon receiving a moderate amount of excitation energy supplied by the neutron binding energy. To describe this exciting new process, Meitner and Frisch proposed the term nuclear "fission" in analogy to the process of division of biological cells. Lise Meitner played an important role in the discovery of fission as she had been a close collaborator of Hahn and head of the Physics sec- tion of the Kaiser Wilhelm Institute for Chemistry until she was forced to leave Germany in 1938. Her contribution to this startling discovery was recognized in 1966 by sharing the U.S. Atomic Energy Commission's Enrico Fermi award with Hahn and Strassmann.

The discovery by Hahn and Strassmann of this new reaction mechanism captured

JOHN R. HUIZENGA

immediately the imagination of chemists and physicists around the world as evidenced by the more than one hundred publications[4] on this subject in 1939 alone. One of these is the comprehensive and classic paper by N. Bohr and Wheeler[10] entitled, "The Mechanism of Nuclear Fission". Bohr had been informed of the early theoretical work of Meitner just prior to his collaboration with Wheeler in Princeton. The new fission process was shown to release an enormous amount of energy and produce a large number of new neutron-rich radioactive species. The possibilities offered by this new process for understanding of basic sciences and for utilization by applied sciences seemed unlimited.

III. SCIENTIFIC HIGHLIGHTS IN NUCLEAR FISSION RESEARCH

Following the large number of papers on nuclear fission in 1939, world events caused all further publications in this field to cease during the first half of the next decade. It was this period that I was initially introduced to nuclear fission in a series of experiments at Oak Ridge with neutrons from a large Ra-Be source. The subject has continued to fascinate me through my entire scientific career.

Nuclear fission is an extremely complex reaction where a cataclysmic rearrangement of a single nucleus occurs yielding two intermediate nuclei and releasing a large amount of energy. In this short review, I can select only a few highlights from the present wealth of literature on nuclear fission. For those interested in a more thorough overview of the field, I recommend the book[11] entitled, "Nuclear Fission" published by Academic Press in 1973.

Nuclear fission is the most dramatic example of collective motion in nuclei. The Bohr-Wheeler liquid-drop model (LDM) is a prototype of nuclear collective models. In this model the nucleus is described as a uniformly charged, constant-density droplet with a sharp sruface. Thus, the liquid-drop model in its simplest from describes the potential energy changes associated with shape distortions in terms of the interplay between surface and Coulomb effects. Only a single parameter is required to characterize the energetics and motions of the droplet. This is the fissionability parameter x.

$$(2) \qquad x = \frac{E_C^\circ \text{ (spherical Coulomb energy)}}{2\ E_S^\circ \text{ (spherical surface energy)}}$$

where $E_C^\circ = k_C\ Z^2/A^{1/3}$ and $E_S^\circ = k_S\ A^{2/3}$. The constants are evaluated by fitting experimental nuclear masses with the semiempirical mass equation. Droplets with x>1 are unstable against small deformations and are expected to fission in a time comparable to a nuclear vibration period. Although the LDM is conceptually simple, detailed calculations of the LDM statics and dynamics are technically difficult except for very small deformations. Important contributions in this field have been made by a number of authors[12-16].

The LDM has been extremely successful in describing the gross features of nuclear reactions including fission. Let me illustrate this point with the following question: Why do we have approximately 100 elements in our periodic table rather than, for example, two or 10,000 elements? With the above definition of the fissionability parameter, one can write immediately an equation for the limiting value of Z, namely

(2)
$$Z^2_{LIMIT} = 2(k_S/k_C)A_{LIMIT}$$

Under the additional assumption that $A_{LIMIT} \approx 2.5 \ Z_{LIMIT}$, eq. (2) can be rewritten as

(3)
$$Z_{LIMIT} \approx 5(k_S/k_C)$$

Hence, the upper bound to the periodic table is dependent upon the ratio of two fundamental coupling constants, the strong or nuclear coupling constant divided by the electromagnetic coupling constant. The ratio of (k_C/k_S) is known from semi-empirical mass formulae to be approximately 20.

It is well known that the liquid-drop model is an inadequate model for predicting many properties of nuclear structure, particularly effects associated with the shell structure of nuclei. On the other hand single-particle models fail to predict reasonable deformation energies at large deformations. Strutinsky[17] first proposed a solution to this dilemma by an ingenious combination of the LD and shell models. In the Strutinsky method, shell effects are considered as small deviations from a uniform single-particle energy level distribution. The deviation is then treated as a correction to the LDM energy which contains the dominant surface and Coulomb effects. The dependence of the pairing strength on deformation can also be treated as a correction in a similar manner. The fission barrier for a heavy nucleus is schematically illustrated in Fig. 1. Nuclear shells introduce significant structure into the fission barrier causing a second minimum in the potential energy surface. This two-humped fission barrier is essential to explain a number of fission phenomena including spontaneously fissioning isomers[18] and sub-barrier neutron-fission resonances[19]. Confirmation of the view that the isomeric state has a much larger deformation than the ground state was first obtained[20] by the identification of the conversion lines of the rotational band build on the shape isomeric state of ^{240}Pu. In recent years, nuclear spectroscopy of states in the second well has developed.

It should be mentioned that although the Strutinsky prescription is intuitively attractive and very successful, there exists no complete theory to justify it. How-

JOHN R. HUIZENGA

ever, the procedure has been reinforced by comparisons[21] of the deformation energy surface from Hartree-Fock calculations with calculations based on the Strutinsky method using Hartree-Fock eigenenergies.

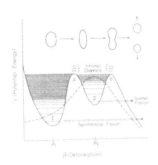

FIGURE 1,   Schematic fission barrier. From Ref. 11

In 1955, A. Bohr[22] first applied transition state theory to explain fission fragment angular distributions. In low-energy fission the quantum numbers of the levels available just above the barrier fix the fragment directions. For example, the angular distributions of fission fragments observed in the $^{238}U(\alpha,\alpha'f)$ reaction for excitation energies up to 0.6 MeV above the barrier are very anisotropic as shown in Fig. 2. These transition states (K=0) require the fission fragments to be

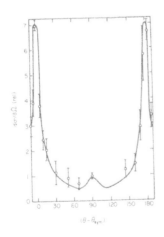

FIGURE 2, $^{238}U(\alpha,\alpha'f)$ angular distributions. From Ref. 11

preferentially emitted along the angular-momentum symmetry axis. Although the above (α,α'f) reaction was first done almost two decades ago, reactions of this type are still now used frequently to probe a variety of fission and nuclear structure problems such as giant resonance states.

One of the earlier observations regarding the fission process was the strong preference for heavy elements at low excitation energy to fission into fragments of unequal mass. Asymmetric mass distributions have proved to be one of the most persistent puzzles in the fission process. Although many suggestions as to the origin of this effect have been offered, no theoretical model has been proposed which has been explored in a complete enough manner or has been sufficiently free of parameter fitting to be generally accepted. The importance of the double shells at Z=50 and N=82 in low-energy fission is illustrated in Fig. 3, where the low-mass side of the heavy peak is rather independent of the fissioning nucleus. In contrast to the asymmetric fission illustrated in Fig. 3, the heaviest fermium[24] isotopes (Z=100, A=258,259) fission symmetrically. Although ^{259}Md(Z=101) spontaneously fissions symmetrically, it appears to be anomalous in its low total kinetic energy[25]. Application of the Strutinsky procedure to heavy nuclei[26] has shown that the second barrier energy is reduced by several MeV for asymmetric distortions. It is important to remember, however, that for heavy nuclei the saddle point is not close to the scission point and it is necessary to consider the dynamics of the descent from saddle to scission.

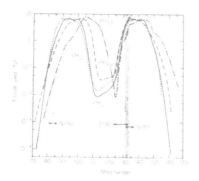

FIGURE 3,   Fission fragment mass distributions. From Ref. 11

Fission dynamics provides a testing ground for nuclear many-body theory and confronts many of the same fundamental questions that are relevant also to heavy-ion reactions. For example, if the motion from saddle to scission is adiabatic with

JOHN R. HUIZENGA

respect to the particle degrees of freedom, the decrease in potential energy appears in collective degrees of freedom at scission primarily as kinetic energy associated with relative motion of the nascent fragments. If, however, the motion is non-adiabatic, there is a transfer of collective energy into nucleonic excitation in a manner analogous to viscous heating. If there is sufficient nonadiabatic mixing of the energy among the single-particle degrees of freedom by the time scission is reached, a statistical model[27] may be a reasonable approximation. Recently, the dynamics of neutron-induced fission of ^{235}U has been studied[28] by time dependent Hartree-Fock methods (TDHF). One important, but disappointing result of these cal-culations is that the appropriate strengths of two-body viscosity and one-body dis-sipation yield similar total kinetic energies in agreement with experiment.

IV.   EARLY MOTIVATION FOR STUDY OF HEAVY-ION REACTIONS

Most of the early proposals for heavy-ion accelerators included in their scien-tific justifications a major section on superheavy elements. These discussions were based on a number of theoretical investigations[29] of the nuclear properties of transfermium elements that revealed an island of relatively stable nuclei due to the influence of shell closures at N=184 and Z=114 (other estimates of the proton shell ranged from Z=110 to 126). Although the uncertainties in the calculated half-lives were very large ($10^{\pm 10}$ for spontaneous fission), the estimated half-lives for some nuclei in the projected island were so long that even with a large error, the feeling prevailed that a good possibility existed for forming superheavy nuclei. An example of a topological map of heavy nuclei is shown in Fig. 4. There is a peninsula of stability for known elements and an island of predicted stability that is centered near Z=110 and N=184. Shown also in Fig. 4 are the landing sites for the fusion reactions $^{248,250}Cm + ^{48}Ca$, two postulated ways to form nuclei near the island of

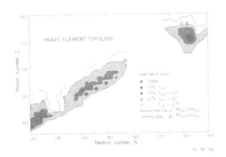

FIGURE 4,   Stability of heavy nuclei. From W. Loveland[29]

HEAVY-ION REACTIONS: A NEW....

stability.   Present experimental attempts to produced superheavy nuclei are utilizing
the strongly-damped or deep-inelastic reaction mechanism.

As discussed in Ref. 29, all experimental searches for superheavy elements both
in nature and at accelerators have failed to date.   However, the new field of heavy-
ion research that has developed in parallel to the super-heavy element searches has
already led to a number of exciting discoveries.   Before discussing these results,
I will show in Fig. 5 the capabilities of several heavy-ion accelerators that are
either running or in construction.   The list is not complete but there is sufficient
information on this figure to give one an impression of the different types of
machines and their capabilities.   The transition region of 10 to 200 MeV per nucleon
is expected to be an exciting and rewarding region for study as one passes through
the sonic, Fermi and mesonic thresholds.   In the lighter projectile region, one sees
from Fig. 5 that the VICKSI accelerator is more powerful than a 25 MV tandem and is
capable of accelerating ions in the mass range of neon up to 20 MeV/u, an energy
well beyond the sonic threshold.

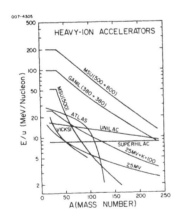

FIGURE 5,   Characteristics of heavy-ion accelerators

V.   SOME RECENT AND CURRENT EXPERIMENTAL RESULTS AND FUTURE DIRECTIONS IN
       HEAVY-ION NUCLEAR SCIENCE

In the 1970's it first became possible to accelerate very heavy ions to relatively high energies.   That is, to energies high enough to overcome the mutual Coulomb repulsion energies for very heavy target masses.   Hence, heavy-ion nuclear
science is that branch of the field that uses nuclei themselves as projectiles to

JOHN R. HUIZENGA

bombard other nuclei. This new field of research has already led to a number of discoveries and is expected to further our understanding of the nuclear system. A
typical nucleus has of the order of 10^2 particles and is neither a few-body system
nor a many-body system in the sense of a gas. Furthermore, in contrast to a classical fluid, a typical constituent nucleon in a nucleus makes only one or two collisions with other nucleons as it travels a distance about equal to the nuclear dia-
meter.

The most important discovery in this field at moderate bombarding energies
(≤ 10 MeV/u) is the unique process now known as deep inelastic (DIC) or strongly
damped (SDC) collisions[30]. Some of the characteristic properties of this reaction
are: (a) its binary nature; (b) the damping of a considerable amount of the initial
kinetic energy and orbital angular momentum into internal energy and spin of each of
the colliding nuclei - sometimes the final kinetic energy is even below the Coulomb
energy of spheres, indicating that the nuclei are strongly deformed on leaving the
interaction zone; (c) the average charge and mass of the reaction products are close
to those of the target and projectile although during the interaction time there is
an exchange of nucleons between the colliding nuclei, the magnitude of which is correlated with the kinetic energy loss; and, (d) angular distributions similar to those
of a relatively fast reaction. In some sense these properties are a mix of those
expected for few-nucleon transfer and fusion reactions.

The division of heavy-ion total reaction cross sections into grazing, fusion
and damped collisions is one of the goals of heavy-ion nuclear science. Some of the
features observed in heavy-ion collisions were expected on the basis of our knowledge
of light-ion reactions. Hence, it was certainly predictable that grazing encounters
would occur by the excitation of surface modes through the nuclear and Coulomb fields
and by the transfer of one or two nucleons. Likewise, the more central collisions
for light-ion reactions are known to fuse. It was indeed surprising to discover
that for very heavy ion systems, the more central collisions led to the damped or
deep-inelastic process at the expense of fusion. In fact for krypton and xenon
induced reactions on very heavy targets, this new process makes up essentially all of
the reaction cross section.

Considerable progress has been made in recent years in understanding the mechanisms operating in damped nuclear reactions. The exchange of many nucleons and
the dissipation of large amounts of kinetic energy are the most significant features
of these reactions. At low bombarding energies of a few MeV/u above V_{COUL}, the
nuclear temperature is very small compared to the Fermi energy and the occupation
probabilities of the single-particle levels are close to those of a degenerate Fermi
gas. Hence, the mean free path of the nucleons is long and of the order of nuclear

HEAVY ION REACTIONS: A NEW....

radii. The nucleus then reacts as a whole to small perturbations such as the transfer of one particle, and one-body dissipation is expected to be valid. Today I wish
to comment briefly on the experimentally well-established correlation[31] between the
energy dissipation and nucleon exchange. These comments follow closely a recent
discussion on this subject[32]. In a phenomenological approach use has been made of
the microscopic time scale provided by the exchange mechanism to give the dissipated
energy as a function of the number N_{ex} of exchanged nucleons[33]

(4)
$$E_{LOSS} = (E_{cm} - V_{COUL}) \{1 - \exp[-(m/\mu)\alpha N_{ex}]\}$$

In eq. (4), the coefficient α conveys information on the character of the exchange
process, m is the nucleon mass and μ is the reduced mass of the dinuclear system.
It is not in general possible to derive a simple and unique relation between N_{ex} and
experimental observables such as the variances σ_A^2 and σ_Z^2 of fragment-A and -Z distributions. However, a precise relation is not essential in the application of Eq. (4)
provided the same relation holds for a given system at different bombarding energies.
For simplicity, it is assumed that $N_{ex} = \sigma_A^2$ or, if only σ_Z^2 is available by
$N_{ex} = (A/Z)^2\sigma_Z^2$ where A and Z apply to the total system. Experimental information on
the relationship between σ_A^2 and σ_Z^2 as a function of energy loss is known for a few
systems as illustrated[34] in Fig. 6.

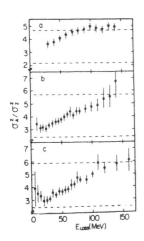

FIGURE 6,   (a) ^{56}Fe + ^{56}Fe; (b) ^{165}Ho + ^{56}Fe; (c) ^{209}Bi + ^{56}Fe.
From Ref. 34

JOHN R. HUIZENGA

Figure 7 shows the resulting fits of Eq. (4) to the experimental data for the reaction ^{209}Bi + ^{136}Xe at 1130 and 940 MeV. The value of α increases as the bombarding energy is reduced. A similar behavior has been observed for other reaction systems. The dependence of α on both bombarding energy and projectile-target asymmetry (Fig. 8) is inconsistent with models based principally on classical kinematic considerations. It is then conjectured that the insufficiency of the classical model to describe the data is due to the neglect of the quantal character of the exchange and dissipation mechanisms[32].

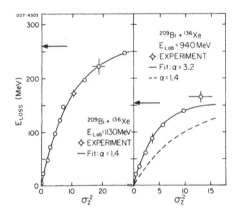

FIGURE 7,   See text.   From Ref. 32

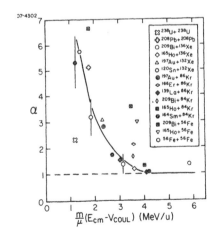

FIGURE 8,   See text.   From Ref. 32

HEAVY-ION REACTIONS: A NEW....

In the following, a recently developed model[35] is applied describing the energy dissipation associated with the exchange of nucleons between two Fermi-Dirac gases in slow relative motion characterized by a relative velocity $\vec{U}$. The two gases have a common temperature τ, and their Fermi energies T_F differ by an amount F_A which is the static driving force for the mass-asymmetry degree of freedom represented by the mass number A of the projectile-like fragment. For ordinary damped collisions, $|\vec{U}|$ is small compared to the Fermi velocity v_F, $|\vec{U}| \ll v_F$, and it is also true that $|F_A| \ll T_F$ and $\tau \ll T_F$ such that the two gases remain nearly degenerate. It then follows from the model[35] that the rate of change of A and the energy dissipation rate can be expressed as

(5)
$$dA/dt = F_A N'(\varepsilon_F)$$

$$dE_{LOSS}/dt = <\omega^2>_F N'(\varepsilon_F)$$

Here $N'(\varepsilon_F) = \partial N(\varepsilon_F)/\partial T_F$ is the differential current of nucleons exchanged between the gases calculated neglecting the Pauli blocking effect. The quantity $\omega = F_A - \vec{U}\cdot\vec{p}$ is the amount of intrinsic excitation produced by the exchange of a nucleon with the intrinsic momentum p, and the brackets denote an average over the orbitals in the Fermi surface, the only ones participating. The two quantities in Eq. (5) can both be represented in terms of one-body operators and may, therefore, be calculated without taking explicit account of the Pauli exclusion principle.

This, however, is not true for the particle-number dispersion σ_A^2, a quantity depending explicitly on the correlations present, such as those imposed by the Fermi-Dirac statistics of the nucleons. The rate of growth of σ_A^2 is, in this model, equal to the total rate of actual exchanges, as long as the system has not evolved too far towards equilibrium. It is given by

(6)
$$d\sigma_A^2/dt = 2\tau^* N'(\varepsilon_F)$$

Here, $\tau^* = <(1/2)\omega\coth(\omega/2\tau)>_F$ is a measure of the energy interval around the Fermi level contributing to exchange processes. In the limit $|\omega| \ll \tau$, it approaches the nuclear temperature $\tau^* \approx \tau$, whereas in the case of $|\omega| \gg \tau$, $\tau^* \approx (1/2)<|\omega|>_F$ may be considerably larger than τ, due to a larger relative displacement of the two Fermi spheres. In any case, the appearance of τ^* in Eq. (6) ensures that proper account is taken of the quantum statistics at all temperatures.

Inspection of the general behavior of the coefficient α is particularly simple for symmetric systems ($F_A \approx 0$) and peripheral collisions, where $\vec{U}$ is almost tangential.

JOHN R. HUIZENGA

Then $<\omega^2>_F \approx (1/4)p_F^2 u^2 = (1/2)mU^2 T_F$, and by comparing the ratio $(dE_{LOSS}/dt)/(d\sigma_A^2/dt)$ from Eqs. (5) and (6) with Eq. (4) it follows that

$$(7) \qquad\qquad \alpha = (\mu/mE)dE_{LOSS}/d\sigma_A^2 \approx T_F/(2\tau^*)$$

The above very approximate estimate suggests that α should typically be substantially larger than unity and decrease as the bombarding energy is increased. Furthermore, according to Eq. (7), a certain degree of universality is expected for α: the α-values for different but still nearly symmetric projectile-target combinations should lie on the same "universal" curve when plotted against $(1/2)mU^2$ essentially determining τ^*. It is striking to observe that these general features, absent in a classical picture, are indeed borne out by the experimental results in Fig. 8.

Since the rough estimate represented by Eq. (7) relies on a number of idealizations, a more refined approach has been taken by performing dynamical calculations of collision trajectories in a coordinate space including the fragment-mass and -charge asymmetries. The dinuclear complex is parameterized by two spherical nuclei joined by a cylindrical neck. Conservative forces are calculated from droplet-model masses, the Coulomb repulsion and the surface and proximity energies of the neck region. Energy dissipation is provided by the nucleon exchange mechanism (cf. Eq. (5)), together with the damping due to the neck motion approximated by a wall-type dissipation formula. Inertial forces are calculated for two sharp rigid spheres. Energy loss and the accumulated variances σ_A^2 and σ_Z^2 are obtained from integrating along the trajectory the dissipation function and the diffusion coefficients, respectively, as given by the model.

Typical results of the calculations are compared to experiment in Fig. 9 for the reactions $^{209}Bi + ^{136}Xe$ [Ref. 36] and $^{209}Bi + ^{56}Fe$ [Ref. 34] at E_{Lab} = 940 and 465 MeV, respectively, that are associated with α-values far in excess of the classical limit. A good reproduction of the data is achieved for $^{209}Bi + ^{136}Xe$ at 940 MeV, and also for the 1130 MeV data for energy losses up to about 150 MeV. However, systematic discrepancies remain between theory and experiment, in particular for high energy losses. These discrepancies are not too surprising in view of the various uncertainties and simplifications of the model. However, explicit account of the quantal nature of these processes has essentially lead to a consistent understanding of the data, the interpretation of which would have been totally misleading otherwise.

HEAVY-ION REACTIONS: A NEW....

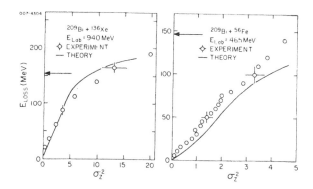

FIGURE 9,   See text. From Ref. 32

As discussed above, nucleon exchange appears to be the major dissipation mechanism in heavy-ion nuclear reactions. However, the presence of other dissipation mechanisms cannot be excluded, such as for example, the excitation of collective sur-face modes. Broglia, et al.[37] have suggested the possible importance of giant resonance excitation and even though this mechanism of energy loss is probably much less important than nucleon exchange, no semi-quantitative experimental information is available on the magnitude of energy loss by the giant resonance process. Dif-ferent theoretical views on the importance of the various dissipation processes is the subject of the afternoon session. Frascaria, et al.[38] first suggested that the observed structure in the inclusive energy spectra of particular product nuclei pro-duced in a heavy-ion damped collision process for symmetric systems is evidence for the excitation of giant resonances. However, I wish to emphasize that before one interprets this observed structure as due to high-energy giant resonances, one must eliminate the expected structure due to statistical processes[39] associated with the de-excitation phase of the damped reaction process. Coincidence techniques may lead to further elucidation of these processes, however, such methods are not likely to give definitive information on the reaction mechanism.

The interaction time for a dissipative heavy-ion collision is defined as the time interval between the formation and rupture of the di-nuclear system. This time is not a directly measurable quantity and must be deduced from a comparison of a reaction model to experimental data. One method[40], based on the nuclear rotation angle $\Delta\Theta$ (ℓ), is illustrated in Fig. 10. For a typical heavy reaction system, the angular momentum dependent interaction times for damped collisions vary from 10^{-22}

JOHN R. HUIZENGA

to 5 x 10^{-21} sec. depending, of course, on the system and bombarding energy. Relaxation times[41] for various degrees of freedom can be estimated from appropriate data. The relaxation times for energy and angular momentum dissipation are comparable and of the order of 3 x 10^{-22} sec., while the relaxation time of the N/Z degree of freedom may be even shorter. The question whether the distribution in Z for fixed A is classical or quantal is still an open one. Some information is available also about the relaxation times of other degrees of freedom, including the mass asymmetry mode.

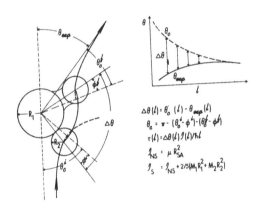

FIGURE 10,   Interaction time for heavy-ion collision. From Ref. 40

During the damped collision process the intermediate di-nuclear complex may acquire a large amount of both spin and excitation energy. Direct measurements[42-44] of the neutron spectra from the fragments show that the excitation energy is shared between the two fragments in proportion to their masses. More importantly as shown in Fig. 11 each fragment has the same nuclear temperature over a range of energy losses[43]. These results indicate that the energy thermalization of the di-nuclear system occurs on a time scale short compared to the interaction time. Furthermore, for low bombarding energies (i.e., < 5 MeV/u above the Coulomb barrier) the pre-equilibrium neutron emission is found to contribute less than 5%, although there is evidence for a small yield of pre-equilibrium charged particles[45]. The charged particle yield increases substantially with higher bombarding energy as will be discussed at this symposium. The parameter marking the boundary between the equilibrium and pre-equilibrium processes appears to be the kinetic energy per nucleon above the Coulomb barrier in the entrance channel rather than the total excitation

HEAVY-ION REACTIONS: A NEW....

energy[43]. Using, then, the kinetic energy per nucleon above the Coulomb barrier as
a characteristic parameter, allows one to compare directly the yields of pre-equilibrium emission from light and heavy-ion reactions. More detailed information on pre-
equilibrium particle emission, especially neutrons, is needed as a function of bombarding energy up to much higher energies in order to search for "Fermi-jet" nucleons.

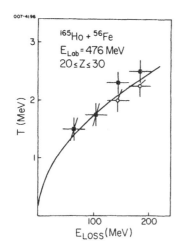

FIGURE 11,   See text. From Ref. 43

Angular momentum dissipation in the damped collision process produces primary
fragments with sizable intrinsic spins as shown by sequential-fission-fragment angular correlation[46] and γ-ray multiplicity[47,48] measurements. The alignment and
magnitude of the spins of the fragments is described by a nucleon exchange model[49,50]
incorporating both the effects of the relative motion of the target and projectile
and the internal motion of the constituent nucleons.

In this conference a number of papers will be presented on the topic of fusion
reactions with heavy ions. Fusion data now exist for a number of systems over a
wide mass range. Although most of the data is for lighter mass systems in the
energy range obtainable with tandem accelerators, limited information is available
at higher energies and for heavier systems. The need for more experimental measurements, especially for heavy systems is obvious.

An example of experimental and theoretical fusion excitation functions for a
light system is shown in Fig. 12 for the $^{16}O + ^{16}O$ reaction. The experimental data
have been measured by six different groups (see Ref. 51). As can be seen, the exper-

JOHN R. HUIZENGA

imental data of different groups are in qualitative but are not in quantitative agreement. This illustrates the problems associated with the experimental identification of the products from fusion and absolute cross section measurements. The solid line in Fig. 12 is a theoretical fusion excitation function calculated with a classical dynamical model[51] based on the nuclear proximity potential and one-body dissipation. An example of experimental and theoretical fusion excitation functions for a much heavier system is shown in Fig. 13 for the ^{109}Ag + ^{40}Ar reaction. The fusion cross section is now composed of both evaporation residues and fission fragments. Again the experimental problems associated with the identification of the fission fragments are sometimes severe, especially when the target and projectile masses are similar. The lines are theoretical curves calculated with different conservative potentials and one-body dissipation.

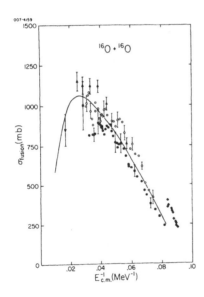

FIGURE 12,   See text.   From Ref. 51

HEAVY-ION REACTIONS: A NEW....

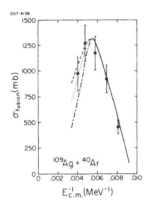

FIGURE 13,   See text. From Ref. 43

A recent comparison of all currently available data on fusion excitation functions for heavy-ion induced reactions with results from the above mentioned classical dynamical model shows overall good agreement between theory and experiment[51].
Although this agreement gives support to the proximity potential and one-body dissipation, it is not possible from the above comparisons to isolate in an unambiguous
way the individualized effects of the conservative and dissipative forces. Individual discrepancies between theory and experiment do exist and can be remedied sometimes by altering the theoretical input parameters such as the conservative potential
or the nuclear radii. However, such changes do not seem warranted at this time for
those cases where the experimental data are uncertain.

A number of fundamental problems arise in any comparison of experimental and
theoretical fusion cross sections. For example, which events are part of the
experimental fusion cross section? How is the theoretical fusion cross section
defined? Are these two cross sections to be compared directly? As the bombarding
energy is increased these questions become even more difficult to answer.

In the context of the classical dynamical model employed by Birkelund,
et al.[52,51], it is possible to determine whether a trajectory leads to fusion of the
target and projectile. This model assumes that all trajectories caught behind a
barrier in the internuclear potential lead to fusion. It is important to note that
whether a trajectory will lead to fusion cannot be determined from inspection of the
effective potential alone, as is the case for some of the so-called 'friction-free'
models. This situation arises primarily because of the effect of tangential friction
which leads to a continuous transfer of orbital to intrinsic angular momentum during

JOHN R. HUIZENGA

the interaction. Thus, a trajectory which shows no barrier and pocket in the effective potential for the asymptotic entrance channel value of angular momentum may experience such a barrier in the potential at some stage during the interaction because the initial value of the angular momentum is reduced during the interaction and the repulsive centrifugal potential is also reduced. It is this reduction in entrance channel angular momentum due to the tangential energy dissipation which is necessary to reproduce the magnitude of the fusion cross section for some of the heavier systems at high energies where simple inspection of the proximity potential predicts much lower cross sections.

It is important to emphasize that trajectories which lead to fusion within the context of the above model may not produce fusion in the commonly understood sense of a system in statistical equilibrium in all its intrinsic degrees of freedom, with a long lifetime, and with deformations which are _inside_ the saddle point for fission. The fusion cross section calculated with the classical dynamical model depends only on whether the system is caught, at least temporarily, behind a barrier in the internuclear potential, and there is no consideration of the question of the subsequent evolution of the trapped system.

The classical dynamical model with dissipation predicts a rather sharp demarkation between fusion and deep-inelastic trajectories. This results from the fact that the fusion trajectories are caught behind a potential barrier and this feature abruptly lengthens the lifetimes of fusion trajectories over deep-inelastic trajectories. Hence, on the basis of the present model, one expects to be able to separate the two types of trajectories rather well except, perhaps, for a narrow angular momentum window between the two types of trajectories where there may be a merging of lifetimes. Hence, for heavier systems, the fusion-fission trajectories with their much longer lifetimes are distinguishable from the deep-inelastic trajectories by their angular distribution and the equilibration of the mass asymmetry degree of freedom (i.e. the final fragments have average masses shifted from the target and projectile).

The measured number of trajectories which lead to fusion-fission-like fragments for heavy systems exceeds, sometimes substantially, the liquid-drop model limit. Although in some cases there is a question about the number of angular momenta to include in the fusion cross section, I think the evidence is rather good that the fusion cross sections for some heavy systems exceeds considerably the liquid-drop model prediction. Such results are consistent also with expectations from a model based on realistic conservative and dissipative forces. It should be emphasized again that these large angular momenta do not lead to what is normally called a compound nucleus, but due to temporary trapping these trajectories have lifetimes sufficiently long for equilibration of the mass asymmetry degree of freedom and a

HEAVY-ION REACTIONS: A NEW....

1 /sin θ angular distribution in distinction from deep-inelastic collisions. Hence,
further separation of the fusing angular momenta into bins below and above the LDM
limit is at best a difficult and challenging experimental task. Some indication
exists that the mass widths of these bins are different[53].

Models of the type referred to above have many limitations, and these short-
comings are expected to become even more important at higher energies where two-body
dissipation becomes important. This subject will be discussed further in a later
session. I do wish, however, to raise one other point about fusion. There is by
now considerable evidence that heavy-ion reactions, including reactions leading to
<u>fusion-like</u> products, may emit fast or non-statistical light particles[54]. The yield
of these particles increases with bombarding energy and their origin may be associated with quite different stages of the reaction. Hence, the presence of these non-
statistical particles should not be taken to preclude <u>necessarily</u> the possibility of
fusion. For example, if non-statistical particles arise on the trajectory after the
system is trapped, or if they arise from processes which are part of the energy loss
mechanism, then any resulting composite nucleus may be regarded as fused. On the
other hand, fast light particles arising from projectile breakup in the entrance
channel are known to lead to "incomplete fusion"[55] or "massive transfer"[56] and are
not to be considered part of the fusion cross section. These fast particles have
about the beam velocity and are expected to experience some shadowing along the beam
direction. Thus, their angular distribution may possibly decrease near 0°. As
more results on fusion become available at higher energies, new insights will be
gained on a number of the above questions dealing with fusion and fast light
particles.

The time-dependent Hartree-Fock method (TDHF) represents a natural first
approximation to a microscopic understanding of heavy-ion reactions at low energies,
including fusion and various damped collision processes. Since TDHF is essentially
a parameter-free theory once a nuclear interaction is chosen, its application to
heavy-ion reactions should teach us about nuclear dynamics. A basic assumption of
this theory is that at any moment, the motion of any one particle is affected by
all the other particles in the system, and that the effects of all the other particles can be approximated by one central field. The extend to which heavy-ion
reaction processes are governed by one-body effects and can be simulated by this
mean field approximation is a subject under active investigation.

The first large scale calculations with an axially symmetric time-dependent
mean field and a Skyrme-type effective interaction were reported[57] in late 1978 for
Kr-induced reactions on Pb and Bi. The calculated fragment energies, mean masses
and scattering angles for strongly damped collisions are in good agreement with

JOHN. R. HUIZENGA

experiments but the calculated mass distribution widths are too small. Fusion cross sections for several systems have been calculated[58] also to different degrees of approximation and with different effective interactions. In general, reasonable agreement with data is obtained, however, in no case is the quality of agreement with data any better than that obtained with the simple classical dynamical model mentioned earlier. It is interesting to note that the TDHF model underestimates considerably the reported fusion cross sections for the ^{40}Ca + ^{40}Ca system, just as the classical dynamical model does. There appears to be something anomalous about this reaction. The TDHF and classical dynamical models disagree on the magnitude of the low-ℓ window. This feature of some of the TDHF calculations may be associated with imposed restrictions on certain degrees of freedom because it has been shown[58] that non-axial modes, for example, are significant in the dissipation of energy at higher bombarding energies. In addition, TDHF includes only one-body dissipation and may underestimate the energy loss for central collisions.

In the latter part of this talk, I have tried to give some impression of the present status of heavy-ion collisions. Considerable progress has been made in recent years, however, there are still many unanswered questions. The availability of new higher-energy heavy-ion beams will open up new vistas and the future of the field looks very exciting as the sonic, Fermi and mesic thresholds are crossed.

The staff of the Hahn-Meitner Institute is to be congratulated on the completion of VICKSI, a verstile and powerful new heavy-ion accelerator. The world community of heavy-ion scientists anticipate many exciting experimental and theoretical discoveries by the staff here in the years ahead.

I wish to close by reminding you that heavy-ion nuclear science is built in part on the foundation blocks erected by two scientific giants working here in Berlin, Otto Hahn and Lise Meitner, and it is fitting that this symposium on heavy-ion reactions recognize their pioneering work in macroscopic nuclear science.

This research was supported by the U.S. Department of Energy. The author acknowledges many helpful discussions with J. R. Birkelund, D. Hilscher, J. Randrup, W. U. Schröder and W. W. Wilcke.

HEAVY-ION REACTIONS: A NEW....

1.   E. Fermi, Nature 133, (1934) 898.
2.   E. Amaldi, et al. Proc. Roy. Soc. A146, (1934) 483.
3.   I. Noddack, Zeit. f. Angew Chemie 37, (1934) 653.
4.   L.A. Turner, Rev. Modern Phys. 12, (1940) 1.
5.   L. Meitner, O. Hahn and F. Strassmann, Zeits. f. Physik 106, (1937) 249.
6.   I. Curie and P. Savitch, J. de Phys. [7] 8, (1937) 385; 9 (1938) 355.
7.   O. Hahn and F. Strassmann, Naturwiss. 27, (1939) 11.
8.   O. Hahn and F. Strassmann, Naturwiss. 27, (1939) 89.
9.   L. Meitner and O.R. Frisch, Nature 143, (1939) 239.
10.   N. Bohr and J.A. Wheeler, Phys. Rev. 56, (1939) 426.
11.   R. Vandenbosch and J.R. Huizenga, "Nuclear Fission", Academic Press, Inc.
      New York (1973).
12.   S. Frankel and N. Metropolis, Phys. Rev. 72, (1947) 914.
13.   S. Cohen and W.J. Swiatecki, Ann. Phys. (NY) 19, (1962) 67; 22, (1963) 406.
14.   D.L. Hill and J.A. Wheeler, Phys. Rev. 89, (1953) 1102; D.L. Hill, Proc. U.N.
      Int. Conf. Peaceful Uses Atomic Energy, 2nd, 15, (1958) 244, United Nations-
      Geneva.
15.   V.M. Strutinsky, Sov. Phys. JETP 18, (1964) 1298.
16.   J.R. Nix and W.J. Swiatecki, Nucl. Phys. 71, (1965) 1.
17.   V.M. Strutinsky, Nucl. Phys. A95, (1967) 420; A122, (1968) 1.
18.   S.M. Polikanov, Usp. Fiz. Nauk. 94, (1968) 93.
19.   E. Migneco and J.P. Theobald, Nucl. Phys. 112, (1968) 603; H. Weigmann, Z. Phys.
      214, (1968) 7.
20.   H.J. Specht, J. Weber, E. Konecny and D. Heunemann, Phys. Lett. B41, (1972) 43.
21.   W.A. Bassichis, A.K. Kerman, C.F. Tsang, D.R. Tuerpe and L. Wilets, In "Magic
      Without Magic: John Archibald Wheeler" (1973), Freeman, San Francisco.
22.   A. Bohr, Proc. U.N. Int. Conf. Peaceful Uses Atomic Energy, 1st, 2, (1955) 151,
      United Nations-Geneva.
23.   B.D. Wilkins, J.P. Unik and J.R. Huizenga, Phys. Lett. 12, (1964) 243.
24.   K.L. Flynn, J.E. Gindler and L.E. Glendenin, Phys. Rev. C12, (1975) 1478;
      E.K. Hulet, et al., Phys. Rev. [in press].
25.   E.K. Hulet, private communication.
26.   P. Möller and S.G. Nilsson, Phys. Lett. B31, (1970) 283.
27.   P. Fong, Phys. Rev. 102, (1956) 434.
28.   J.W. Negele, S.E. Koonin, P. Möller, J.R. Nix and A.J. Sierk, Phys. Rev. C17,
      (1978) 1098.
29.   For references to the sizable literature in this field see the following
      review articles: C.E. Bemis, Jr. and J.R. Nix, Comments on Nuclear and
      Particle Physics, 7, (1977) 65; G.T. Seaborg, W. Loveland and D.J. Morrissey,
      Science 203, (1979) 711; and G. Herrmann, Nature 280, (1979) 543.
30.   References to the early literature are contained in the review article by
      W.U. Schröder and J.R. Huizenga, Ann. Rev. Nucl. Sci. 22, (1977) 425.
31.   J.R. Huizenga, J.R. Birkelund, W.U. Schröder, K.L. Wolf and V.E. Viola, Phys.
      Rev. Lett. 37, (1976) 885.
32.   W.U. Schröder, J.R. Birkelund, J.R. Huizenga, W.W. Wilcke and J. Randrup,
      preprint (1979).
33.   W.U. Schröder, J.R. Birkelund, J.R. Huizenga, K.L. Wolf and V.E. Viola, Physics
      Reports 45, (1978) 301.
34.   H. Breuer, B.G. Glagola, V.E. Viola, K.L. Wolf, A.C. Mignerey, J.R. Birkelund,
      D. Hilscher, A.D. Hoover, J.R. Huizenga, W.U. Schröder and W.W. Wilcke, Phys.
      Rev. Lett. 43, (1979) 191.
35.   J. Randrup, Nucl. Phys. A307, (1978) 319; A327 (1979) 490.
36.   W.W. Wilcke, J.R. Birkelund, A.D. Hoover, J.R. Huizenga, W.U. Schröder,
      V.E. Viola, Jr., K.L. Wolf and A.C. Mignerey, to be published in Phys. Rev. C.
37.   R.A. Broglia, C.H. Dasso and Aa. Winther, Phys. Lett. B61, (1976) 113;
      R.A. Broglia, C.H. Dasso, G. Pollarolo and Aa. Winther, Phys. Rev. Lett. 41,
      (1978) 25.

JOHN. R. HUIZENGA

38. N. Frascaria, C. Stéphan, P. Colombani, J.P. Garron, J.C. Jacmart, M. Riou and L. Tassan-Got, Phys. Rev. Lett. 39, (1977) 918.
39. D. Hilscher, J.R. Birkelund, A.D. Hoover, W.U. Schröder, W.W. Wilcke, J.R. Huizenga, A.C. Mignerey, K.L. Wolf, H.F. Breuer and V.E. Viola, Phys. Rev. C20, (1979) 556.
40. J.P. Bondorf, J.R. Huizenga, M.I. Sobel and D. Sperber, Phys. Rev. C11, (1975) 1265; W.U. Schröder, J.R. Birkelund, J.R. Huizenga, K.L. Wolf and V.E. Viola, Phys. Rev. C16, (1977) 623.
41. C. Riedel, G. Wolschin and W. Nörenberg, Z. Physik A290, (1979) 47.
42. Y. Eyal, et al., Phys. Rev. Lett. 41, (1978) 625.
43. D. Hilscher, et al., Phys. Rev. C20, (1979) 576.
44. B. Tamain, et al., preprint (1979).
45. J.M. Miller, et al., Phys. Rev. Lett. 40, (1978) 1074.
46. P. Dyer, et al., Phys. Rev. Lett. 39, (1977) 392.
47. A. Olmi, Phys. Rev. Lett. 41, (1978) 688.
48. M.M. Aleonard, Phys. Rev. Lett. 40, (1978) 622.
49. R. Vandenbosch, Phys. Rev. C20, (1979) 17.
50. G. Wolschin and W. Nörenberg, Phys. Rev. Lett. 41, (1978) 691.
51. J.R. Birkelund, L.E. Tubbs, J.R. Huizenga, J.N. De and D. Sperber, Physics Reports, 56 (1979) 107.
52. J.R. Birkelund, J.R. Huizenga, J.N. De and D. Sperber, Phys. Rev. Lett. 40, (1978) 1123.
53. C. Lebrun, et al., Nucl. Phys. A321, (1979) 207; B. Heusch et al., Z. Physik, A288, (1978) 391; M. Lefort, Proc. of this conference.
54. See proceedings of this conference.
55. K. Siwek-Wilczyńska, E.H. DuMarchie van Voorthuysen, J. van Popta, R.H. Siemssen and J. Wilczyński, Phys. Rev. Lett. 42, (1979) 1599.
56. D.R. Zolnowski, H. Yamada, S.E. Cala, A.C. Kahler and T.T. Sugihara, Phys. Lett. 41, (1978) 92.
57. K.T.R. Davies, V. Maruhn-Rezwani, S.E. Koonin and J.W. Negele, Phys. Rev. Lett. 41, (1978) 623.
58. S.J. Krieger and K.T.R. Davies, Phys. Rev. C20, (1979) 167; Phys. Rev. C18, (1978) 2567 and references therein.

SYMPOSIUM ON DEEP-INELASTIC AND FUSION REACTIONS WITH HEAVY IONS
HAHN-MEITNER INSTITUT FUR KERNFORSCHUNG. BERLIN
October 23-25, 1979.

MASS DISTRIBUTION IN DISSIPATIVE REACTIONS.
THE FRONTIER BETWEEN FUSION AND DEEP INELASTIC TRANSFERS.

Marc LEFORT
Institut Physique Nucléaire Orsay and GANIL Caen. France.

For many years, two classes of dissipative reactions have been proposed for describing the result of hard collisions between two complex nuclei.

i)Those leading to complete fusion where a single excited ensemble of nucleons is formed and lasts for a duration much longer than a single rotation period. The subsequent decay may follow various channels amongst which fission processes.

ii) Those corresponding to higher ℓ-waves, where an incomplete fusion occurs and the intermediate system keeps a two center potential so that disruption into two fragments follows shortly after an exchange of nucleons and a large kinetic energy loss. The life time of this composite system is long enough for applying statistical transport concepts.

Considering the mass distribution of the products, without any particular restriction, the general feature is expected to show three peaks corresponding to : 1) the light fragments, 2) the heavy fragments and 3) the evaporation residues. Schematically this is typically sketched in figure 1 a) for an asymmetric system $^{16}O + ^{92}Mo$. As a matter fact, most of the time, only the light fragments resulting from the deep ine-

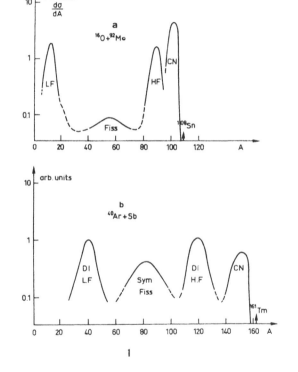

Fig. 1 : Schematic representation of mass distributions for all the reaction products in dissipative collisions.
a. Between a light projectile and a medium mass target, for example $^{16}O + ^{92}Mo$
b. Between a medium mass projectile, ^{40}Ar, and a heavier target, ^{121}Sb, at sufficient energy (E = 225 MeV).

lastic collisions are detected at various angles, while the evaporation residue yield is measured in the very forward direction (fig. 2). Also, when the compound nucleus is heavy enough, symmetric fission occurs in a fourth broad peak. This already happens for the compound nucleus ^{108}Sn, although the fission yield is only of the order of a few millibarns[1].

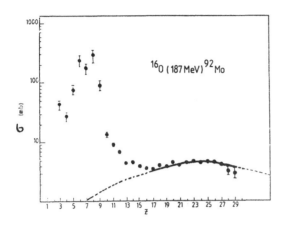

Fig. 2 : The Z distribution measured for the product of the collision (^{16}O + ^{92}Mo) by Agarwal et al[1]. The curve indicates the upper limit of estimates for fission cross section. The evaporation residues do not appear on this picture.

We would like to discuss the various aspects of mass distribution and to show that there are a number of cases for which the distinction between deep inelastic fragments and fission fragments issued from complete fusion is not so easy. Morever, there might be a third category which ressembles experimentally fission after fusion in many respects, although the usual statistical criteria are not strictly fulfilled.

Let us begin with a number of other typical mass distributions corresponding to different systems. In figure 1 b), the yield of fragments corresponding to a symmetric splitting of the compound nucleus is much greater than in figure 1 a), because of the higher angular momentum and the heavier mass. However, it has been shown [2-3] that the fission process appears only above E = 120 MeV, when an angular momentum greater than 35 $\hbar$ can be reached.

Figure 3 corresponds to a rather different situation. The entrance channel in the example (^{52}Cr + ^{56}Fe) is nearly symmetric, and the result is that the two deep inelastic components in the exit channel are mixed-up around the symmetric splitting. Furthermore, if there are fission fragments from the compound nucleus, they contribute also to the symmetric peak. This is indeed indicated in the actual data obtained by Agarwal et al[1] and presented in figure 4. There, the angular distribution follows a forward peaking which is characteristic of deep inelastic reactions only for 22 < Z < 27, whereas all the other products (Z< 22 and Z > 27) present a constant differential cross section dσ/dθ versus the emission angle, θ, in the

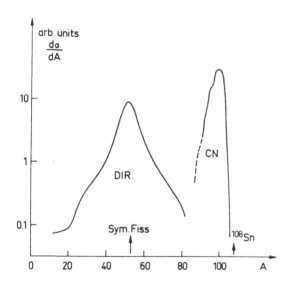

Fig. 3 : Schematic representation of
the mass distributions for a symmetric system in the entrance
channel (medium masses). Deep
inelastic reaction products and
fission fragments from the compound nucleus are mixed up with
a maximum around mass 53-54, at
symmetry.

center of mass system (Fig. 5). However, if all these events are summed
up, the cross section reaches 430mb,
nearly equal to the evaporation residue cross section (513 ± 80 mb).
If they were attributed to symmetric
fission following complete fusion,
the complete fusion cross section
would reach nearly 1 barn corresponding to a critical angular momentum
ℓ_{cr} = 72, a value a bit higher than
various theoretical estimates between 63 and 72, depending on the critical distance parameter (r_{cr} = 0.95
to r_{cr} = 1.05 fermi). However it is
difficult to believe that the compound nucleus ^{108}Sn de-excites
through fission in nearly half of
the cases, since ($\Gamma_{f}/\Gamma_{f}+\Gamma_{p}$), the calculated ratio of fission width over
total width, is equal to 0.5 only for
an angular momentum $\ell\hbar$ = 62 and the
ℓ population in our particular example extends between 0 and 85 units.
Furthermore, the same compound nu-

cleus with the same ℓ population was made with an asymmetric entrance (^{16}O + ^{92}Mo) and
the measured cross section for all symmetric splittings does not exceed 50 mb, as expected from calculations.

The preliminary conclusion that may be drawn is that most of the events
corresponding to $d\sigma/d\theta$ = cst present some intermediate feature between true complete
fusion and deep inelastic transfer reactions. If one admits a critical value around
63 $\hbar$, for the system (^{52}Cr + ^{56}Fe), the evaporation residue contribution corresponds
to the range 0 < ℓ < 53, and those fission-like events would be attributed to ℓ waves
on both sides of ℓ_{cr}, i.e. between 53 and 72. Another possibility could be that complete fusion occurs in the range 25 < ℓ < 63, while the fission-like phonomenon would
correspond to 0 < ℓ < 25 . And, of course, one has to explain why they are not found
in the asymmetric system (^{16}O + ^{92}Mo).

Let us shift now to heavier systems. In figure 6, corresponding to ^{40}Ar+^{197}Au
the evaporation residues have totally disappeared, as one would expect since the compound nucleus ^{237}Bk has a very great fission probability. Three typical peaks are

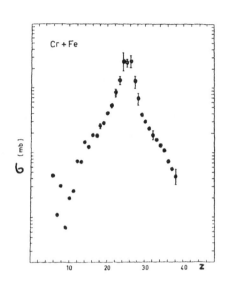

Fig. 4 : The actual Z distribution as it was measured[1]
for the products of the reaction (^{52}Cr + ^{56}Fe) at a center of mass energy E = 135 MeV.
The angular distribution follows dσ/dθ = cst for Z < 22
and Z > 27. The forward peaking appears only for
22 < Z < 27.

shown and one may believe that all the events which are seen in the broad symmetric mass distribution are due to fission after complete fusion.

Figure 7 indicates[4] the well known contour plot $\frac{d^2\sigma}{dEdA}$, where the separation between fission fragments and deep inelastic light and heavy products seems rather well defined. There are several characteristics of the fragments belonging to the symmetric mass distribution which suggest that they are indeed true fission fragments. As it was shown many years ago, they correspond to a full momentum transfer[5] follo-wed by a repulsion after scission with coulomb energies. Moreover, they exhibit a very typical angular distribution, $d\sigma/d\Omega$ proportional to $1/_{\sin\theta}$. This has been demonstrated very clearly[6] by collecting the recoiling fragments upon foils lo-cated at all angles between 0 and 180°, and by counting characteristic x rays.

However, a question arises, which is even more crucial for targets heavier than Au, like bismuth or uranium. Is it possible to form a well defined compound nu-cleus when the fission barrier has disappeared, like in a nucleus of Z equal to 110 (Ar + U) ? And therefore, is there any meaning to admit the usual concept of deforma-tion, saddle point and fission for such a compound nucleus ? Furthermore, the effect of high angular momentum decreases strongly the fission barrier, and it has been clai-med that complete fusion cannot occur for ℓ waves which correspond to a rotating liquid drop[7] for which B_{fR} = 0.

Recently the three peaks have also been observed[8-9] for the system Fe+Xe and for the system ^{238}U + ^{48}Ca (Fig. 8 and 9). However, in the last case, the medium peak (Fission like symmetric mass distribution) appears only for an excitation energy of E^* = 60 MeV, whereas it is not present at E^* = 18 MeV, corresponding to a bombar-ding energy just above the Coulomb barrier. Since the compound nucleus is Z = 112, the

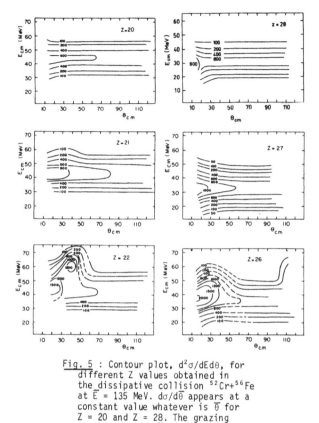

Fig. 5 : Contour plot, $d^2\sigma/dEd\theta$, for
different Z values obtained in
the dissipative collision $^{52}Cr+^{56}Fe$
at $\bar{E}$ = 135 MeV. $d\sigma/d\bar{\theta}$ appears at a
constant value whatever is $\bar{\theta}$ for
Z = 20 and Z = 28. The grazing
angle is around $\bar{\theta}_{gr}$ = 40 degrees
(from ref.([1])).

fission process should be already
dominant at the lowest energy.

The last set of typical mass
distributions corresponds to collisions between projectiles heavier
than copper and heavy targets.
There, only two peaks are clearly
seen (Fig. 10), one for the light
fragments, the other for complementary heavy fragments, and the domain of masses around symmetric
splitting of the compound nucleus
is totally missing([10]). This was
the great surprise of the year 1973
and the origin of the denomination
([11]) of quasi-fission(1974). Those
cases are now quite well defined.
The entire set of ℓ waves from 0 to
ℓ_{max} contribute to the deep inelastic process. Then the obvious
question which is raised is the
following :
Let us take two systems leading
to the same Z for the composite
system, for which, because of $B_{fr}=0$,
the compound nucleus formation is supposed impossible. Then why does the symmetric
splitting of the composite system in a fission-like process appear in the case of
($^{40}Ar + ^{238}U$), whereas nearly no fission-like fragments appear in the case of ($^{84}Kr
+ ^{186}W$), althoug the same value, Z = 110, is obtained for the composite system. Figure
11 extracted from reference 12 illustrates nicely this difference.

"TO FUSE OR NOT TO FUSE"
FISSION-LIKE SPLITTING AND THE ROTATING LIQUID DROP FISSION BARRIER.

We may draw a list of systems for which the three peaks have been observed
and where the question is open   to know whether or not the symmetric component is
due to a fission process issued from a compound nucleus (Table 1). For nearly all of
them the critical angular momentum was calculated according the method of Ngô et al.
([13]) by using the concept of critical distance introduced by Galin et al([14]). Nearly
the same values are also obtained with the Bass model([15]). Also the angular momentum
for which the rotating liquid drop fission barrier vanishes has been estimated, accor-

ding to the well known expression : $B_{f_R} = B_f - (E_{Ro} - E_{RS})$ (1)

where E_{Ro} and E_{RS}, the rotational energies of the spherical and deformed shapes respectively, are calculated according to the angular momentum $E_R = \dfrac{\ell(\ell + 1)\ \hbar^2}{2\ \mathcal{J}}$ (2)
with moments of inertia $\mathcal{J}_0$ and $\mathcal{J}_S$ corresponding to the spherical and the deformed
shape respectively.

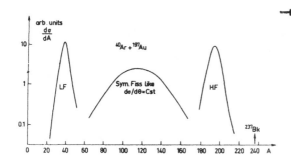

Fig. 6 : Schematic picture of the
mass distribution of reaction
products in the collision of a
medium mass projectile (A < 50)
and a heavy target. The exam-
ple corresponds to $^{40}Ar + ^{197}Au$
at energy close to the barrier.

Fig. 7 : Typical two dimensional plot $d^2\sigma/dEdA$ ob-
tained([4]) at a given detection
angle in the case of the reaction $^{40}Ar + ^{197}Au$. The symmetric fission events appear
in the central part of the
picture and the separation
from deep inelastic ordinary
products is quite well defined.

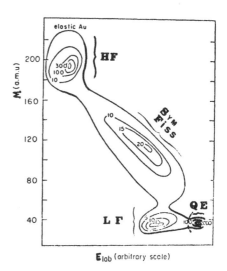

In the table 1, one may notice that the fission barrier always vanishes at
angular momenta smaller than the critical limit. It has been argued([7]) that when
a composite system is made with a fission barrier equal to zero, it will disintegrate
without passing the stage of a definite nucleus so that the resulting products are
not fission fragments but correspond already to deep inelastic reactions. However, the
concept of fission barrier concerns the exit channels whereas the concept of critical
angular momentum deals with the entrance channel and is mostly governed by dynamical approach in a sudden approximation.One dimensional models present the total effective
potential (coulomb + nuclear + rotational) as a function of internuclear distance.

TABLE 1

Reaction	$\overline{E}$(MeV)	CN	ℓ_{cr}	ℓ for $B_f = 0$	Ref
$^{52}Cr + {}^{56}Fe$	135	^{108}Sn	73	72	1
$^{40}Ar + {}^{92}Mo$	208	^{132}Nd	114	87	16
$^{40}Ar + {}^{107,109}Ag$	120	$^{147,149}Tb$	90	76	17.
$^{86}Kr + {}^{65}Cu$	160	^{151}Tb	100	78	21
$^{134}Xe + {}^{56}Fe$	225	^{190}Hg	115	82	9
$^{40}Ar + {}^{121}Sb$	225	^{161}Tm	138	87	2 - 3
$^{35}Cl + {}^{141}Pr$	160	^{176}Os	120	83	20
$^{40}Ar + {}^{165}Ho$	239	^{205}At	120	86	18
Ar + Au	250	^{237}Bk	118	74	4
" "	188	^{237}Bk	88	74	4
Ar + Th	210	$^{272}108$	78	64	19
Ar + U	250	$^{278}110$	119	63	2 - 3

Table 1 : Some systems for which ℓ_{cr} is larger than ℓ corresponding to $B_f = 0$.

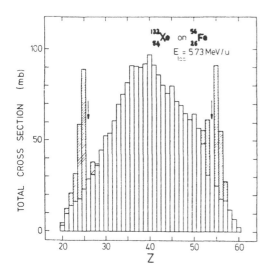

Fig. 8 : Reaction product cross-section measured[8] in the system $^{132}Xe + {}^{56}Fe$. A very important fission like Z distribution is observed around Z = 40, which corresponds to equal splitting of the compound nucleus ^{188}Hg.

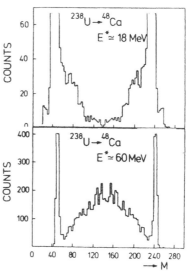

Fig. 9 : Z distributions for the coincident binary products in the reaction $^{238}U + {}^{48}Ca$, obtained[9] at two different bombarding energies. For the lowest one, there is no symmetric splitting at E*=17 MeV, whereas at E*=60 MeV, the fission like bump is very important.

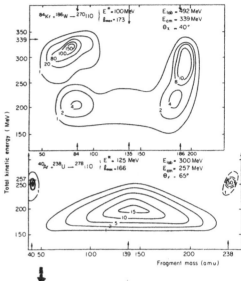

Fig. 10 : Schematic mass distribution for heavy systems. When the projectile mass exceeds A around 50 amu, fission fragments issued from a compound nucleus are not observed, as it is typically found for ^{63}Cu + ^{197}Au or ^{84}Kr + ^{209}Bi.

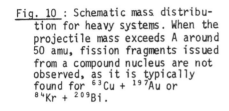

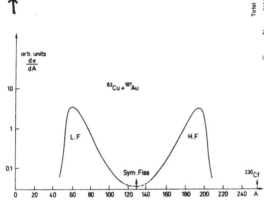

Fig. 11 : Contour plots $d^2\sigma/dEdA$ for the two systems leading to the same Z value of the composite (Z=110) Fission like fragments appear as usual around A=139 in the case of ^{40}Ar projectiles, and only deep inelastic events (quasi-fission) are observed with ^{84}Kr projectiles (from ref.([12])).

It generally shows a minimum(pocket).When dissipative forces act strongly between the two nuclei the classical trajectory may become trapped within the pocket, because of kinetic energy loss as well as angular momentum transfer. Generally, a rapid relaxation into the compound equilibrium configuration occurs so that the scheme of figure 12 changes from the sudden approximation (for example proximity potential) into an adiabatic transformation towards a single nucleus potential.

For high ℓ waves, the large centrifugal potentials suppress the pocket and quasi-elastic or deep inelastic events occur. For smaller impact parameters, even when the pocket has disappeared, complete fusion may occur if the distance of approach is closer than the critical distance, because adiabatic effects modify the potential curve.

When the compound nucleus decays through fission, the separation occurs along an axis which is different from the separation distance between the two approaching nuclei. This means that even when $B_f = 0$, there might appear a pocket, like in figure 12, but the system is open towards another direction. Because the shape evolution between entrance and exit may be hindered by viscosity effects (one-body dissi-

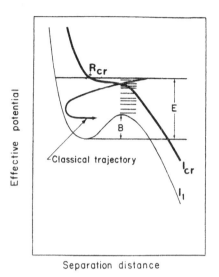

Fig. 12 : Schematic one-dimensional
potential energy curves versus
separation distance for two ℓ
values. For ℓ_1 a pocket is obtained in the sudden approximation calculations and a given
trajectory may be trapped. For
ℓ_{cr}, the pocket has disappeared
and complete fusion occurs only
if the critical distance ℓ_{cr} is
reached (ref.([14])).

pation), one can indeed, even bet-
ween $\ell_{Bf = 0}$ and ℓ_{cr}, conceive of
a quasi-compound nucleus. In a two
dimensional plot of potential energies, the system may be reached
after overcoming a ridge, and then
survives long enough before decaying along an open valley, so that
some equilibrium would be attained.
Particularly, the remembrance of
mass distributions close to the
projectile and target masses might
disappear in favour of a symmetric
splitting. In a way, we assume that,
in addition to the usual fission
passing through a saddle point, a
out of barrier fission-like process
exists.

MAIN CHARACTERISTICS OF THE FISSION-LIKE PHENOMENA

There are three sets of data in favour of our intermediate mechanism.

i) Enhancement of the apparent fusion-fission yield.

These are the results quoted in table 1. For all the systems the cross section for symmetric splitting and fragments emitted within an angular distribution
$d\sigma/d\theta$ = cste, is greater than expected from complete fusion calculations and $B_f = 0$
limits. For ℓ waves above $\ell_{(B_f = 0)}$, there are still reactions which furnish symmetric scissions, clearly separated from deep inelastic mass distributions, and with a
complete energy relaxation.

ii) Excitation functions have been drawned([21]) for fission-like events resul-
ting from 5 different systems : ^{20}Ne, ^{35}Cl, ^{40}Ar projectiles and ^{63}Ni, Ag, ^{116}Sn and
^{141}Pr targets. Neither the threshold nor the cross section magnitude at several MeV
above the threshold can be reproduced in a single way by classical statistical calculations including the usual Γ_f/Γ_{tot} ratio. The only possibility for an agreement

was to lower the liquid-drop fission barrier by 40 %. As an example, B_f has been measured for ^{186}Os at an experimental value of 24.3 MeV, the effective heigth necessary
for fitting the data on (^{35}Cl + ^{141}Pr) where ^{176}Os is the compound nucleus, is only
11.6 MeV. Another explanation would be that part of these fission fragments are due
to a fission without barrier process, i. e., above $B_{f_{Rot}} = 0$.

iii) The last type of results concerns the width of the mass distribution
around the symmetric splitting, measured typically on figure 13.

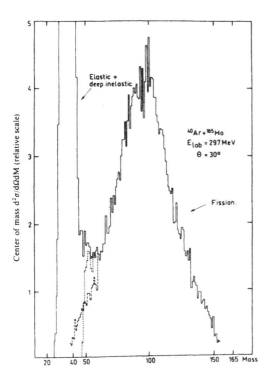

Fig. 13 : Measured mass distribution
in the system ^{40}Ar + ^{165}Ho at a
bombarding energy of 297 MeV and
at an angle of 30°. The value of
the width, Γ, can be easily measured (ref.([18])).

It has been established experimentally that the increase of nuclear
temperature (or intrinsic excitation
energy) of the compound nucleus has
a positive effect on the FWHM of
the mass distribution([16]). This has
been predicted in the frame of the
liquid drop model([23]). On the other
hand, the influence of angular momentum is poorly known and experimental
data are contradictory.The recent comparison made by Tamain et al([18-22])
between the fragment mass distributions of (^{40}Ar + ^{165}Ho) and
(^{20}Ne + ^{nat}Re), indicates a rather
strong broadening of the width,
from 30 units to 56 units when the
calculated limit, ℓ_{cr}, increases
from 49 up to 120. A correction
has to be made for the excitation
energy effect since E^* varies at
the same time from 68.5 to 153 MeV
(T varies from 1.6 to 2.2 MeV). The
result after correction is shown
in figure 14. The rather striking
finding is that the FWHM increases
very slowly for all the range where ℓ is lower than $\ell(B_{f=0})$ and there is a much more
marked broadening when the composite nucleus, ^{205}At is able to reach ℓ values higher
than $\ell_{Bf=0}$. A wider compiling (^{27}) has been made for all the systems included in Table
1. In order to make the comparison amongst rather different compound nucleus masses,
a reduced width has been defined Γ/A. Figure 15 shows very clearly that, after correction for temperature effects, Γ/A is more or less constant when the fission barrier
still exists, but increases dramatically when $B_f \leq 0$. Then it seems rather straight

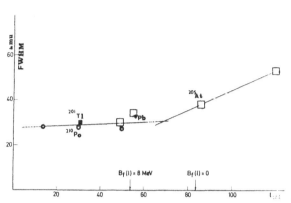

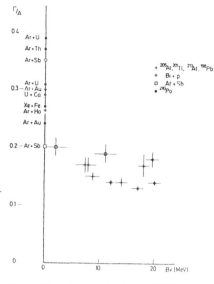

Fig. 14 : Full width at half maximum for
the fission-like mass distributions
measured for various compound systems
sharing various angular momenta.
○ ^{210}Po, ■ ^{201}Tl, ▼ ^{198}Pb, □ ^{205}At
(from ref.$(^{22})$).

Fig. 15 : Reduced width, Γ/A, versus
calculated fission barriers for
various systems. When B_f is zero
or negative, values of Γ/A are put
on the ordinate axis (from $(^{27})$).

forward to suggest that all ℓ-waves higher than $\ell_{(B_f = 0)}$ produce a reaction a bit
different from the ordinary fission, in a sense that the mass distribution is wider.
This is quite understandable since there is not any more the constraint of the saddle
point path.

WHAT IS ORIGIN OF FISSION-LIKE PROCESS WITHOUT FISSION BARRIER ?

A number of qualitative suggestions have been made in order to explain this
broad symmetric mass distribution. The first one, due to Mathews and Moretto[17] is to
consider that it is the result of fission after fusion, the enhancement being due
to thermal barrier penetration towards the exit channel of a classically trapped trajectory. They calculated the width :

$$\Gamma_{out} \simeq \hbar \omega \exp (- B/_T) \tag{3}$$

where $\hbar\omega$ is the vibrational phonon energy and T the temperature. Then they apply
the diffusion model which gives the charge distribution, but instead of a life time
around a gaussian distribution, they take it from Γ_{out} in equation (3). Also, because
the trapped trajectories are trapped during a rather long time as compared to the
mass equilibration time, they find a charge distribution :

$$\sigma_Z = \pi \lambda^2 \sum_\ell (2 \ell + 1) \frac{\rho_Z (E_Z)}{\sum_Z \rho_Z (E_Z)} \tag{4}$$

The level densities ρ_Z apply to the scission point of the two ions.

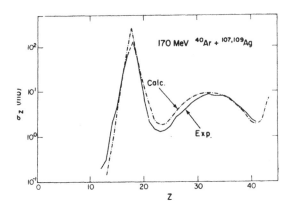

Fig. 16 : Cross sections for different Z products in the reaction ^{40}Ar + $^{107, 109}$Ag. Experimental data are compared to calculations made assuming penetration of the barrier and the diffusion model (ref.([17])).

The result is quite satisfactory as shown in figure 16 for Ar + Ag.

However, the absence of fission-like peak for ^{238}U + ^{48}Ca at the lowest excitation energy is not explained. Also we don't know why the fission like cross section is much larger for Cr + Fe than for O + Mo, although the compound nucleus is exactly the same, and the pocket is even deeper for Cr + Fe than for O + Mo.

Finally it is not clear to see how does the system evolve from the sudden approximation two centers potential towards the single potential of a compound nucleus.

The second type of explanation has been given by Nörenberg([24]) in the frame of the diffusion model applied to dissipative collisions. However at the first glance a continuous evolution in time between the deep inelastic products focused around projectile mass and the fission-like component focused around symmetrical division should lead to a continuous mass distribution, and not to three distinct peaks. Nevertheless, Nörenberg and Riedel introduce the formation of a doorway configuration which is formed during the fast approach following diabatic states. This doorway configuration decays into statistical equilibrium, but the density of available states is reduced during the approach. The degree of coherence with respect to the entrance channel is introduced in the transport theory as an additional macroscopic variable. Therefore, they use a time-dependent dynamical potential made of an adiabatic part and a diabatic part ([24]):

$$U (\chi(t) = U_{ad} (1 - \chi(t)) + U_{dia} \cdot \chi(t) \tag{5}$$

An interesting consequence of such a treatment is shown in figure 17 for the system Ar + Pb, where different trajectories are drawn in the (r, α)-plane, r describing the distance between fragments and α the mass asymmetry ratio. The trajectory with ℓ = 51 leads to a compound nucleus. The trajectory ℓ = 104 corresponds to a deep inelastic collision which evolves in 2.10^{-21} seconds. The trajectory ℓ = 102 is captured for a while (2.10^{-20} seconds) so that the system develops towards mass

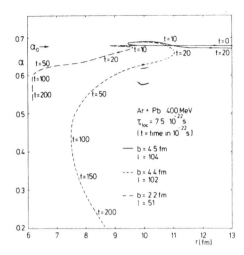

Fig. 17 : Trajectories for three ℓ-
values. In the asymmetry separation distance plane, the asymmetry α may change very much as
time evolves (from ref.([17])).

symmetry ($\alpha = 0$) and finally splits
into two fission like fragments
since the time is much longer than
a rotational period ($1/\sin \theta$ angular distribution). In order to decide if this mechanism works as
an extension of the ordinary dissipative collisions to a long living
component, it would be interesting
to know how does the model predict
the mass distribution. Still open
is the question why the fission
like products are not observed when
the projectile mass reaches around
50 (nearly no symmetric fragments
for ^{63}Cu induced reactions).

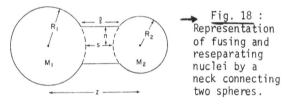

Fig. 18 :
Representation
of fusing and
reseparating
nuclei by a
neck connecting
two spheres.

The third approach has been
proposed by Swiatecki([25]) at the
International School of Nuclear
Physics in Erice, in a simplified
treatment of the dynamics of nucleus-nucleus collisions using the
wall-plus-window formula([26]). As
usual, the shapes of fusing or
reseparating nuclei are described
by two spheres connected by a cylindrical neck. For a given asymmetry, the two degrees of freedom describing the system are σ, the separation distance between the
two surfaces and ν, the neck radius, with a unit of length equal to $2 \bar{R} = 2R_1R_2/R_1+R_2$.
The length of the neck is then $\lambda \approx \sigma + \nu^2$ (Fig. 18).

A neck in the half-density contour of the spheres will form when quarter-
density points have touched, i. e., for $\sigma = \sigma_1$, and the boundary of the configuration space due to the fact that the neck length cannot be negative is :

$$\sigma \geqslant \sigma_1 - \nu^2 \tag{6}$$

The potential energy landscape is built up in coordinates s and n, starting with the
coulomb energy :

$$E_{coul} = \frac{3}{5} (\frac{(Z_1e)^2}{R_1} + \frac{(Z_2e)^2}{R_2}) + \frac{Z_1Z_2e^2}{R_1+R_2+s} \quad \text{for } s > 0 \tag{7}$$

and the nuclear energy calculated with a proximity correction

$$E_{nucl} = 4 \pi \gamma (R_1^2 + R_2^2) + 2 \pi \gamma (n (\ell - \ell_f) - n^2) \tag{8}$$

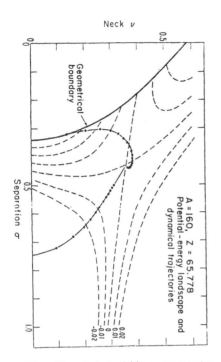

Neck ν

Geometrical boundary

Separation σ

A = 160, Z = 65.778
Potential-energy landscape and
dynamical trajectories

<u>Fig. 19</u> : Swiatecki's representa-
tion([25]) of potential energy
map. The right side trajectory
corresponds to fission path
from saddle to scission. The
left side trajectory is labelled 0.136 in figure 20.

where γ is the surface energy per
unit area and ℓ_f is a length of the
neck corresponding to zero surface
energy. Finally in units $8\,\pi\gamma\bar{R}^2$,
the potential energy is expressed
as a function of ν, σ, χ the Coulomb parameter $Z_1 Z_2 e^2 / 16\pi\gamma\bar{R}^{-3}$,
and of the asymmetry parameter
$\Lambda = (R_1 + R_2)/2\bar{R}$.

Figure 19 shows such a potential
energy map, where two regions of
low energy are separated by a saddle
point path. Except in the case of
a reflection symmetric system, e-
nergy is not stationnary with respect to asymmetry and the saddle
point is not unconditional.

The equations of motion reduce
to expressions

$$\mu\ddot{\sigma} + \nu^2\dot{\ddot{\sigma}} = \chi / (\Lambda + \sigma)^2 \quad \text{for } \sigma > 0 \qquad (9)$$

$$\mu\ddot{\sigma} + \nu^2\dot{\ddot{\sigma}} = \frac{\chi}{\Lambda^2} - \nu \quad \text{for } \sigma < 0 \qquad (10)$$

and
$$4\,\nu\,(\sigma + \nu^2)\,\dot{\nu} = (\sigma_1 - \sigma) + 2\,(1 - \sigma_1)\,\nu - 3\,\nu^2 \qquad (11)$$

When an angular momentum L is present, the disruptive tendency is increased by
centrifugal forces $E = \frac{L^2}{2\mathfrak{J}}$.

Swiatecki has shown that one can define a generalized rotational parameter

$$\chi = \frac{Z_1 Z_2 e^2}{16\pi\gamma\bar{R}^3} + \frac{L^2 \, M_r \, (R_1 + R_2)^3}{16 \, \pi \, \gamma \, \bar{R}^3 \, \mathfrak{J}_0^2} \qquad (12)$$

where $\mathfrak{J}_0$ is the rigid body moment of inertia and M_r the reduced mass.

The result is that two di-nuclear systems with different asymmetries, i. e., different Z_1 and Z_2, different size and different angular momenta can be characterized
by the same disruptive parameter χ, which is made of a fissility part x, and a centrifugal part Y in the proportions

$$\chi = \frac{10}{3} \, (\, x + \frac{48}{49} \, 2^{1/3} \, Y) \qquad (13)$$

For example χ has the same value in a symmetric system $A_1 = A_2 = 80$, $Z_1 = Z_2 = 32.89$
with L = 0, and in an asymmetric system $Z_1 = 12$, $Z_2 = 54$ with L around 50 $\hbar$.

Figure 20 shows seven dynamical trajectories for such a system where χ = 2.1 units. These evolutions are taking place on the topography of the potential energy illustrated in figure 19.

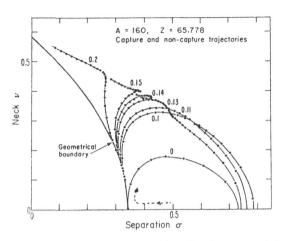

Fig. 20 : Seven trajectories in the potential energy landscape of two colliding nuclei [25]. The composite system is A = 160, Z = 65.778. Captured trajectories correspond to a peeling-off point where the neck radius ν is equal to 0.2, 0.15 and 0.14. Non-captured trajectories correspond to 0, 0.1, 011 and 0.13. The time evolved between two points on a trajectory is 10^{-22}sec.

The trajectories are derived from the equation of motion (10-11). The approach of two nuclei is represented by a point moving from right to left along the σ-axis. At the point $\sigma = \sigma_1$, the superposition of the diffuse densities of the two nuclei is at the origin of the neck growing along the geometrical locus

$$\nu = \sqrt{\sigma_1 - \sigma} \qquad (14)$$

The rate of neck growth, $\dot{\nu}$, is related to the speed of approach $\dot{\sigma}$ by :

$$\dot{\nu} = \frac{-\dot{\sigma}}{2\nu} \qquad (15)$$

The collision proceeds along the geometrical locus (14) until the neck growth velocity given by equation (11) has attained the geometrical velocity given by (15).

At this point the trajectory in (σ-ν) space leaves the geometrical locus. Swiatecki calls it the "peeling-off" point for which the conditions are :

$$\dot{\sigma} = -\nu(1 - \sigma_1 - \nu)/\sigma_1$$
$$\sigma = \sigma_1 - \nu^2 \qquad (16)$$

For the trajectory labelled ν = 0, the pealing-off point is on the σ axis, i.e., the neck radius increases when the two nuclei are in contact, but the neck elongation occurs soon after and is followed by a prompt splitting. For greater velocities, $\dot{\sigma}$, the trajectories follow the geometrical locus for a while, and the pealing-off point is obtained at ν = 0.1, ν = 0.11, etc... They come more and more into the neighborhood of the saddle point configuration, where the potential energy is stationary. Nevertheless they lead to re-separation. For a higher injection velocity, $\dot{\sigma}$, and a greater neck growth, the pealing-off is obtained at ν larger than 0.136 and the trajectories lead to capture on the inside of the saddle point (north east region). The critical trajectory dividing capture and non capture is shown on figure 19. Now in the region around the saddle, two classes of dynamical trajectories appear in the first stage of a given asymmetry. They depend on whether they are captured or not on the inside of the "conditional saddle point", in the configuration space σ-ν. Amongst those trajectories which are not captured, the possibility of kinetic energy loss exists because of interplay of collective deformations and one-body dissipation.

But, later on, for the called "captured" trajectories, the asymmetry degree of freedom is not frozen all the time and a change in this mode modifies the place of the saddle point.

In a purely symmetric system, conditional and true saddle points are the same. But for an asymmetric system, the conditionnal saddle point is on the outside of the true saddle point. For a captured trajectory (inside the conditional saddle) a change in the asymmetry degree of freedom towards symmetry (due to the driving potential) will shift the conditional saddle point closer to the true saddle point. Therefore, the trajectory will travel in a changing landscape. The captured trajectory along which the symmetry increases will, after a while, find itself on the right side of the saddle and therefore will lead to re-disintegration. The original compacting force has changed of sign. A relatively long time is spent during this "trip" in a moving landscape, so that, at the end the splitting occurs with a symmetry mass distribution. This composite system is perhaps not really a compound system since it is still described as two nuclei connected by a neck. However, it looks also like a deformed single nucleus when the neck radius becomes large and the separation distance is rather short, like it appears for example near the saddle region for the trajectory labelled $\nu = 0.13$ in figure 20, where $\nu_s = 0.38$ and $\sigma_s = 0.4$ units.

Now suppose a potential energy map where the low energy region on the left side of the saddle corresponds to a very shallow valley, because the influence of large angular momentum ($B_f = 0$). The trajectories with large ν values will still travel in a plateau above the pealing-off point. The time evolved might be long enough so that symmetric splittings will be reached, and the system will behave quite similar to a compound nucleus.

Finally let us quote Swiatecki's conclusion : "Thus a new type of reaction is suggested on theoretical grounds by the notion of conditional saddle point shapes at fixed asymmetry and by the fact that with growing symmetry a large region of configuration space opens up (for heavy systems) between the conditional saddle and the true saddle. These composite nucleus reactions, which are captured inside the conditional saddle but not inside the true saddle, should be characterized by time scales intermediate between deep inelastic and compound nucleus times and should have mass and angular distributions approximatively, but only approximatively, similar to compound nucleus reactions."

We believe that this class has been found experimentally since the very first work of Sikkeland, and the results quoted in table I confirm this discovery. Also, this type of reaction could not exist for very asymmetric systems where the driving potential is not towards symmetry but on the contrary, is towards asymmetry and therefore the division between captured and non-captured on inside the saddle point, is not affected by the changed asymmetry. It could neither exist for very hea-

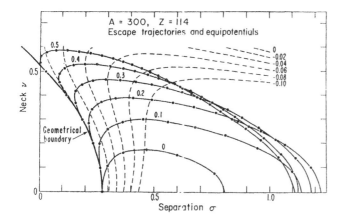

Fig. 21 : Potential energy map ([25]) and trajectories for a very heavy composite system, A=300, Z=114. There is no valley on the left hand side, and no saddle path. There is no captured trajectory. Interaction times for the most relaxed events do not exceed 3.10^{-21} sec.

vy projectiles when the potential energy map in the (σ-ν) configuration does not show any valley, so that all trajectories lead to re-disintegration, as shown in figure 21.

Summarizing, the symmetric mass-distributions which appear very typically in reactions induced by medium mass projectiles on medium and heavy targets in the approximate range ($200 < Z_1Z_2 < 2000$) may be classified well as a third dissipative mechanism in addition to the compound nucleus formation and to the deep inelastic process. It might also be considered as a particular aspect of complete fusion for all those nuclei produced in a ℓ-wave range between the value of ℓ for which the rotating liquid drop fission barrier is equal to zero and the dynamical critical angular momentum ($\ell_{B_f=0} < \ell < \ell_{cr}$).

I should like to thank W.J. Swiatecki for very enlightening discussions at the Erice School of nuclear physics and for making me available his preprint and figures. I thank also very warmly Bernard Tamain who has done a lot in Orsay for this third mechanism concept.

REFERENCES

(¹) S. Agarwal, J. Galin, D. Guerreau, X. Tarrago, unpublished and
     S. Agarwal, Thesis, Orsay (1979).
(²) M. Lefort, Y. Le Beyec, J. Péter, Rivista del Nuovo Cimento 4, 79   (1974).
(³) F. Hanappe, C. Ngô, J. Péter, B. Tamain, Physics and Chemistry of Fission,
     IAEA-SM-174/42, Vienna (1974), II, 289.
(⁴) C. Ngô, J. Péter, B. Tamain, M. Berlanger, F. Hanappe, Z. Phys. A283, 161 (1977).
(⁵) T. Sikkeland, Phys. Lett. 31B, 451 (1970).
(⁶) C. Ngô, J. Péter, B. Tamain, M. Berlanger, F. Hanappe, Z. Phys. A283, 161 (1977).
(⁷) S. Cohen, F. Plasil, W.J. Swiatecki, Ann. of Phys. 82, 577 (1974),
     F. Plasil, Proc. Nashville Conf. between complex nuclei, North Holland vol. 2,
     107, (1974).
(⁸) B. Heusch, C. Volant, H. Freiesleben, R.P. Chestnut, K.D. Hildebrand,
     F. Pühlhofer, W.F.W. Schneider, Z. Phys. A288, 391 (1978).
(⁹) A. Olmi, H. Sann, U. Lynen, V. Metag, S. Björnholm, D. Habs, H.J. Specht,
     R. Bock, A. Gobbi, H. Stelzer, 15th meeting on Nuclear Physics, Bormio 724 (1978)
(¹⁰) M. Lefort, C. Ngô, J. Péter, B. Tamain, Nucl. Phys. A216, 166 (1973).
(¹¹) F. Hanappe, M. Lefort, C. Ngô, J. Péter, B. Tamain, Phys. Rev. Lett. 32, 738
     (1974).
(¹²) J. Péter, C. Ngô, B. Tamain, Nucl. Phys. A250, 351 (1975).
(¹³) C. Ngô, B. Tamain, J. Galin, M. Beiner, R.J. Lombard, Nucl. Phys. A240, 353
     (1975).
(¹⁴) J. Galin, D. Guerreau, M. Lefort, X. Tarrago, Phys. Rev. C9, 1018 (1974).
(¹⁵) R. Bass, Nucl. Phys. A231, 45 (1974).
(¹⁶) B. Borderie, F. Hanappe, C. Ngô, J. Péter, B. Tamain, Nucl. Phys. A220, 93 (1974).
(¹⁷) G.J. Mathews, L.G. Moretto, L.B.L. 8972 (1979).
(¹⁸) C. Lebrun, F. Hanappe, J.F. Lecolley, F. Lefebvres, C. Ngô, J. Péter, B. Tamain,
     Nucl. Phys. A231, 207 (1979).
(¹⁹) Io. Oganecian, Dubna Report, E2 3942 (1968).
(²⁰) M. Beckerman, M. Blann, Phys. Lett. 68B, 31 (1977).
(²¹) M. Beckerman, M. Blann, Phys. Rev. Lett. 38, 272 (1977) and Phys. Rev. C17, 1615
     (1978).
(²²) F. Hanappe, J. Péter, B. Tamain, C. Lebrun, J.F. Lecolley, F. Lefebvres, C. Ngô,
     Physics and Chemistry of Fission, Julich, IAEA.SM/241 D3, May (1979).
(²³) J.R. Nix, Nucl. Phys. A130, 241 (1969).
(²⁴) W. Nörenberg, C. Riedel, Z. Phys. A290, 335 (1979).
(²⁵) W.J. Swiatecki, Lecture at Erice School on Nuclear Physics, to appear in
     Pergamon Press. D. Wilkinson Editor (1979).
(²⁶) J. Blocki, Y. Boneh, J.R. Nix, J. Randrup, M. Robel, A.J. Sierk, W.J. Swiatecki,
     Ann. Phys. 113, 330 (1978).
(²⁷) B. Tamain, personnal communication.

Collective Motion in Deep Inelastic Collisions

C.H. Dasso

NORDITA, Blegdamsvej 17, DK-2100 Copenhagen Ø, Denmark

Abstract

A brief review of a program to analyze deep inelastic collisions with enphasis in the collective response of the participant ions is given. Recent results of the approach valid for near-grazing collisions are presented.

This contribution is just one of several scheduled for this after-noon on the subject of "Different Views of Deep Inelastic Collisions". Given the short time available I will not be able to discuss, not even present, details of the particular approach we have followed in Copen-hagen. Those interested in a more complete description of the subject may refer to the Lecture Notes of the last Varenna Summer School [1] which represent the most comprehensive presentation of this effort so far.

Ishall limit myself to sketch the main elements of our model, to list different problems where it has been used and to review a recent application of these ideas which is not included in the reference just quoted. The example in question is of special interest since it allows for a direct check on basic ingredients of the picture.

The motivation behind this work was the hope, if nothing else, that one could use the rather complicated collisions between heavy ions to test our knowledge on the structure of the nuclei involved. Or, if you want, the wish to learn to what extent some aspects of the experimental evidence could be understood in terms of gross features of the response function of the reacting partners. We had been, of course, aware of the important work being done emphasizing the statistical aspects of the process. However, the usual arguments about time scales were not convincing enough to exclude some amount of coherent response and we decided to investigate this possibility. Thus evolved a different way of looking at deep inelastic collisions which, in a sense, is complementary to some of the work that will be covered in the following contributions.

I stated that the aim was to establish a connection between the experimental evidence and the structure of the ions involved. When one thinks that information about the population of specific nuclear states will be needed it is quite natural to cast the problem in the framework of a coupled channel calculation. If $|\varphi_\gamma\rangle$ denotes the state vector of the reaction channel γ the idea is to construct the vector $|\psi(t)\rangle$ representing the intrinsic state of the nuclear system as a superposition

$$|\psi(t)\rangle = \sum_\gamma C_\gamma(t) |\varphi_\gamma\rangle \quad . \tag{1}$$

Given the coupling matrix elements between the channels, which are functions of $\vec{r}$ (and hence of t in the approximation that the coordinate of relative motion can be treated clasically) one can set up coupled channel equations of the form

$$\dot{C}_\gamma(t) = \sum_{\gamma'} M_{\gamma\gamma'} (\vec{r}(t)) C_{\gamma'}(t) \tag{2}$$

where $M_{\gamma\gamma'}$ is a matrix in the channel index. Following (2), every channel amplitude evolves in time reflecting the characteristics of the couplings and the population of all the different channels.

Surely in a heavy ion collision a lot of reaction channels are open. Thus, it appears to be a tremendous task to carry out an explicit coupled channel calculation which realistically includes the relevant degrees of freedom. However, there is a specific case in which taking care of an infinite number of channels becomes trivial. That happens when we consider states of a pure harmonic system, coupled to a time-dependent external field which is linear in the amplitude of the mode. In such case, the full complexity of (2) reduces to solving the time evolution of the amplitude of the mode in the equivalent classical problem. Out of this information one can construct the probabilities of occupation for all the different channels.

This realization per se would not be of great help, if one could not single out, in the nuclear spectrum, states of this type. They should, in addition, play an important role in the process of excitation. Degrees of freedom which are ready-made for our purpose are the surface vibrational modes. Their collective nature makes it possible to represent by just a few of them most of the sum rule for the different multipolarities. Both the Coulomb and nuclear fields derived from the

ion-ion interaction couple strongly to them, linearly in a first approximation.

We collect an added bonus from this choice. By carrying out explicitly the excitation of deformation degrees of freedom one can follow the evolution at all times of the nuclear shapes. This has undis-
putable advantages insofar as defining properly the ion-ion potential
and corresponding form factors. The need to preserve during the collision enough coherence to produce rather deformed surfaces in the exit
channel has long been recognized.

The simplicity of the harmonic modes that allows us to treat infinite channels with one classical equation can be exploited further.
In our calculations we have often represented the distribution of multipole strength by low-lying collective states and high modes in the
so called giant resonances region. The coupling of the latter to a
background of two-particle two-hole and more complicated states results
in the spreading of the strength over many states within a certain
energy interval. The coupling to all these channels can be incorporated in the scheme by introducing a damping term in the equation for the
otherwise purely harmonic forced vibration representing the mode. This
prescription has full validity provided that multiple excitation of
giant resonances does not take place to an appreciable extent. The
energy (and angular momentum) taken away from the modes through damping
is supposed to be irretrievably lost to the other degrees of freedom.

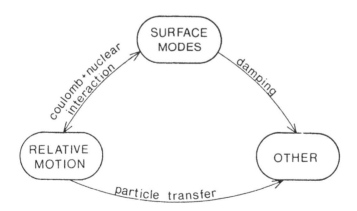

Fig. 1. Pictorial representation of the calculation scheme described
in the text.

The picture that emerges from these considerations is schematically illustrated in fig. 1. As can be inferred from the drawing we have later added in our calculations another one-way street connecting the variables of relative motion with the "other" degrees of freedom. This coupling represents the reduction in relative momentum due to particle transfer and is treated in the one-body dissipation proximity approximation [2]. I shall not talk about this subject partly because I am not quoting today any results and partly because it has been covered in detail this morning by J. Huizenga [3]. In connection with his presentation, however, I would like to point out that one should not conclude from the fact that particle transfer can account for the observed energy losses that there are no other competing mechanisms. Actually, in many instances either the upper or lower branches in fig. 1. could account by themselves for the total kinetic energy loss in relative motion. It is only by considering both mechanisms that one may begin to answer questions about their relative importance.

Let me just list a few problems (and the corresponding references) in which the model has been applied with encouraging results.
- Calculation of fusion cross sections [4].
- Analysis of polarization experiments [5].
- Estimation of non-planar scattering and angular momentum alignment [6].
- Calculation of energy and angular distributions for different reactions [1].

In addition, an illustrative comparison with results of a different theoretical approach (TDHF) was carried out in reference [7].

Two persistent features are uncomfortably present in most calculations, i) the spread of the mass distributions is largely underestimated, ii) angular distributions show too much cross section at backward angles. This last result may be, to some extent, related to the fact we have consistently used the proximity potential of reference [8]. The specification of the ion-ion potential is not essential to the application of the method. On the other hand, the whole chapter of the mass transfer is under scrutiny and a major revision may be needed.

In our calculations, the large values of energy loss associated with deep inelastic processes result from multiple excitation of collective modes and from particle transfer. In this regime it is rather difficult to relate features of the observed energy and angular distributions to the excitation of any specific degree of freedom. On the other hand, when the average energy loss is of the order of 10-20 MeV

we expect single-phonon excitation to take place with a considerable
probability. As this happens for grazing collisions, the study of this
class of peripheral reactions provides with a unique opportunity to put
in evidence the spectra of collective modes of the colliding ions.

I would like to close this brief review by presenting the results
of the analysis of peripheral collisions in the reaction $^{16}O + ^{208}Pb$
at the bombarding energies of 160, 315 and 640 MeV. Particle transfer
was taken into account in these calculations as a depopulation of the
initial mass partition in the way described in reference [9]. In fig.2
we can see how the optimal conditions for observation of high-lying
modes occur, at a given energy, near or at the grazing angle ($\theta_g = 15^O$
at 315 MeV). As we move away from the grazing angle, the contribution
of multiple excitation to the energy spectrum becomes predominant.

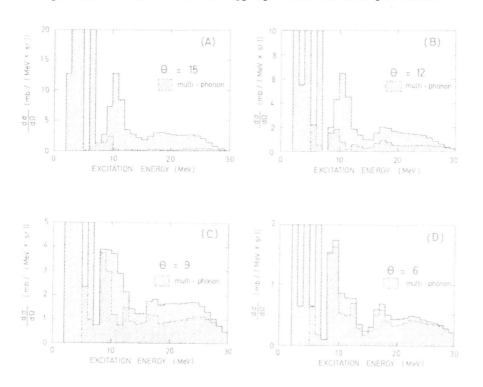

Fig. 2. Distribution of cross section as a function of excitation
energy for the reaction $^{16}O + ^{208}Pb$ at 315 MeV. The multiple-phonon
background becomes predominant as the observation angle θ moves away
from the grazing value $\theta_g = 15^\circ$.

The qualitative changes in the contributions to the energy spectrum which arise from increasing bombarding energies were also studied.
The results of the model calculation are shown in fig. 3. In the

upper part of the figure a significant increase in the contribution of
the Coulomb excitation is observed. This may be of interest given the
fact that the electromagnetic coupling favors the excitation of quadrupole and octupole modes. A concrete possibility of resolving the different multipolarities which contribute to the structures about 10 MeV
and 18 MeV is thus suggested. In the lower part of the figure it is
possible to appreciate a substantial decrease in the relative importance of the multiple-phonon excitations for higher bombarding energies.

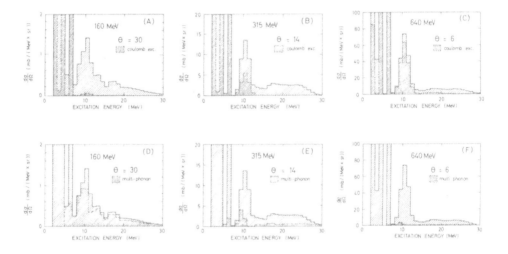

Fig. 3. Contribution of Coulomb excitation (A-C) and multiple-phonon
excitation (D-F) for the reaction $^{16}O + ^{208}Pb$ at the bombarding energies of 160, 315 and 640 MeV. In all cases an observation angle θ
close to the grazing was chosen.

The qualitative trends indicated in fig. 3 make it appealing to
search for excitation of high-lying modes in peripheral collisions for
energies in the range of 40-80 MeV/nucleon. However, one should not
forget that for higher energies the cross section concentrates more
and more in forward angles. Besides, the very violent nature expected
for the more central collisions at these energies may impose severe experimental restrictions.

References:

[1] R.A. Broglia, C.H. Dasso and A. Winther, Proceedings of the
     International School of Physics Enrico Fermi, Course LXXVII,
     to be published (available in preprint).
[2] J. Randrup, Nucl.Phys. A307(1978)319.
[3] J.R. Huizenga, these proceedings.
[4] R.A. Broglia, C.H. Dasso, G. Pollarolo and A. Winther, Phys. Rev.
     Lett. 40(1978)707.
[5] R.A. Broglia, S.Afr. Tydskr. 1(1978)173.
[6] R.A. Broglia, C.H. Dasso, G. Pollarolo and T. Døssing, Phys. Rev.
     Lett. in press.
[7] R.A. Broglia, C.H. Dasso, A.K. Dhar, H. Esbensen and B. Nilsson
     Phys. Lett. 83B(1979)301.
[8] R. Blocki, J. Randrup, W.H. Swiatecki and C.F. Tsang, Ann. Phys.
     105(1977)427.
[9] R.A. Broglia, C.H. Dasso, H. Esbensen, G. Pollarolo, A. Vitturi
     and A. Winther, NBI-79-22 and Phys. Lett. in press.

TRANSPORT THEORY AND DOORWAY CONFIGURATION IN DISSIPATIVE

HEAVY-ION COLLISIONS

W. Nörenberg

Gesellschaft für Schwerionenforschung (GSI)
Darmstadt, F.R. Germany

The evolution of the collision complex can be divided roughly into three stages: mutual approach of the nuclei, local equilibration and slow relaxation of macroscopic degrees of freedom. The last stage has been extensively studied within the framework of transport theories. The main steps in the derivation and application of a Fokker-Planck equation for the slowly varying macroscopic (collective) variables are discussed. Mass transfer and dissipation of relative angular momentum are well described by the theory. In order to bridge the gap between the initial stage and the third stage an explicitly time-dependent dynamical potential in the relative motion is introduced. This potential is due to the initial correlations which arise from the ground-state configurations of the approaching nuclei. The generalized Fokker-Planck equation is applied to compound-nucleus formation and dissipative collisions. The possible existence and some characteristic features of a long-living component of dissipative collisions are studied.

1. Introduction

As illustrated in fig. 1, the total process can be roughly divided into three stages.

(1) The initial stage is characterized by the mutual approach of the nuclei in their ground states. The motion of the nucleons during this first stage of the collision is expected to be governed by their self-consistent single-particle potential which evolves slowly in time. This stage should therefore be well described by the time-dependent Hartree-Fock theory (TDHF).

(2) By residual interactions the Slater determinant of time-dependent single-particle states of TDHF decays to more complex configurations. This decay leads to a local statistical equilibrium. At the end of this decay the system occupies the total phase space (total configuration space) which is available for fixed values of the macroscopic (collective) degrees of freedom.

(3) During the third stage the slow macroscopic (collective) variables relax towards

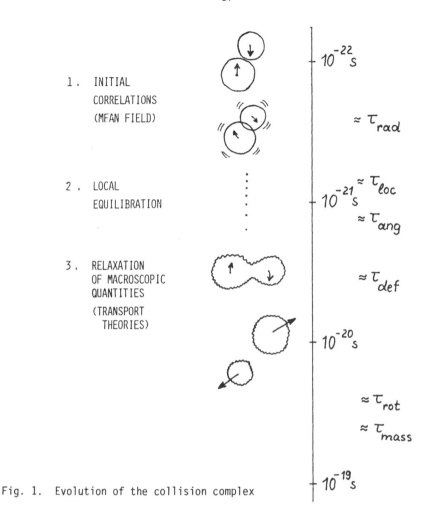

1. INITIAL
   CORRELATIONS
   (MEAN FIELD)

10^{-22} s

$\approx \tau_{rad}$

2. LOCAL
   EQUILIBRATION

$\approx \tau_{loc}$
10^{-21} s
$\approx \tau_{ang}$

3. RELAXATION
   OF MACROSCOPIC
   QUANTITIES
   (TRANSPORT
   THEORIES)

$\approx \tau_{def}$

10^{-20} s

$\approx \tau_{rot}$

$\approx \tau_{mass}$

10^{-19} s

Fig. 1. Evolution of the collision complex

their equilibrium distributions. The complexity of the wave function in this stage of the process suggests to take advantage of random properties of matrix elements as implied by the local statistical equilibrium. This leads naturally to the formulation of transport theories.

2. Derivation of transport equations

Transport theories of dissipative heavy-ion collisions have been derived by several groups [1-5]. A review has been given recently by Weidenmüller [6]. All derivations start from an assumed separation of the degrees of freedom into slow macroscopic (collective) and fast equilibrating "intrinsic" variables. The resulting local sta-

tistical equilibrium is either represented by a heat bath of given temperature [2,4]
or by random properties of the coupling matrix elements [1,3,5].

Let me illustrate the successive steps in the derivation of transport equations.
For simplicity we treat the relative motion classically and assume that the trajectory
$\vec{r}(t)$ and the coupling $V(t)$ between excited states of the interacting nuclei are known.
This semiclassical treatment of relative motion [1,7,8] has been generalized in [5]
by including the relative variables explicitly in the equation of motion.

(1) By dividing the total channel space into subsets $\mathcal{H}_\mu$ we can define the macroscopic occupation probabilities

$$f_\mu(t) \equiv \sum_{m \in \mathcal{H}_\mu} \rho_{mm}(t) \qquad (1)$$

which denote the total occupation probability of the subset $\mathcal{H}_\mu$. Rewriting the Liouville equation for the density matrix $\rho(t)$, yields the generalized master equation
[1]

$$\frac{d}{dt}f_\mu(t) = I_\mu(t,t_0) + \sum_\nu \int_0^{t-t_0} d\tau \, K_{\mu\nu}(t,\tau)d_\mu f_\nu(t-\tau) . \qquad (2)$$

The first term on the r.h.s. represents the contribution from those parts of the
initial density matrix $\rho(t_0)$ which are not contained in $f_\mu(t)$. The second term on
the r.h.s. describes the coupling between different subsets. In analogy to the Boltzmann equation it is referred to as the collision term. The collision term consists of
gain terms ($\nu \neq \mu$) which correspond to transitions from all subsets $\nu \neq \mu$ to the subset μ, and another term $\nu = \mu$ which describes the loss from the subset μ. The kernel
$K_{\mu\nu}(t,\tau)$ is non-local in time. The change of the macroscopic occupation probabilities at time t depends on all earlier times $t-\tau>t_0$. Because of this property, the
kernel and its characteristic decay time are referred to as the memory kernel and
the memory time $\tau_{mem}^{\nu\mu}$, respectively.

(2) The second step in the derivation of a transport equation is the introduction of
random properties of the coupling matrix $V(t)$. This leads to the following structure
of the memory kernel,

$$K_{\mu\nu}(t,\tau) = \hbar^{-2}<V_{mn}(t)V_{nm}(t-\tau)>_{\nu\mu} G_{\nu\mu}(t,t-\tau) + c.c. . \qquad (3)$$

The mean correlation function $<V_{mn}(t)V_{nm}(t-\tau)>_{\nu\mu}$ decays as function of τ within a
characteristic time $\tau_{cor}^{\nu\mu}$ (correlation time). This correlation time is related to a
correlation length σ by $\tau_{cor} = \sigma/|\vec{r}|$. The correlation length σ is typically given
by the mean distance between the nodes of the single-particle wave functions. For

strong overlap, $\sigma \approx 1$ fm, while in the nuclear surface $\sigma \approx (3...4)$ fm [9]. Therefore, $\tau_{cor} \gtrsim 10^{-22}$s. The mean propagator $G_{\nu\mu}(t,t-\tau)$ for a state $n\in\mathcal{H}_\nu$ and a state $m\in\mathcal{H}_\mu$ has a lifetime denoted by $\tau_{dec}^{\nu\mu}$. The relative values of $\tau_{cor}^{\nu\mu}$ and $\tau_{dec}^{\nu\tilde{\mu}}$ determine the two limits of weak and strong coupling. The weak-coupling limit is defined by

$$\tau_{mem}^{\nu\mu} \equiv \tau_{cor}^{\nu\mu} \ll \tau_{dec}^{\nu\mu} \qquad (4)$$

and the strong-coupling limit by

$$\tau_{mem}^{\nu\mu} \equiv \tau_{dec}^{\nu\mu} \ll \tau_{cor}^{\nu\mu} . \qquad (5)$$

Calculations [10] of the decay time for small overlap give values $\tau_{dec} \gtrsim 5\cdot 10^{-23}$s. Thus, in the basis of asymptotic eigenstates $\tau_{dec} \ll \tau_{cor}$ such that the strong-coupling limit is applicable for dissipative collisions. The initial correlations $I_{\mu\nu}(t,t_0)$ have the same structure as the memory kernel (3). It is usually assumed that the resulting transport equation is applied for $t-t_0 \gg \tau_{mem}^{\nu\mu}$ where $I_{\mu\nu}(t,t_0)$ can be neglected. The memory time characterizes the time interval during which the phase correlations are lost.

(3) As discussed at the beginning of this section, the macroscopic variables have to be chosen as the slow modes of the system such that the relaxation times for the macroscopic variables satisfy $\tau_{relax} \gg \tau_{mem}^{\nu\mu}$. This justifies the Markov approxima-tion. It consists in replacing $f_\nu(t-\tau)$ by $f_\nu(t)$ in eq. (2) and yields the master equation for $t-t_0 \gg \tau_{mem}^{\nu\mu}$,

$$\frac{d}{dt} f_\mu(t) = \sum_{\nu\neq\mu} w_{\mu\nu}(t) [d_\mu f_\nu(t) - d_\nu f_\mu(t)] \qquad (6)$$

with the transition probability $w_{\mu\nu}(t) \equiv \int_0^\infty d\tau K_{\mu\nu}(t,\tau)$. Such a master equation was originally introduced by Pauli [11] in order to justify the $\mathcal{H}$-theorem from the quantum-mechanical point of view. Moretto and Sventek [12] have directly applied such a master equation to the diffusion of nucleons between the colliding nuclei.

3. Fokker-Planck equation, transport coefficients and relaxation phenomena

A particularly convenient method of describing the solutions of the master equation (6) has been introduced by transforming the integral equation into a second-order differential equation, the Fokker-Planck equation [7,13]

$$\frac{\partial f(\vec{y},t)}{\partial t} = -\sum_{i=1}^{g} \frac{\partial}{\partial y_i} \{v_i(\vec{y},t)f(\vec{y},t)\} + \sum_{i,j=1}^{g} \frac{\partial^2}{\partial y_i \partial y_j} \{D_{ij}(\vec{y},t)f(\vec{y},t)\}. \qquad (7)$$

Here we have replaced the discrete variables μ by the continueous variable $\vec{y}$ with g components. The transport coefficients v_i (drift coefficients) and D_{ij} (diffusion coefficients) are completely determined by the second moments of the transitions probabilities and the density of states. Drift and diffusion coefficients are related by (generalized) Einstein relations [8, 13-15].

The transport coefficients have been calculated for three macroscopic variables: the mass A_1 of one fragment (mass-asymmetry variable), the z-component M of the total intrinsic angular momentum and the total excitation energy E^* [14]. On the other hand experimental values for the transport coefficients have been obtained by analyzing data on various reactions [8, 15]. Figure 2 shows as an example the fit (solid curve) of the mass-transport coefficients to the data for Kr (5.99 MeV/u) + Er. Using the theoretical values for the transport coefficients we obtain the dashed line.

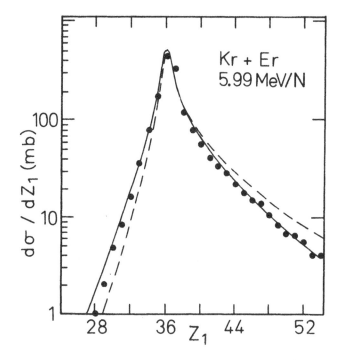

Fig. 2. Calculated element distributions for the reaction [86]Kr (5.99 MeV/u) + [166]Er compared with experimental data [18]. The dashed curve corresponds to theoretical values of drift and diffusion coefficients, the solid curve to a fit by adjusting only the drift coefficient. From [19].

Figure 3 summarizes experimental and theoretical values of the diffusion coefficient for various reactions [15]. We realize that the dependence of the diffusion coefficient on bombarding energy, total mass and mass asymmetry is well described theoreti-

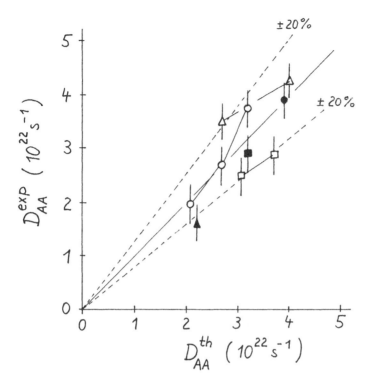

Fig. 3. Comparison of experimental and theoretical values for the mass diffusion
coefficient. The reactions are: (o) ^{86}Kr (515, 619, 703 MeV) + ^{166}Er, (□)
^{136}Xe (900, 1130 MeV) + ^{209}Bi, (△) ^{208}Pb (1456, 1560 MeV) + ^{208}Pb, (■)
^{84}Kr (712 MeV) + ^{209}Bi, (▲) ^{132}Xe (779 MeV) + ^{120}Sn (●) ^{208}Pb (1560 MeV)
+ ^{238}U.

cally. Figure 4 compares the calculated angular momenta of the fragments with results
from γ-multiplicity measurements. The dip around the projectile charge is understood
as follows: The fragments close to the projectile are predominantly produced in collisions with large l-values where the interaction time and hence, the dissipated angular momentum is small. Sufficiently far away from the projectile-charge number, the
dissipated angular momentum saturates. These fragments are mainly populated in collisions with l-values where the interaction time is large enough to reach sticking.

Returning to the time evolution of the collision complex let us discuss the
characteristic times envolved in the process (cf. fig. 1). From the analysis of
angular distributions it is possible to deduce nuclear interaction times [16]. These
interaction times range from about 10^{-22}s for grazing collisions up to several 10^{-21}s
for close collisions (with small impact parameters b << b_{gr}). From a fit to experimental energy spectra the relaxation times for the loss of radial kinetic energy

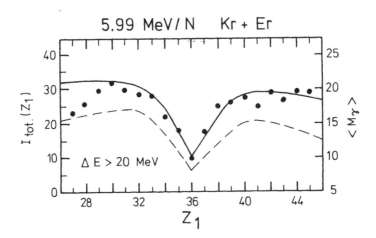

Fig. 4. The total angular momentum $I_{tot}(Z_1)$ of the fragments as function of the ragment charge number and corresponding γ-multiplicity data from [18] for the reaction ^{86}Kr (5.99 MeV/u) + ^{166}Er. The dashed curve is obtained by neglecting fluctuations. From [19].

(τ_{rad}) and for the evolution of fragment deformations (τ_{def}) have been determined [17]. In this analysis the theoretical relaxation time τ_{ang} for the dissipation of relative angular momenta has been used (cf. fig. 4). The values for the relaxation times ($\tau_{rad} \approx 0.3 \cdot 10^{-21}$ s, $\tau_{ang} \approx 1.0 \cdot 10^{-21}$ s, $\tau_{def} \approx 4 \cdot 10^{-21}$ s) imply that the fast loss of radial kinetic energy is followed by the dissipation of relative angular momentum and finally by the evolution of fragment deformations. The analysis of mass distributions (cf. fig. 2) show that no equilibrium is reached in the mass-asymmetry coordinate for typical interaction times. The corresponding equilibration time ($\tau_{mass} \approx 2 \cdot 10^{-20}$ s) is larger by an order of magnitude.

We may summarize the transport-theoretical results obtained so far by the statement that we understand best the slow processes. The success of transport theories can be attributed to the short duration of the initial stages as compared to the third stage. For a more complete understanding of dissipative heavy-ion collisions, however, it is necessary to bridge the gap between the treatment of the initial and the final stages. In the following we outline some basic considerations which lead to the introduction of an explicitly time-dependent potential in the relative motion.

4. Dynamical potential and local equilibration

During the fast approach of the nuclei the individual nucleons cannot follow the

lowest possible (adiabatic) levels. The nucleonic wave functions stay essentially unchanged (conservation of the number of nodes). Such a 'diabatic' behaviour is encountered also for electrons in atom-atom collisions. The motion of nucleons on diabatic
levels gives rise to a large potential energy in addition to the adiabatic potential.
This additional potential can be estimated from the inspection of two-center shell-
model calculations. From a more schematic consideration of correlation diagrams we
obtain

$$(\Delta U)^o_{diab} \simeq 30 \cdot A_1^{1/3} \text{ MeV} \tag{8}$$

for the additional repulse potential at the compound-nucleus shape and for $A_1 = A_2$.
The occurrence of an additional repulsive potential has been recognized also from
TDHF calculations [20, 21].

The doorway configuration which is formed during the fast approach of the two
nuclei can be considered as a highly correlated state of (n particle - n hole) excitations with respect to the adiabatic configuration. This correlation is effectively
lost by the decay via residual interactions. The time which is necessary to obtain
local statistical equilibrium between all excited states at a given shape, is denoted
by τ_{loc}. Estimates for this local equilibration time can be obtained from the decay
time of one of the particle-hole states or from the corresponding time in precompound
reactions. These considerations lead to $\tau_{loc} \simeq 10^{-21}$s.

5. Correlation parameter and consequences for transport theories

For describing the effect of the doorway configuration and its decay, we introduce an order parameter χ which measures the degree of correlation with respect to
the entrance channel. It is defined to be one initially and approaches zero for
$t \gg \tau_{loc}$. This parameter is an additional macroscopic variable which has to be taken
into account explicitly in the transport theory [5]. Only if τ_{loc} would turn out to
be much smaller than all other characteristic times of the collision process we could
neglect χ. As a consequence the relative motion of the nuclei would be determined by
the adiabatic potential. Since we expect τ_{loc} to be of the order of 10^{-21}s, such an
approximation seems not to be justified.

We assume in the following that the correlation parameter $\chi(t)$ is given by

$$\chi(t) = \exp \left[-\frac{1}{\tau_{loc}} \int_{t_o}^{t} f(r(t'))dt' \right] \tag{9}$$

where we regard τ_{loc} as an unknown parameter. The integral over the form factor $f(r)$

smoothly switches on the equilibration. The quantities $r(t)$ and t_0 denote respectively the mean trajectory and a time well before the collision. As compared to the transport theory formulated in [2, 3, 5], the essential new feature is the introduction of an explicitly time-dependent dynamical potential,

$$U_{dyn}(r,t) = U_{ad}(r) \; [1-\chi(t)] + U_{diab}(r) \tag{10}$$

where U_{ad} and U_{diab} denote the adiabatic and diabatic potentials, respectively. The relative motion of the nuclei and the transfer of nucleons is described by the Fokker-Planck equation

$$\partial f \; (r,p; \; \theta,l; \; \alpha; \; t) \; / \partial t \;\; =$$

$$- \frac{p}{\mu} \frac{\partial f}{\partial r} + \frac{\partial U_{dyn}}{\partial r} \frac{\partial f}{\partial p} - \frac{\partial}{\partial p} (v_p f) + \frac{\partial^2}{\partial p^2} (D_{pp} f)$$

$$- \frac{1}{\mathcal{J}_{rel}} \frac{\partial f}{\partial \theta} - \frac{\partial}{\partial l} (v_1 f) + \frac{\partial^2}{\partial l^2} (D_{11} f) + \frac{\partial^2}{\partial l \partial \alpha} (D_{1\alpha} f)$$

$$- \frac{\partial}{\partial \alpha} (v_\alpha f) + \frac{\partial^2}{\partial \alpha^2} (D_{\alpha\alpha} f) \tag{11}$$

with the reduced mass μ. This equation is written in the variables of radial momentum p, relative angular momentum l and mass asymmetry $\alpha = (A_1-A_2)/(A_1+A_2)$. In addition, the Fokker-Planck equation (4) includes the relative distance r, the angle of rotation θ and the effect from the force $-\partial U_{dyn}/\partial r$ on the distribution function f. The transport coefficients are calculated from the expressions of [14] with modifications arising from the treatment of the radial motion. In particular, a form factor is introduced in accordance with the numerical calculations of [9]. For the adiabatic potential we use the results of Möller and Nix [22]. The diabatic potential is given by adding to the adiabatic potential the central value (8) with an adequate form factor. The proximity form for the mass-asymmetry dependence is used for calculating $(\Delta U)^0_{diab}$ for different projectile-target combinations,

$$(\Delta U)^0_{diab} \simeq 60 \; \frac{A_1^{1/3} \; A_2^{1/3}}{A_1^{1/3} + A_2^{1/3}} \; \text{MeV} \; . \tag{12}$$

For Ar + Pb this gives 130 MeV. Effects from the dynamical potential are present directly in the relative motion of (11) but also in the transport coefficients. Here, the dynamical potential enters via the effective exitation energy

$$E^*_{eff} = E - U_{dyn} - E^{rel}_{kin} \tag{13}$$

which is available as heat. Whereas the transport coefficients v_p, D_{pp}, v_1, D_{11}, $D_{1\alpha}$,

D_{pp} are only affected via the effective temperature $T_{eff} \equiv (E^*_{eff}/a)^{1/2}$ with the level-density parameter a, the mass-drift coefficient v_α becomes explicitly proportional to the dynamical force $-\partial U_{dyn}/\partial \alpha$ and hence should directly show the local equilibration. An indication of such an effect is shown in fig. 5 where the mean values of the element distribruion for ^{86}Kr (5.99 MeV/u, 8.18 MeV/u) + ^{166}Er are plotted as functions of the total kinetic energy loss ΔE and the interaction time τ_{int}, respectively. At both bombarding energies a local equilibration time of the order 10^{-21}s is indicated.

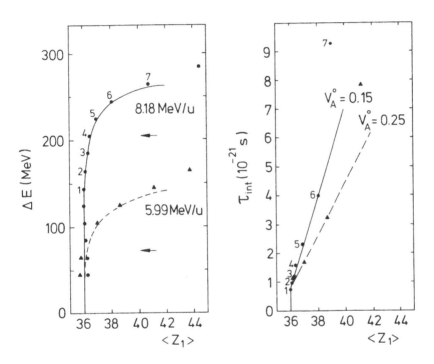

Fig. 5.   Correlation of the mean values $<Z_1>$ of the element distributions with the total kinetic energy loss ΔE and the interaction time τ_{int}, respectively, for the reactions ^{86}Kr (5.99 and 8.18 MeV/u) + ^{166}Er. From [17].

6.   Numerical results for the compound nucleus ^{248}Fm

    For the numerical treatment of the Fokker-Planck equation we introduce a moment expansion. Figure 6 shows three different trajectories for ^{40}Ar (400 MeV) + ^{208}Pb in the (r,α)-plane for a local equilibration time 0.75 $\cdot$ 10^{-21}s. The trajectory with the smallest l-value (l=51) leads to the formation of a compound nucleus. The largest l-value (l=104) corresponds to a fast process known as dissipative (or deeply inelastic) collision. The trajectory with l = 102, although similar to that with l = 104 for

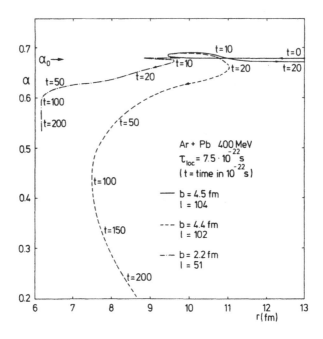

Fig. 6. Trajectories for three l-values in the collision ^{40}Ar (400 MeV) + ^{208}Pb. From [23].

small times, is captured. In contrast to the trajectory with l = 51 it does not lead to a compound nucleus because l > $l_{crit} \approx 70$ (largest l-value for the existence of the compound nucleus ^{248}Fm). Instead, the system develops towards mass symmetry (α=0) and is expected to split into two fragments if deformations of the fragments are allowed. This long-living component of dissipative collisions results after the system has rotated several times and hence, exhibits essentially a 1/sinθ angular distribution like the fragments from compound-nucleus fission. In contradistinction to compound-nucleus fission, the mass distribution of this long-living component of dissipative collisions is expected to be broader because it is not limited by the saddle-point shape.

We can divide the reaction cross-section schematically according to the l-values. For 0 < l < l_{crit} we have compound-nucleus formation if the critical l-value l_{cap} for capture (here $l_{cap} \approx 103$) is larger than l_{crit}. For $l_{cap} < l_{crit}$ compound-nucleus formation is limited by l_{cap}. The long-living component of dissipative collisions is expected for $l_{crit} < l < l_{cap}$. Figure 7 shows the capture cross-section in units of

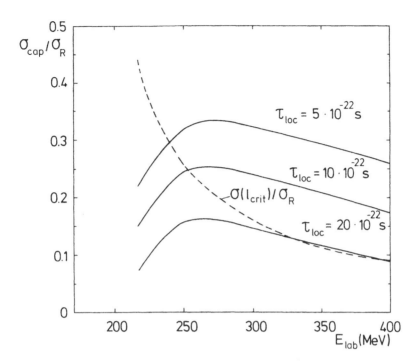

Fig. 7. Capture cross-sections for different local equilibration times τ_{loc}. From [23].

the reaction cross-section as function of the bombarding energy and for different values of the local equilibration time. For $\sigma_{cap} < \sigma(l_{crit})$ the capture leads to compound-nucleus formation. For small bombarding energies the compound-nucleus formation is strongly suppressed. It reaches the maximal value at the crossing point with the dashed line which corresponds to l_{crit}. Above this cross-over a long-living dissipative collision occurs. The threshold for this component is well above the interaction barrier. According to the preliminary results of fig. 7 this long-living component occurs for $l_{loc} < 2 \cdot 10^{-21}$ s.

Experimental evidence for the existence of the long-living dissipative component have been reported for ^{238}U (5.7 MeV/u) + ^{48}Ca [24], ^{132}Xe (5.9 MeV/u) + ^{56}Fe [25], ^{20}Ne + ^{nat}Re and ^{40}Ar + ^{165}Ho [26]. A threshold somewhat above the reaction barrier as indicated in fig. 7 has been observed.

7. Concluding remarks

Dissipative collisions play a dominant role in heavy-ion reactions and reveal new nuclear properties which are connected with mass transfer, kinetic-energy loss and angular-momentum dissipation. These processes represent an interesting many-body problem at rather high excitation energies. In contrast to nuclear spectroscopy dissipative heavy-ion collisions supply information about relaxation phenomena in nuclei. Such phenomena have also been observed in precompound reactions and in fission, but only in heavy-ion collisions a rich variety of such cooperative phenomena has been discovered. In this respect the study of dissipative collisions has opened a new field of nuclear research.

The transport theories as formulated up to now for dissipative heavy-ion colli-sions are subject to three major restrictions:

(i) Restriction due to the choice of collective and macroscopic variables: For the sake of simplicity one takes into account only a few collective degrees of freedom (for example, the relative distance between the centers of the fragments) and sóme macroscopic variables which characterize the observable intrinsic properties of the fragments (for example, excitation energy, mass asymmetry and intrinsic angular mo-mentum). It is clear that further collective coordinates like the deformations of the fragments, and maybe also additional macroscopic variables are necessary for a more complete understanding of the process.

(ii) Restriction to local statistical equilibrium: For fixed values of collective and macroscopic variables it is assumed that all intrinsic degrees are populated accor-ding to their statistical weight.

(iii) Restriction due to the choice of basis: The derivations of transport equations assume complete randomness of the coupling between different basis states. This is only partly a consequence of the assumption (ii) and goes beyond it because the ran-domness assumption within a particular basis (for example the eigenstates of the separated nuclei) artificially eliminates correlations which might be important.

The restriction (ii) to local statistical equilibrium limits the applicability of present formulations of transport theory to the later stages of the total process. We have introduced a dynamical potential which is explicitly time-dependent. This dynamical potential smoothly connects in time the initial repulsive-core potential and the adiabatic potential which governs the motion after local equilibrium is established. The inclusion of such a dynamical potential in the microscopic transport theory extends the applicability of the theory to the initial stage. The mechanism of energy loss becomes two-fold. Part of the kinetic energy is lost directly by friction. Another part is first stored as potential energy (dynamical potential) and then transformed into heat via residual interactions. Thus we expect the relative motion to be quite different from the treatment by friction forces and the adiabatic potential.

This concept has been applied to heavy-ion collisions describing both compound-nucleus formation and dissipative collisions. A long-living component of dissipative collisions (DIS II) is found to exist beside the well established fast component (DIS I). The DIS II component is characterized by similarities to compound-nucleus fission. It differs from compound-nucleus fission by a broader mass distribution and by a threshold in bombarding energy which lies well above the interaction barrier.

References

1. W. Nörenberg, Z. Physik A 274, 241 (1975) and 276, 84 (1976)
2. H. Hofmann and P.J. Siemens, Nucl. Phys. A 257, 165 (1976) and 275, 464 (1977)
3. D. Agassi, C.M. Ko and H.A. Weidenmüller, Ann. Phys. (N.Y.) 107, 140 (1977)
4. J. Randrup, Nucl. Phys. A 307, 319 (1978) and Nucl. Phys. A 327, 490 (1979)
5. S. Ayik and W. Nörenberg, Z. Physik A 288, 401 (1978) and in preparation
6. H.A. Weidenmüller, Progress in Nuclear and Particle Physics (in press)
7. W. Nörenberg, Proc. of the Predeal International School on Heavy Ion Physics, Predeal, Romania, 1978, ed. by A. Berinde et al. (Central Institute of Physics, Bucharest, 1978) p. 825 and GSI-Report 79-5
8. W. Nörenberg and H.A. Weidenmüller, Introduction to the Theory of Heavy-Ion Collisions, Lecture Notes in Physics, second edition (Springer-Verlag, Berlin - Heidelberg - New York) in press
9. B.R. Barrett, S. Shlomo and H.A. Weidenmüller, Phys. Rev. C 17, 544 (1978)
10. S. Ayik, B. Schürmann and W. Nörenberg, Z. Physik A 277, 299 (1976)
11. W. Pauli, in "Probleme der modernen Physik", Festschrift zum 60. Geburtstag A. Sommerfelds, ed. by P. Debye (Hirzel-Verlag, Leipzig, 1928) p. 30
12. L.G. Moretto and J.S. Sventek, Phys. Lett. 58B, 26 (1975); J.S. Sventek and L.G. Moretto, Phys. Lett. 65B, 326 (1976)
13. W. Nörenberg, Phys. Lett. 52B, 289 (1974)
14. S. Ayik, G. Wolschin and W. Nörenberg, Z. Physik A 286, 271 (1978)
15. A. Gobbi and W. Nörenberg, in Heavy-Ion Collisions vol. II, ed. by R. Bock (North-Holland, Amsterdam, 1979) in press
16. G. Wolschin and W. Nörenberg, Z. Physik A 284, 209 (1978)
17. C. Riedel, G. Wolschin and W. Nörenberg, Z. Physik A 290, 47 (1979)
18. A. Olmi, H. Sann, D. Pelte, Y. Eyal, A. Gobbi, W. Kohl, U. Lynen, G. Rudolf, H. Stelzer and R. Bock, Phys. Rev. Lett. 41, 688 (1978)
19. G. Wolschin and W. Nörenberg, Phys. Rev. Lett. 41, 691 (1978)
20. H. Flocard, S.E. Koonin and M.S. Weiss, Phys. Rev. C 17, 1682 (1978) P. Bonche, B. Grammaticos and S.E. Koonin, Phys. Rev. C 17, 1700 (1978)
21. A.K. Dhar, private communication (March 1979)
22. P. Möller and J.R. Nix, Nucl. Phys. A 281, 354 (1977)
23. W. Nörenberg and C. Riedel, Z. Physik A 290, 335 (1979)
24. H. Sann, private communication
25. B. Heusch, C. Volant, H. Freiesleben, R.P. Chestnut, K.D. Hildenbrand, F. Pühlhofer, W.F.W. Schneider, B. Kohlmeyer, W. Pfeffer, Z. Physik A 288 (1978) 391
26. C. Lebrun, F. Hanappe, J.F. Lecolley, F. Lefebvres, C. Ngô, J. Péter and B. Tamain, Nucl. Phys. A 321 207 (1979)

LINEAR RESPONSE THEORY OF DEEPLY INELASTIC COLLISIONS

by

Helmut Hofmann

Physik-Department, Technische Universität München
8046 Garching, Federal Republic of Germany

Abstract

We present the linear response approach as to be
consistent with a model in which collective motion
of large amplitude is approximated locally by a harmonic one. In this model the transport coefficients
are not constants but depend on local collective
quantities like the mean value of the coordinates
and the frequency, as well as on a time dependent
temperature. We describe how the transport coefficients can be calculated within a realistic microscopic model. In this model we account properly
for the quantum behavior of the nucleons inside
the nucleus and the presence of residual two-body
forces. In the whole picture the relative motion
of the fragments is treated on equal footing with
the dynamics of other collective degrees of freedom
of both the fragments and the composite system.

I. Introduction

In this contribution I want to give a brief review on a theory of DIC
on which we have been working since a few years.[+1] The main emphasis
shall be on the discussion of the basic physical picture, the justi-
fication of the approach and its internal consistency.

The essential features of our procedure can be characterized as
follows:

i) We aim at the computation of multidimensional cross sections. This
   requires a proper treatment of relative motion.

ii) We do this by means of introducing a set of collective coordinates
   Q_μ and momenta P_μ which are related to measured quantities like:
   kinetic energy of relative motion, angles, spins, mass and charge
   asymmetry degrees of freedom etc.

iii) We claim that the transport coefficients (which enter the equation
   of motion for the $\{Q_\mu\}$ can be calculated within realistic, microscopic models.

Our theory is meant to take account of the following constraints:

i) We want to describe an experimental situation which we encounter
   in reactions where the kinetic energy of relative motion is typically
   of the order of a few MeV per nucleon. As the most striking features
   of these experiments appears to be the almost complete dissipation
   of this kinetic energy into high intrinsic excitation and, in many
   cases, the considerable transfer of mass and charge.

ii) It is good to remember that the scattered objects are nuclei and
   to face the problems which arise from this fact: 1) We shall have
   to deal with problems of non-equilibrium thermodynamics of a small
   system. 2) The nucleons inside the nuclei behave like quantum objects and not like classical particles. 3) In some cases quantum
   features are important even for collective degrees. (As an example
   we mention high frequency modes for which the application of classical statistical mechanics is inadequate (see below)). 4) The introduction of collective coordinates underlies a self-consistency requirement.

[+]At this opportunity I would like to thank A.S. Jensen, C.Ngô and P.J.
Siemens for their close collaboration.

iii) As the last, but not least important, constraint appears the
feasibility of the approach: If not prevented by the conditions
above, we should try to get transport equations which can be
solved and the (transport) coefficients of which can be calculated within realistic models.

II. Basic Concepts of the Theory

To develop the theory we have to consider first the constraints. Let
us begin with the feasibility. Even if we restrict ourselves to an
explicit treatment of the dynamics of the collective degrees only:
their full, exact equation will be tremendously complicated. It would
definitely be a integro-differential equation[2] with a complex structure of all the different terms. But even if we are able to reduce
it to a partial differential equation this might still be too complicated to allow for a feasible solution. As an example, think of a Fokker-
Planck equation of one variable, excluded the conjugate momentum.
Suppose the force is a non-linear function of the collective coordinate.
For DIC as well as for fission, this case is certainly very realistic.
Already in this simple case a solution of the Fokker-Planck equation
which is close to the correct one requires special and complicated
techniques, and some of them are applicable only in special cases.

What is feasible? We certainly can solve classical equations of motion,
even for non-linear forces. But this is insufficient; we also want to
know the fluctuations in order to be able to calculate the measured
variances. Let us, therefore, adopt the following approximation scheme:
We describe the complicated collective motion by means of the mean
values (first moments) and the second moments of the Q_μ , and P_μ . This
amounts to say: The first moments we shall obtain by solving the
classical equations of motion, the second moments we shall obtain by
approximating the complicated dynamics locally by a harmonic motion.

Conditions for this approximation scheme

Any dynamics can be approximated locally by a harmonic motion if the
time interval δt is sufficiently short. This statement is almost evident for classical Newtonian dynamics. But it is equally correct for
the time evolution of a quantum system. This observation can be most
easily inferred from Feynman's method of calculating the evolution
operator: for infinitesimal times this operator is determined by the
classical Lagrangian[3] .

In our case there is a lower limit for a δt, in which we can study the collective motion. This limit is dictated by our desire that we want to get rid of memory effects coming from the intrinsic, nucleonic de-grees: We wish our collective equation of motion to be local in time, i.e. to be a differential equation and not a integro-differential equation. Suppose this intrinsic time is given by τ, so our condition will read:

$$\tau \ll \delta t \ll \tau_{coll} , \tag{1}$$

where τ_{coll} is a typical time for the collective motion.

There is a second condition. The fluctuations should not grow too large. Otherwise we need to take into account moments in $(Q-\langle Q \rangle_t)$ and $(P-\langle P \rangle_t)$ of an order higher than two. (This condition is probably not as re-strictive as the first one. We may be able to use further tricks (see ref. 1)).

<u>To which extent are the conditions fulfilled</u> ?

The typical intrinsic relaxation time τ is of the order of 10^{-22} sec. We find this value from calculations of the relevant response functions [4,5] (from which a unique definition of τ can be deduced). Typical times for collective motion are one order of magnitude larger. So the condition is moderately well fulfilled.

As for the second condition: We know from our experience with the de-scription of heavy ion collisions[6] that the fluctuations do not grow very large within the time of contact provided we start with zero width before the reaction: This is to say the distribution function $d(Q,P,t)$ in the collective phase space is given by

$$d(Q,P,t) = \delta(P-\langle P \rangle_t) \, \delta(Q-\langle Q \rangle_t) \quad \text{for} \quad t \to -\infty \tag{2}$$

This ansatz defines fully the initial distribution if quantum effects can be neglected. In the case the latter ones are important we must change our interpretation. We may then take the solutions of the equations for the first and second moments to define the <u>propagation</u> of the distribution. Say the initial distribution function is given by $f(Q,P,0)$. Let us pick out a single point Q_0,P_0. For time 0 its re-presentation in the phase space can be described as:

$$d(Q,P,0) = \delta(P-P_o)\,\delta(Q-Q_o)$$

(3)

For later times it will develop within the tube shown in the figure. To obtain the full distribution at time t we only have to fold f(Q,P,0) with d(q,P.t):

$$f(Q,P,t) = \int dQ_o\,dP_o\;f(Q_o,P_o,0)\,d(Q,P,t;Q_o,P_o,0)$$

(4)

So it is for this propagation where we claim the locally harmonic approximation to be adequate. Since d always starts with zero width the fluctuations for this d do not grow large.

It should be stressed that our approximation scheme does not involve classical arguments beyond those of treating the motion locally harmonic. Therefore, all quantum effects which we encounter in systems with linear forces are properly included. (This is true even for the penetration through an inverted oscillator).

In order to avoid too much scepticism about this inclusion of quantum effects, I want to recall again Feynman's path integration method. From there one knows[7] that the inclusion of harmonic vibrations around the classical trajectory leads to a solution of the time dependent Schrödinger equation within the WKB approximation. Different to the case mentioned before, this solution is now meant to describe the dynamics for large time intervals. A description of the large scale motion within the WKB limit should be sufficient for the examples we are interested in.

Let us now come back to the special situation we want to describe:Which kind of collective coordinates do we need for a description of typical DIC? We certainly have to include the vector of relative motion and its conjugate momentum. Furthermore, we will need angles to describe rotations of the fragments. Degrees of this kind are related more or less to the two well separated fragments.

On the other hand we do expect large mass transfer. This most likely implies that the two fragments loose entirely their identity during the reaction. But this means that we have to face the problem of collective motion of the composite system. In this sense the situation is then similar to fission.

For all these degrees of freedom we have to find the proper coordinates. It is here where the aforementioned problem of self-consistency appears: The nuclear mean field has to follow the density deformation and vice versa. I do not want to go into these problems today. Let me just mention that we are presently working on some method to account for this problem of self-consistency in the presence of dissipative phenomena. Also this method exploits strongly our locally harmonic approximation (see 1)).

III. Definition of the perturbation approach

Disregarding the self-consistency problem we may assume the Hamiltonian of the total system to have the following form:

$$\hat{\mathcal{H}}(\hat{x}_i, \hat{p}_i, \hat{Q}, \hat{P}) = \hat{H}(\hat{x}_i, \hat{p}_i) + \hat{V}(\hat{x}_i, \hat{Q}) + \hat{H}_{Coll}(\hat{Q}, \hat{P}) \qquad (5)$$

(We restrict the discussion to one collective degree Q and its conjugate momentum P).

Let us now use our conjecture that the locally harmonic approximation is sufficient to describe the time propagation of the collective system. Suppose we want to study the motion around a time t_0. We then expand

$$\hat{\mathcal{H}}(\hat{x}_i, \hat{p}_i, \hat{Q}, \hat{P}) \quad \text{around}$$

$$Q_0 = \langle \hat{Q} \rangle_{t_0} \quad \text{and} \quad P_0 = \langle \hat{P} \rangle_{t_0}$$

to second order in $\hat{Q}-Q_0$ and $\hat{P}-P_0$:

$$\hat{\mathcal{H}} \cong \hat{H}(\hat{x}_i, \hat{p}_i) + \hat{V}(\hat{x}_i, Q_0) + (\hat{Q} - Q_0)\frac{\partial \hat{V}}{\partial Q_0} + \frac{1}{2}(\hat{Q} - Q_0)^2\frac{\partial^2 V}{\partial Q_0^2} +$$

$$+ \quad H_{Coll}^{harm} \left(\hat{Q}, \hat{P}, Q_0, P_0 \right) \tag{6}$$

Here, H_{coll}^{harm} describes the bare collective motion around Q_0, P_0 in harmonic order.

In this approximation the coupling between collective and intrinsic degrees is given by:

$$\delta \hat{V} = (\hat{Q} - Q_0) \frac{\partial \hat{V}}{\partial Q_0} + \frac{1}{2}(\hat{Q} - Q_0)^2 \frac{\partial^2 \hat{V}}{\partial Q_0^2} \tag{7}$$

During $\delta t = t - t_0$ $\delta \hat{V}$ will be small provided our conditions of slow collective motion and small fluctuations are fulfilled. Therefore we can treat this $\delta \hat{V}$ by means of perturbation theory. To perform this perturbation approach it is very convenient to use linear response theory. This will enable us, for instance, to distinguish unambiguously between dissipative and conservative forces[1].

This special way of treating the coupling between collective and intrinsic degrees of freedom is one of the places where our theory differs strongly from those of Groß[8], Nörenberg et al.[9] (see also Ayick[11]), and Weidenmüller et al.[10]. In their work the full $\hat{V}$ is used as coupling operator. This $\hat{V}$, I agree, should not be treated by perturbation methods. The two approaches are difficult to compare as far as the line of reasoning is concerned. In my opinion, every statement against the linear response treatment which does not refer to this renormalization of the interaction must be taken with great caution.

To actually perform the perturbation or the linear response approach it is necessary to define the unperturbed state. We describe it by means of the following density operator

$$\hat{\rho}_0 = \hat{\rho}_{int} \left(\hat{H}_{int}(\hat{x}_i, \hat{p}_i, Q_0), T_0 \right) \cdot \hat{d}(t_0) \tag{8}$$

with

$$\hat{H}_{int}(\hat{x}_i, \hat{p}_i, Q_0) = \hat{H}(\hat{x}_i, \hat{p}_i) + \hat{V}(\hat{x}_i, Q_0) \tag{9}$$

Here, $\hat{d}(t_0)$ is the unperturbed density operator for collective motion at t_0.

The density operator for the intrinsic degrees, $\hat{\rho}_{int}(\hat{H}_{int})$, we assume to represent a statistical equilibrium defined by the renormalized intrinsic Hamiltonian $\hat{H}_{int}$ and a temperature T_0. Both these assumptions are not necessary for the application of linear response theory, but very convenient. They can be justified as follows:

a) The reason for using an equilibrium distribution are the comparatively short intrinsic relaxation times of 10^{-22} sec as obtained for
instance from the calculation of response functions[1]. Furthermore,
as you all know, the equilibrium assumption does not contradict the
experimental results about n-emission (see the paper by D.Hilscher at
this conference).

b) The reason for using the temperature is again a matter of convenience.
As an example, we then need to specify only the mean intrinsic excitation energy and not its distribution. The temperature concept would,
however, be out of place if the transport coefficients would be sensitive to the width of this energy distribution. This does not seem to
be true since the corrections can be shown to be of second order in
$\Delta E_{int}/\langle E_{int}\rangle$(at least for not too small temperatures, see ref.5). It is
essential, of course, that the temperature is calculated from the
mean energy of the redefined intrinsic Hamiltonian, i.e. the equation
for the energy balance reads

$$\langle H_{coll}\rangle_{t_0} + \langle \delta V\rangle_{t_0} + \langle \hat{H}(\hat{x}_i,\hat{p}_i) + \hat{V}(\hat{x}_i,Q_0)\rangle_{t_0} = E_{tot} = const. \quad (10)$$

In this way the energy is shifted- off-shell if one wishes - from
the value $\langle \hat{H}(\hat{x}_i,\hat{p}_i)\rangle_{t_0}$ by the amount $\langle \hat{V}(\hat{x}_i,Q_0)\rangle_{t_0}$. (The small coupling
$\hat{\delta V}$ can be neglected in this energy balance).

In this whole discussion we presume, of course, that we are interested
finally in the distribution of the collective energy only; for example
that of the fragments kinetic energy. We do not want to calculate explicitly the distribution of the intrinsic energy.

IV. Features of the transport equation

We may now proceed to derive an equation of motion for the density
operator $\hat{d}(t)$ for the collective phase space. This is done by treating
$\hat{\delta V}$ consistently to second order and averaging over the nucleonic degrees
of freedom.

The word consistently does mean here that all those orders in this
coupling are neglected which would destroy the harmonic behavior of
the motion around the mean values.

This equation has the following features:

1) It is a quantal equation which may be solved by transforming to an
equation for the Wigner-transform d(Q,P,t). This latter equation then

turns out to be identical to the classical Fokker-Planck equation. The reason for that, of course, is our restriction to the locally harmonic approximation. But - besides this approximation - no classical argument has been used.

2) The effective forces in this equation depend on $(Q-\langle Q\rangle_t)$ and $(P-\langle P\rangle_t)$ at most to first order. This means that the conservative force, for instance, appears as:

$$K(Q) = K(\langle Q\rangle_t) - C(\langle Q\rangle_t)(Q - \langle Q\rangle_t) \tag{11}$$

It again shows that with respect to the mean values of Q and P the equation of motion accounts for the full nonlinearity of the force. (The same is true for the friction force). For the fluctuations in $(Q-\langle Q\rangle_t)$, on the other hand, the conservative potential acts by means of the effective stiffness coefficient C ,calculated at the center of the distribution, i.e. at $\langle Q\rangle_t$.

It is this particular approximation for the nonlinear forces which finally enables the simple solution d of the Fokker-Planck equation. Choosing the form (3) as initial condition, d(Q,P,t) will be a Gaussian centered at $\langle Q\rangle_t$ and $\langle P\rangle_t$. Had we treated the coupling $\hat{\delta v}$ to higher order, the effective force K(Q) would definitely be a non-linear function of Q. No simple solution of the transport equation would be possible in this case. This shows clearly the internal consistency of solving the transport equations by means of first and second moments on the one hand and using locally the weak coupling limit on the other hand.

3) All the transport coefficients and in particular those representing the friction and fluctuating forces will also depend on the (unperturbed) collective frequency defined locally. These transport coefficients can be expressed in terms of response and correlation functions.

For the friction coefficient we obtain

$$\gamma(\omega_{coll} ; \langle Q\rangle_t) = \frac{\chi''(\omega_{coll})}{\omega_{coll}} \tag{12}$$

with χ'' being the dissipative part of the relevant response function. Here, I have indicated that χ depends on the mean value of the coordinates. The origin of this, of course, is our adjustment of the intrinsic Hamiltonian to the collective motion.

This definition of the friction coefficient is different to the one in our early publications[1,4,5]. There we always used the picture of extremely slow collective motion. This limit can be regained by letting ω_{coll} go to zero:

$$\gamma(\omega_{coll} \to 0) = \lim_{\omega_{coll} \to 0} \frac{\chi''(\omega_{coll})}{\omega_{coll}} = \frac{\partial \chi''}{\partial \omega_{coll}}\bigg|_{\omega_{coll} = 0} \tag{13}$$

The latter equation is obtained since $\chi''(\omega)$ is an odd function of ω[1].

In keeping with our approximation scheme described above there is no need for this extreme limit. To avoid confusion: For a description of a motion underlying nonlinear forces the frequency ω_{coll} cannot be unlimited. We still have to fulfill the condition $\tau \ll \delta t < \tau_{coll}$.

The diffusion coefficient can be expressed by a correlation function. This correlation function is related to the response function by the fluctuation-dissipation theorem. We may therefore expect a generalized Einstein relation between the diffusion and the friction coefficient. Indeed, for real frequencies (of unperturbed motion) we obtain:

$$D(\omega_{coll}) = \gamma(\omega_{coll}) \cdot T^*(\omega_{coll}) \tag{14}$$

Here, T^* is given by:

$$T^*(\omega_{coll}, T) = \frac{\omega_{coll}}{2} \cotgh \frac{\omega_{coll}}{2T} \tag{15}$$

and represents the mean energy of an oscillator coupled to a heatbath with temperature T.

We observe that the usual Einstein relation is obtained in the high temperature limit only; i.e. for $\omega_{coll} \lesssim T$:

$$D = \gamma T \left(1 + \frac{1}{12} \frac{\omega_{coll}^2}{T^2} + \cdots \right) \tag{16}$$

Already in our first paper on the Fokker-Planck equation more than two years ago[1] we have stressed that $D = \gamma T$ is obtained only in the high temperature limit. There, we also pointed out that this limit is very much related to going from quantum statistics to classical statistics. This is well known for equilibrium thermodynamics. But it seems to hold true also for our case of nonequilibrium. The Fokker-Planck equation describes the relaxation process. If quantum features of this relaxation are to be retained one needs to have T^* instead of T.

The first correction term to the Einstein relation, as given in eq.(16), has a structure similar to a result obtained by Agassi, Ko and Weidenmüller[12]. In our notation their result can be written as:

$$D = \gamma T \left(1 + \frac{1}{3} \gamma_0^2 \right) \tag{17}$$

with

$$\gamma_0 = \frac{\hbar^2 k^2}{2\mu} \cdot \sqrt{\frac{a}{E_{int}}} \cdot \frac{1}{\sigma k} = \frac{1}{\sigma} \cdot \frac{1}{\sqrt{2\mu}} \cdot \sqrt{\frac{E_{kin}}{T^2}} \tag{18}$$

On the right hand side of eq.(18) we have used $E_{int}=aT^2$, in accordance with ref.11, and introduced the kinetic energy $E_{kin}=\frac{\hbar^2 k^2}{2\mu}$. For a vibration with frequency ω_{coll} and amplitude q_0 we could estimate this kinetic energy by $E_{kin} \sim \omega_{coll}^2 q_0^2$. This implies the correction term in eq.(17) to be proportional to

$$\frac{1}{3} \gamma_0^2 \sim \frac{q_0^2}{\sigma^2} \frac{\omega_{coll}^2}{T^2}$$

showing the similarity to the second term in eq.(16).

Let us finally comment on the case that the unperturbed frequency is purely imaginary, i.e. that our local oscillator is an inverted one. Such a situation arises close to the top of the barrier. In this case a generalized Einstein relation is more delicate. It involves a thorough discussion of a continuation of the fluctuation-dissipation theorem into the complex plane. This is possible under certain conditions. Then one sees that the correction term to the Einstein relation becomes negative, different to the case for an ordinary oscillator and different to the result (17) of ref. 12. But I do not wish to go into these details here. They will be published elsewhere[1].

V. Comments on microscopic computation of the transport coefficients

Let us now discuss the computation of the response and correlation functions. For this computation we need the matrix elements of the coupling operator $\hat{F}(\hat{x}_i, Q_o) = \frac{\partial \hat{V}(\hat{x}_i, Q_o)}{\partial Q_o}$ with the eigenstates of the intrinsic Hamiltonian as well as the eigenenergies of the latter:

$$\hat{H}_{int}(\hat{x}_i, \hat{p}_i, Q) \, |m, a\rangle = E_m(Q) \, |m, a\rangle \qquad (19)$$

These quantities will depend on the model we are willing to accept for $\hat{H}_{int}$:

$$\hat{H}_{int} = \hat{H}(\hat{x}_i, \hat{p}_i) + \hat{V}(\hat{x}_i, Q) \qquad (20)$$

There is a general agreement that the coupling $\hat{V}(\hat{x}_i, Q)$ is determined by single particle fields. This is similar to the conventinal model for nuclear collective motion. It is for $\hat{H}(\hat{x}_i, \hat{p}_i)$ where different theories make different assumptions. Let us for the following discussion split $\hat{H}(\hat{x}_i, \hat{p}_i)$ into a pure single particle Hamiltonian plus two-body residual interactions

$$\hat{H}(\hat{x}_i, \hat{p}_i) = \hat{H}_o(\hat{x}_i, \hat{p}_i) + \hat{V}_{res}(\hat{x}_i, \hat{p}_i) \qquad (21)$$

In the random matrix model of DIC[10-12] it is assumed that this residual interaction plays a dominant rôle. The structure of the eigenstates and thus of the matrix elements of $\hat{F}$ is then so complicated that a detailed calculation is meaningless. On the basis of the central limit theorem one can then argue that the matrix elements are random numbers following a Gaussian distribution.

We want to approach the computation from the opposite point of view: We argue that in the first approximation one may try to neglect $\hat{V}_{res}$ completely. As the next step one may then incorporate different orders of $\hat{V}_{res}$. As we shall describe below there does indeed exist a systematic way of doing this.

When neglecting $\hat{V}_{res}$ completely, the computation of the response function is identical to the one of mass parameters in the cranking model - with one important exception. This exception, however, is a matter of principle and basic to all computations of transport coefficients for dissipative phenomena: It is the question of why and how irreversibility can be incorporated into the macroscopic equations

of motion. This problem appears whenever one has to deal with a discrete excitation spectrum. In nuclear physics the problem may appear
in a more critical fashion than it does for a solid: the level spacing
of simple (elementary) excitations of a nucleus is comparatively broad.
This is one of the places where we feel the problem of dealing with a
small system.

The answer to the problem is simple. It is very much related to the
question of relaxation of the intrinsic excitations. For a discrete
spectrum a relaxational behavior can be expected to happen at most
within a limited time say τ_{obs}, which must be much shorter than the
Poincaré recurrence time, τ_{poinc}. Only if a measuring device cuts off
the times $t > \tau_{obs}$ can we talk about relaxation. Translating to ener-
gies we see that we need an energy uncertainty $\delta E \simeq \frac{\hbar}{\tau_{obs}}$ much larger
than the level spacing D: $\delta E \gg D$. (This unequality is obtained after
estimating the Poincaré recurrence time by $\hbar/D$).

Only if this fact is used is a calculation of friction coefficients
meaningful. However, since τ_{obs} and thus δE, cannot be known precisely the friction coefficient should be independent of δE. Such situations are possible - even if one takes into account only single
particle excitations (see ref.4). We may then talk about one-body
dissipation in the proper sense: The friction coefficient is determined entirely within the one-body picture: One-body matrix elements,
single particle (and hole) energies. It should be stressed, however,
that we deal here with the quantal version of one-body dissipation.

For one-body dissipation in the sense of the wall and window formula,
on the other hand, one has to make further assumptions: One has a)
to assume that the nucleonic motion may be described by classical
dynamics and b) that the size of the system tends to infinity.
It has been shown by Koonin et al.[13] that under these conditions
the quantal formula for one-body dissipation approaches the wall
formula. The second condition is necessary to get proper relaxation-
and probably does not imply serious deficiencies. But the first
condition certainly does. For the examples we want to calculate the
friction force the nucleonic motion certainly cannot be adequately
described by means of classical mechanics. Indeed, calculations for
fission as presented by A.S. Jensen at Jülich[14] show the values for
γ obtained with the quantal formula to be lower by about a factor 3
or more. (This factor may be influenced by the inclusion of pairing
into the quantal expression). But more important than this overall

factor are shell effects which influence strongly the coordinate dependence of the friction coefficient. Such shell effects are not included in the wall formula. These deficiencies of the wall and window models also shed some doubts on the use of these models for calculation of drift and diffusion coefficients for mass exchange.

The somewhat delicate questions of level smearing (uncertainty in the energy) mentioned above appear in the most dramatic way, of course, only for this one-body picture. As soon as more complicated states play a rôle the level density gets much larger and the smearing can be handled with much more ease. Most likely the importance of the pure one-body mechanism for nuclear dissipation has been overestimated in the past. Indeed, the spreading width of single particle states easily reaches values of the order of a few hundred keV as soon as the energy of the states is a few MeV away from the Fermi surface. These spreading widths just cannot be neglected when calculating a response function.

The simplest way to account for this decay into more complicated configurations is by means of complex self-energies. This is to say one replaces the p.h-energies appearing in the response functions by self-energies which contain an imaginary part. The imaginary part of the particle-hole excitation is assumed to split into two terms:

$$\Gamma_{k\ell} = \Gamma_k + \Gamma_\ell \tag{22}$$

representing the independent decay of the particle and the hole. Both will depend on the energy. The simplest ansatz one can make is to use the "on shell" expression, which around the Fermi energy reads:

$$\Gamma_k = \Gamma_k(\epsilon_k) = \Gamma_0 \left[(\epsilon_k - \lambda)^2 + (\pi T)^2 \right] \tag{23}$$

The first term is well-known from nuclear matter calculations. The second term accounts for the larger phase space being available at finite temperature T. This prescription has been used to calculate transport coefficients within a two-center shell model. The configuration of the shapes corresponded to those encountered in fission up to the second saddle. (A.S. Jensen has presented these calculations at the Jülich meeting of fission[14]).

This first ansatz, as given in eq.(23), may be too crude. This can be seen if the residual interaction $\hat{V}_{res}$ is taken into account in a

systematic way when calculating the response and correlation functions[14].
The method one may use is the continued fraction approach of Mori and
Götze to calculate response functions. E. Werner has introduced[15] this
method to problems of nuclear physics at zero temperature, for instance
for a description of giant resonances.

The formula for $\Pi'_{k\ell}$ has to be changed in two ways: firstly, the energy
dependence has to account for off-shell effects and secondly, there
will be additional factors involving the occupation probabilities of
the states. These factors do depend on the collective frequency, the
energies of the levels as well as the temperature.

These changes do not complicate the numerical evaluation of the response
functions. We expect to have such computations ready within the next
few months, for fission configurations at least. These calculations are
not simple, but feasible and realistic. They are realistic in the sense
that they treat the nucleonic motion in a realistic two center shell
model. A somewhat more critical quantity will be the parameter for the
spreading width. But this quantity can be related to either nuclear
matter calculations or even to optical model parameters. Please note
that here we can greatly benefit from known results in the temperature
zero case, namely, as mentioned before, for instance from results on
giant resonances etc.

I will not have time to show you applications of the theory to actual
heay ion collisions. As you may know we calculated multidifferential
cross sections - mainly in collaboration with Christian Ngô[6] - by
assuming phenomenological friction coefficients, taken from models
of trajectory calculation. These applications have been and will be
presented elsewhere. I may just note that most recently we made a
simple application to the charge exchange process[17]. We solved a
quantal Fokker-Planck for this particular mode using constant coefficients. We found good agreement with experiment when plotting
the variance as a function of energy loss. Brosa and Groß[18] have
argued against this paper in raising the point that the inertia for
this mode will depend on the shape configuration. It will thus change
the variance in the exit channel. In this model the inertia even
extends to infinity. In my opinion the question they raise is an important one. It has to do with a proper description of the dynamics
around the scission point. This dynamics requires that around scission
the exchange processes should cease to happen. However, it seems to
me that a) this should be possible in different ways - not only by

letting the inertia go to infinity and b) this singularity in the
inertia is probably due to the liquid drop model adopted. A treatment
within the cranking model will probably show very different results.
So further research is required in this important problem.

Let me summarize: I tried to explain to you the basic points of our
theory of DIC. We exploit the linear response theory to achieve a
locally suitable description of the complicated dynamics. I tried to
show to you the intimate connection of this local linearization of the
coupling and a description of the large scale collective motion by
means of its first and second moments. Such a description leads to
dynamical equations of motion which are easy to solve numerically.
The input to these equations can be computed within microscopic
but nevertheless realistic models.

The author would like to thank K. Hartmann and F. Scheuter for
reading the manuscript.

References

1) Hofmann, H., Siemens, P.: Nucl.Phys. A257, 165(1976) and A275,
   464 (1977)
   for reviews see ref. 5) as well as: Hofmann, H., Jensen, A.S.,
   Ngô, C., Siemens, P.J.: Proc. IVth Balaton Conference on Nuclear
   Physics, Keszthely, Hungary, June 1979 and to appear in Physics
   Reports.
2) Nakajima, S.: Prog.Theor.Phys. 2o, 948 (1958)
   Zwanzig, R.: J.Chem.Phys. 33, 1338 (1960)
   and Lect.Theor.Phys. (Boulder)Vol.3, 16o (196o).
3) Feynman, R.P., Hibbs, A.R.:"Quantum Mechanics and Path Integrals",
   McGraw-Hill Book Company, New York, 1965
4) Johansen, P.J., Siemens, P.J., Jensen, A.S., Hofmann, H.: Nucl.
   Phys. A288, 152 (1977)
5) Hofmann, H.: Int.Symp. on Nuclear Collisions and their Microscopic Description, Bled, Yugoslavia, Sept. 1977; FIZIKA 9,
   Suppl.4, 441 (1977)
6) Ngô, C., Hofmann, H.: Z.Physik A282, 83 (1977)
   Berlanger, M., Grangé, P., Hofmann, H., Ngô, C., Richert, J.:
   Phys.Rev. C17, 1495 (1978); Z.Physik A286, 2o7 (1978)
   Berlanger, M., Ngô, C., Grangé, P., Richert, J., Hofmann, H.:
   Z.Physik A284, 61 (1978)
7) see ref. 3), for instance p.63.
8) D.H.E. Gross, Z.Physik A291, 145 (1979) and references therein;
   see also proceedings of this conference
9) Nörenberg, W.: Z.Phys. A274, 241 (1975) and A276, 84 (1976);
   see also proceedings of this conference and references therein.
1o) Agassi, D., Ko, C.M., Weidenmüller, H.A.: Ann.Phys. (N.Y.) 107,
    14o (1977)
    Ko, C.M., Agassi, D., Weidenmüller, H.A.: Ann.Phys. (N.Y.) 117,

237 (1977); see also: Weidenmüller, H.A.: "Transport theories of heavy-ion reactions", to appear in "Progress in particle and nuclear physics" ed. D. Wilkinson

11) Ayik, S.: Z.Physik, in print

12) Agassi, D., Ko, C.M., Weidenmüller, H.A.: Ann.Phys.(N.Y.) 117, 407 (1979)

13) Koonin, S.E., Hatch, R.L., Randrup, J.: Nucl.Phys. A283, 87 (1977)
    Koonin, S.E., Randrup, J.: Nucl.Phys. A289, 475 (1977)

14) Jensen, A.S., Reese, K., Hofmann, H., Siemens, P.J.: Physics and Chemistry of Fission, Jülich 1979, IAEA, Vienna, SM/241-H4

15) Wio, H., Werner, E., Hofmann, H.: to be published

16) Werner, E.: Z.Physik A276, 265 and 275 (1976)
    Theis, W., Werner, E.: Z.Physik A287, 323 (1978)

17) Hofmann, H., Grégoire, C., Lucas, R., Ngô, C.: Z.Physik, in print.

18) Brosa, U., Gross, D.H.E.: to be published.

DEEP INELASTIC COLLISIONS VIEWED AS BROWNIAN MOTION

D.H.E. Gross

Hahn-Meitner-Institut für Kernforschung, Berlin, and
Fachbereich Physik der Freien Universität Berlin
D-1000 Berlin 39, Glienicker Straße 100

Abstract: The theory of Brownian motion is presented in a form applicable for deep inelastic nuclear collisions. It is illustrated by simple classical models.

Introduction

Deep inelastic collisions of heavy nuclei are known since about 8 years. Attempts to describe these events by statistical theories are as old. One may ask, why should one try a new theory instead of exploring the existing ones. On the other hand non-equilibrium transport processes like Brownian motion are studied since perhaps 100 years and one should ask why does one not use these theories to explain deep inelastic collision data. Moreover, these theories have reached a high standard of sophistication, experience, and precision [1, 2] that I believe them to be very usefull for our problem. I will try to sketch a possible form of an advanced theory of Brownian motion [2] that seems to be suitable for low energy heavy ion collisions.

Before going into details I should just make a link between Brownian motion of a heavy particle and heavy ion collisions. The relative motion of the two nuclei and perhaps their surface vibrations constitute the degrees of freedom corresponding to the Brownian particle, whereas the internal nucleon motion is the bath system. The nuclei are about 100 times heavier than the bath particles (nucleons). In that respect the situation is similar to the conventional Brownian motion where, however, the observed Brownian particle is of course much more heavy ($> 10^8$) than the bath molecules. This also clearly shows the difference between nuclear collisions and Brownian motion. This problem, however, is well known. It exists already in the application of equilibrium statistics, made for $\sim 10^{23}$ particles, to nuclear problems.

Introductory Example

We consider a heavy particle M, Q, P moving through a harmonic bath m_i, q_i, p_i, ω_i. The total system is governd by the Hamiltonian:

$$H = \frac{P^2}{2M} + V(Q) + \sum_i \left\{ \frac{p_i^2}{2m_i} + \frac{m_i \omega_i^2}{2} (q_i + f_i(Q))^2 \right\} , \tag{1}$$

where $q_i f_i(Q)$ is the coupling of the heavy particle and the bath. This leads to the coupled equations of motion:

$$M\ddot{Q} + \frac{dV}{dQ} + \sum_i m_i \omega_i^2 (q_i + f_i(Q)) \frac{df_i(Q)}{dQ} = 0 \tag{2a}$$

$$\ddot{q_i} + \omega_i^2 (q_i + f_i(Q)) = 0 . \tag{2b}$$

If we take Q(t) as given, equation (2b) can be solved:

$$q_i(t) = \hat{q}_i(t) - \int_0^t ds \, f_i(Q(s)) \, \omega_i \, \sin\{\omega_i(t-s)\} , \tag{3}$$

$$\hat{q}_i(t) = q_i(0) \cos \omega_i t + \frac{p_i(0)}{m_i \omega_i} \sin \omega_i t .$$

Substituted into eq. (2a) yields the equation of motion for the heavy particle:

$$M\ddot{Q} + \frac{dV(Q)}{dQ} + \int_0^t ds \sum_i m_i \omega_i^2 f_i'(t) \cos\{\omega_i(t-s)\} f_i'(s) \dot{Q}(s) = F^+(t) \tag{4}$$

with $$F^+(t) = -\sum_i m_i \omega_i^2 f_i'(t) \hat{q}_i(t) ;$$

$$f_i'(t) = \frac{df_i(Q(t))}{dQ}$$

We see that the induced force on the left side of eq. (4) is quadratic in f_i' (Q) and proportional to $\dot{Q}$. It is a generalized <u>retarded friction</u> force. Let us assume for the rest of this chapter, that the Q-dependence of eq. (4) is linear also. The left hand side of (4) does not depend on the initial condition of the bath and can be calculated without refering to the bath system at all. The unmeasured information about the initial condition of the bath appears only in the force $F^+(t)$, which we may call the <u>random</u> force. The initial values $q_i(0)$, $P_i(0)$ may be distributed according to some probability. The most natural choice is any function $\varrho(e_i....)$ of the initial energy of the bath particles

$$e_i = \frac{P_i^2(0)}{2m_i} + \frac{m_i\omega_i^2 q_i^2(0)}{2} \qquad \text{e.g. } \varrho(e_i...) \propto \delta(e_i - e),$$

the classical microcanonical ensemble, or $\varrho(e_i...) \propto \exp\left\{-\sum_i e_i/T_i^*\right\}$
for a quantum or classical canonical ensemble at temperature T with

$$T_i^* = \frac{\hbar\omega_i}{2} \coth\frac{\hbar\omega_i}{2T}. \tag{5}$$

Equation (4) may be called a retarded Langevin equation for the heavy particle trajectory Q(t). It can be solved repeatingly for various possible choices of $q_i(0)$, $p_i(0)$ i.e. of the random force $F^+(t)$. By this a whole bunch of trajectories is generated. It describes the time evolution of the distribution d(Q,P,t) in an ensemble of many similar experiments that differ in the unmeasured initial values $q_i(0)$, $p_i(0)$ of the bath particles (like in a nuclear scattering experiment).

An equivalent method to solve the stochastic Langevin equation is to solve a partial differential equation for the distribution d(Q,P,t), the Fokker Planck equation. Without going into details we mention that there is a diffusion term $\partial_p^2 Dd$ which describes the spreading of the distribution. The diffusion coefficient D is given by the mean value of $\langle F^+(t)\, F^+(s)\rangle$, the force-force correlation function:

$$\langle F^+(t) \quad F^+(s)\rangle = \int dq_i\, dp_i \ldots \varrho(e_i...)\, F^+(t)\, F^+(s)$$
$$= \sum_i m_i \omega_i^2\, f_i'\, \cos\left\{\omega_i(t-s)\right\}\, f_i'\, \langle e_i\rangle \tag{6}$$

which is up to the factor $\langle e_i \rangle$ the same as the kernel of the dissipative force in eq. (4). This is a general form of the fluctuation-
dissipation theorem. For a classical Boltzmann distribution $g(e_i \ldots)$
$\alpha \exp\left\{ - \Sigma_i e_i / T \right\}$ we have $\langle e_i \rangle = T$ and (6) is the famous Einstein
relation.

If we take $\int_0^\infty d\omega f' f' \cos\{\omega_i(t-s)\}$ for $\Sigma_i f_i' f_i' m_i \omega_i^2 \cos\{\omega_i(t-s)\}$,
i.e. a dense spectrum of frequencies ω_i, we have $\langle F_i^+(t) F_i^+(s) \rangle \propto \delta(t-s)$
and eq. (4) is a proper Langevin equation for a Markoffian process
and a white noise spectrum of the random force, as it is used in
the standard theory of Brownian motion.

Equation (4) is exact. No approximation was needed. The result for
the retarded dissipation force is the same as the one deduced form
parturbation theory in ref. [3]. Here it is valid for arbitrary
coupling strength and arbitrary velocities $\dot{Q}(t)$. It is interesting
to notice that complications arising for more realistic systems
seem not to come from strong coupling or high velocities but from
the anharmonicities in the bath particle motion.

Approximate Treatment of Anharmonic Bathes, Projection Formalism.

There is an extended literature about Brownian motion in genuine
bathes. The most modern ones use the projection operator technique
due to Nakajima and Zwanzig [4]. Suited for scattering problems is
the theory by Mazur and Oppenheim [2]. We will translate it to
quantum motion but follow it as close as possible.

a) Mapping of quantum dynamic on the classical one.

We take the total Hamiltonian as

$$\tilde{H} = \tilde{H}_o (PQ) + \tilde{h}_o(p_i q_i \ldots) + \tilde{V}(q_i \ldots, Q) . \qquad (7)$$

The Liouville equation for any operator A is

$$\dot{\tilde{A}} = \frac{i}{\hbar} [\tilde{H}, \tilde{A}] = i \mathcal{L} \tilde{A}, \qquad (8)$$

where the Liouville operator L may be defined by this equation.
The Wigner transform in the P Q phasespace

$$\hat{A}(PQ) = \int ds \, \langle Q + \tfrac{s}{2} | \tilde{A} | Q - \tfrac{s}{2} \rangle \, e^{-\tfrac{i}{\hbar} sP}$$

has the equation of motion [5]

$$\dot{\hat{A}}(PQ) = i\mathcal{L}_0 \hat{A}(PQ) + i\mathcal{L}_1 \hat{A}(PQ) \qquad\qquad (9)$$

where

$$i\mathcal{L}_0 \hat{A}(PQ) = \tfrac{i}{\hbar} [\hat{h}(Q), \hat{A}(PQ)]_- \, ,$$

$$\qquad\qquad\qquad\qquad (10)$$

$$\hat{h}(Q) = \hat{h}_o(p_i q_i \dots) + \hat{V}(q_i \dots, Q) \qquad \text{(= two centre shell model problem for the bath system)}$$

and

$$i\mathcal{L}_1 \hat{A}(PQ) = \tfrac{P}{M} \partial_Q \hat{A}(PQ) + \partial_P \tfrac{1}{2} [\hat{F}(Q), \hat{A}(PQ)]_+ + O(\hbar)$$

$$\qquad\qquad\qquad\qquad (11)$$

with

$$\hat{F}(Q) = -\partial_Q \{ H_o(PQ) + \hat{V}(q_i \dots, Q) \} \, .$$

If $H_o(P,Q)$ is at most quadratic and $V(q_i, Q)$ linear in Q the correction $O(\hbar)$ vanishes and the quantum dynamic of the P, Q degree of
freedom coincides with the classical one [5]. This is the reason,
why the classical treatment of the previous chapter is sufficient
also for a harmonic quantum bath. Furthermore we restrict ourself
to the classical approximation of the Brownian particle motion and
neglect the correction $O(\hbar)$ in eq. (11).

b) The heavy mass approximation

If the Brownian particle is very much heavier than the bath particles
($\sqrt{m_i/M} = \lambda \ll 1$) the change of momentum $\Delta P \sim 2p$ of the Brownian
particle due to a single collision with a bath particle (typical momentum p) is much smaller than P itself (P $\gg$ p). On the other
hand we assume the velocity $\dot{Q}$ to be small compared to the bath particle velocities $\dot{q}$ (in the heavy-ion case is the relative velocity
of the two nuclei $\dot{R}$ much smaller than the Fermi velocity v_F of the
nucleons inside the nuclei). In this case we approximate the motion
of the bath particles by their motion L_0 in the field of the fixed
Brownian particle. In the heavy-ion case this is the nucleon motion
in the adiabatic two-center shell model. I.e. our approximation is
expected to be a good description for $\lambda = \sqrt{m/M} \ll 1$, P $\gg$ p, $\dot{Q} \ll \dot{q}$.
(p/P = λ^2 $\dot{q}/\dot{Q}$ = λ for $\dot{Q}/\dot{q} = \lambda$). It is evident that low energy
collisions of heavy nuclei fit well into this regime ($\lambda \sim 1/10$, $\dot{Q} \sim$
$v_F/5$).

The split of the equation of motion (9) into the "rapid" part (10)
and a "slow" one (11) allows for a consistent expansion in powers
of λ. The "slow" part L_1 scales with λ as $L_1 = \lambda L_1^*$ where

$$i\mathcal{L}_1^*\hat{A} = \frac{P^*}{M^*} \partial_Q \hat{A} + \partial_{P^*} \frac{1}{2}\left[\hat{F}(Q),\hat{A}\right]_+ \qquad (12)$$

$$P^* = \lambda P \sim p \;,\; M^* = \lambda^2 M \sim m \;,\; \frac{P^*}{M^*} \sim \dot{q}$$

As a good part of the rapid (L_0) motion of the bath is well described
by equilibrium statistical mechanics, it is usefull to split the bath
particle motion into a statistical part and a part that contains its
deviation from equilibrium. One defines the projection operator $\mathcal{P}$ by

$$\mathcal{P}\hat{A}(PQ) = \text{tr} \, e^{-\beta(\hat{h}(Q)-\Omega(Q))}\hat{A}(PQ)$$

$$e^{-\beta\Omega(Q)} = \text{tr} \, e^{-\beta\hat{h}(Q)} \;,\; \mathcal{P}\hat{A} = \langle\hat{A}\rangle \;,\; \mathcal{Q} = 1 - \mathcal{P} \qquad (13)$$

where the trace is be taken over the quantum states of the bath.

Similar to our introductory example one eliminates the "rapid" non-equilibrium motion of the bath to obtain an effective equation for the motion of the Brownian particle (details can be found in [5]:

$$\dot{\hat{P}}(t) = e^{i \mathcal{L} t} \langle F \rangle + F^{+}(t) - \int_{0}^{t} d\tau \, e^{i \mathcal{L}(t-\tau)} \dot{\hat{Q}} \beta \langle \hat{6}_{Q} e^{i \mathcal{L}_{0} \tau} \hat{F}_{Q} \rangle + \mathcal{O}(\lambda^{3})$$

(14)

where:

$$F^{+}(t) = e^{i \hat{Q} \mathcal{L} t} \hat{Q} \hat{F}$$

$$\hat{F}_{Q} = \hat{Q} \hat{F} = \hat{F} - \langle \hat{F} \rangle$$

$$\hat{6}_{Q} = \frac{1}{\beta} \int_{0}^{\beta} d\lambda \, e^{\lambda \hat{h}(Q)} \hat{F}_{Q} \, e^{-\lambda \hat{h}(Q)} \xrightarrow[\hbar \to 0]{} \hat{F}_{Q}$$

This effective equation of motion is a generalized Langevin equation similar to the one we obtained in the introduction. Again there is a random force $F^{+}(t)$, the average of which vanishes $\langle F^{+} \rangle = \hat{P} F^{+} = 0$, and a retarded dissipative force proportional to $\dot{Q}$. There is, however, an important difference. The propagator as well as the averaging itself depend on the adiabatic states of the bath interacting with a fixed Brownian particle. The kernel of the friction force is just Kubo's force-force response function [1]. Besides $\lambda \ll 1$ there is another important condition for the validity of (14) [2, 5]: A correlation function of the type $\langle A \, e^{iL_{0}t} \, B \rangle$ should decay at large times as

$$\langle \hat{A} \, e^{i \mathcal{L}_{0} t} \hat{B} \rangle = \langle A \rangle \langle B \rangle \quad \text{for } t > t_{mem}$$

(15)

where t_{mem} is some finite memory time independent of $\hat{A}$, $\hat{B}$. This is a plausible statement saying that the correlation between the operators $\hat{A}$ and $\hat{B}$ relax to equilibrium in a time t_{mem} that depends on the system but not on the choice of $\hat{A}$ and $\hat{B}$. However, for a realistic system this to prove may be difficult if not impossible. Mazur and Oppenheim call this the ergodic behaviour of the bath [2]. In the next chapter we will study on a simple model some of the aspects of the memory time.

Classical Piston as Brownian Motion.

In ref. [3] I have introduced the piston model as a guide to study nuclear dissipation processes. In order to illuminate the present formalism and to get some insight into the nature of the memory time, I will discuss a modified model where the piston potential has a finite range instead of a sharp edge like in [3]. We consider a big container with vertical walls containing a one dimensional classical ideal gas of N noninteracting particles.

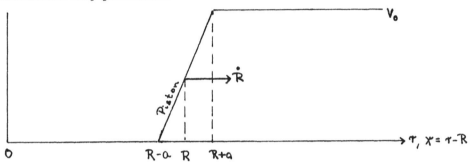

For a single bath particle we have:

potential of piston: $V(x) = V_0 \frac{a+x}{2a} \Theta(a+x)\Theta(a-x) + V_0 \Theta(x-a)$,

force on the piston: $F(x) = -\frac{dV}{dR} = \frac{V_0}{2a}\Theta(a+x)\Theta(a-x)$,

dressed motion of a bath particle in the field of the piston:

$$x(t) = x_0 + \frac{p_0}{m}t - \frac{V_0}{4am}t^2,$$

dressed distribution function for a single bath particle:

$$g(x,p) = \exp\left\{-\beta\left(\frac{p^2}{2m} + V(x) - \Omega\right)\right\}, \quad e^{-\beta\Omega} \xrightarrow[\substack{V_0 \to \infty \\ a \to 0}]{} R\sqrt{2\pi m kT},$$

$$\langle F \rangle = e^{\beta\Omega}\int_{-a}^{a} dx \int_{-\infty}^{\infty} dp\, e^{-\beta\left(\frac{p^2}{2m} + V(x)\right)} \frac{V_0}{2a} \xrightarrow[\substack{V_0 \to \infty \\ a \to 0}]{} \frac{kT}{R} \qquad (16)$$

which is the pressure of the ideal gas, and $\Omega_0 F = F - \frac{kT}{R}$. The Langevin equation (14) reads for our model:

$$\frac{1}{N}\dot{P}(t) = \frac{kT}{R(t)} + F^+(t) - \int_0^t d\tau \, e^{i\hat{x}(t-\tau)} \beta \dot{R} \langle F_Q \, e^{i\hat{x}_o\tau} F_Q \rangle$$

$$= \frac{kT}{R(t)} + F^+(t) - \int_0^t d\tau \, e^{i\hat{x}(t-\tau)} \frac{\dot{R}}{kT} \left\{ \langle F e^{i\hat{x}_o\tau} F \rangle - \left(\frac{kT}{R}\right)^2 \right\} \qquad (17)$$

and

$$\langle F e^{i\hat{x}_o t} F \rangle = \left(\frac{V_o}{2a}\right)^2 \int_{-a}^a dx \int_{-\infty}^\infty dp \, e^{-\beta(\frac{p^2}{2m} + V(x) - \Omega_o)} \Theta(a + x(t)) \, \Theta(a - x(t)) \qquad (18)$$

It is easy to get a imagination of the time structure of $\langle F e^{iL_o t} F \rangle$. A single particle with momentum p outside the piston contributes to $\langle F e^{iL_o t} F \rangle$ a sequence of pulses of the form:

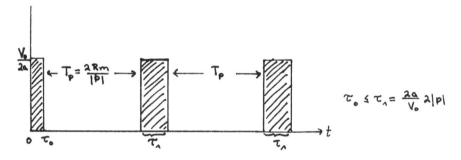

$$\tau_o \leq \tau_\wedge = \frac{2a}{V_o} \, 2|p|$$

The first (direct) pulse is due to the flight of the particle from t = 0 to τ_o within the range 2a of the piston field and the other come from multiple reflections at the opposite container wall. These correspond to the Poincaré recurrences as in general many-body systems. After averaging over x and p, $\langle F e^{iL_o t} F \rangle$ has a narrow direct peak at t = 0 with a width proportional to $T^{1/2} a/V_o$ and a background rising from 0 at t = 0 to $3 (\frac{kT}{R})^2$ at t = ∞. The contribution to the friction force of the sharp direct peak is for $V_o/kT \gg 1$ identical to the friction force calculated from momentum balance for a one-sided

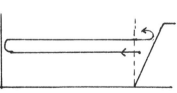

sharp edged piston [3]. The background, being connected to the Poincare recurrences summed over all particles, is for large times equal to $3N(\frac{kT}{R})^2$ and consequently vanishes in the thermodynamic limit $N,R \longrightarrow \infty$ $\frac{N}{R} = \mathbf{g}$ fixed.

We have seen that the memory time splits into two parts. One is long and comes from the recurrences. The long part vanishes in the thermodynamic limit of a large container. In this limit the correlation function does relax quickly, as was required, if the coupling has a short range. The motion of our piston is well described by the Langevin equation (14). For a finite bath the recurrence effects do exist but may be reduced considerably by two-body collisions within the bath <u>if the mean free path</u> of the particles in the bath <u>is less than 2R</u>.

For the case of heavy nuclear collisions, we may expect that the theory of Brownian motion works, if

   a) the reduced mass of the two nuclei is much larger than the
      nucleon mass,

   b) the relative velocity of the two nuclei at contact is small
      compared to the Fermi-velocity,

   c) the coupling of internal and relative motion (e.g. the mutual
      single particle potentials) has a short range,

   d) the mean free path of s.p. motion is considerably less than
      twice the nuclear diameter.

These conditions are moderately fulfilled for low energy deep inelastic collisions, especially in the exit channel.
I am gratefull to S. Grossmann for valuable comments.

<u>References</u>

[1]   R. Kubo, J.Phys. Soc. Japan. Vol. <u>12</u> (1957) 570
[2]   P. Mazur and I. Oppenheim, Physica <u>50</u> (1970) 241
[3]   D.H.E. Gross, Nucl. Phys. <u>A240</u> (1975) 472
                  Phys. Lett. <u>68B</u> (1977) 412
                  and Lectures held at Orsay (1976), IPNO/TH 77-10
[4]   S. Nakajima, Progr. Theor.Phys. <u>20</u> (1958) 948
      R. Zwanzig, J.Chem. Phys. <u>33</u> (1960) 1338
[5]   D.H.E. Gross, Z.Phys. <u>A291</u> (1979) 145

A FAST SPLITTING OF PROJECTILE-LIKE FRAGMENTS
IN THE REACTION ^{86}Kr-^{166}Er AT 12.1 MeV/u

U. Lynen[§]

Max-Planck-Institut für Kernphysik
6900 Heidelberg, Germany

In this report first results of the reaction ^{86}Kr-^{166}Er at an incident energy of 12.1 MeV/u will be presented.

The reason for extending the investigation of deep-inelastic collisions (DIC) to higher energies is twofold: On the one hand, a comparison of the new results with extrapolations of our present understanding
of DIC might help to disentangle the effects of nuclear structure and
of the dynamics of the reaction, and, on the other hand, we hope to
find new phenomena. As an example the high momentum resulting from the
coherent motion of all projectile nucleons when impinging on the target
nucleus should favor collective modes of excitation.

The experiment was performed at the UNILAC accelerator at Darmstadt.
A beam of ∿1 pnA was used to irradiate an ^{166}Er target of 160 µg/cm^2
thickness enriched to 96%. The target was deposited on a carbon backing of 30 µg/cm^2. The projectile-like reaction products were detected
in a large-area position-sensitive ionization chamber ($\Delta\Omega$ = 50 msr,
$\Delta\Theta_{Lab}$ = 21°) [1,2] which due to the small grazing angle of the reaction
Θ_{gr} = 17° covered the angular range 3° ≤ Θ_{Lab} ≤ 24°. For every event
dE/dx, E, Θ, and Φ and by using a parallel-plate avalanche detector in
front of the ionization chamber also the time of flight was recorded.
This latter detector was also used to detect pile-up events. In this
way the ionization chamber could reliably be operated up to 100 kHz.
The center-of-mass transformation was performed event by event and the
effect of neutron evaporation on the reconstructed Q-value has been
taken into account in an iterative way [3]. Absolute cross sections
were determined by a normalization of elastically scattered events to
the Rutherford cross section.

The system ^{86}Kr-^{166}Er has already been investigated quite exten-

[§]In collaboration with M. Dakowski[1], P. Doll[1], A. Gobbi[1], J.B.
Natowitz[2], A. Olmi[2], D. Pelte[3], H. Sann[1], and H. Stelzer[1] ([1]Gesell-
schaft für Schwerionenforschung mbH, 6100 Darmstadt, W. Germany;
[2]Max-Planck-Institut für Kernphysik, 6900 Heidelberg, W. Germany;
[3]I. Physikalisches Institut der Universität, 6900 Heidelberg, W.
Germany).

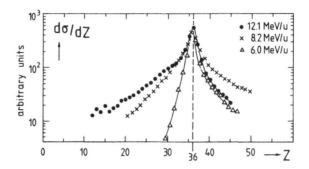

Figure 1   *Element distributions dσ/dZ of projectile-like fragments for 6, 8.2, and 12.1 MeV/u integrated over all inelastic events and over angle. The relative yield has been normalized for Z = 36.*

sively at several lower bombarding energies [3,4,5]. Although a DIC is a continuous evolution in many quantities, we will in the following concentrate on element distributions and their variation with energy loss. The integrated element distributions of projectile-like fragments at incident energies of 5.9 and 8.2 MeV/u are shown in Fig. 1. They are found to be nearly symmetric around Z = 36 with some increase towards a symmetric mass split. These distributions are well reproduced by diffusion model calculations. In striking contrast to this, the element distribution at 12.1 MeV/u (full points) shows a long tail towards very light elements and falls off steeply for elements heavier than Z = 36.

In order to obtain a more detailed view especially at which Q-values the enhanced production of light elements occurs, a contour plot of the double-differential cross section $d^2\sigma/dTKE \cdot dZ$ is given in Fig. 2. Starting from elastic scattering, with increasing energy loss the width

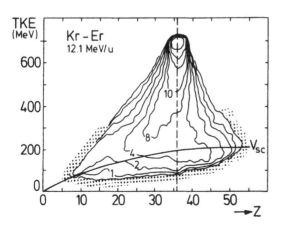

Figure 2
Contour plot of the double differential cross section $d^2\sigma/dTKE \cdot dZ$ integrated over angle ($3^0 \leq \Theta_{Lab} \leq 24^0$) in units of mb/(20 MeV·Z). The dotted background shows the region where the distributions are affected by experimental cuts. V_{sc} indicates the height of the Coulomb barrier at the moment of scission (see also Fig. 6).

of the distribution not only broadens, but at the same time the maximum
of the element distribution is shifted towards lighter elements.   In
Fig. 3 the centroid of the element distribution as a function of energy
loss is shown for the system ^{86}Kr-^{166}Er at 6, 8.2, and 12.1 MeV/u.

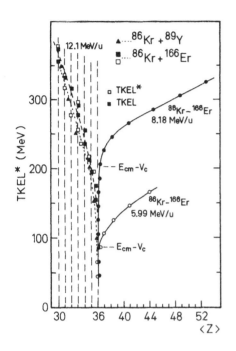

Figure 3
Centroid of the element distributions as a
function of the energy loss for the reaction
Kr-Er at different energies. For comparison,
at 12.1 MeV/u also the results of Kr-Y are
shown. Concerning the difference between TKEL
and TKEL* see Ref. [3].

Whereas at low incident energies the centroid remains at the Z of the
projectile and only for outgoing energies below the Coulomb barrier a
drift towards symmetry is observed, at 12.1 MeV/u the drift occurs much
earlier and is leading towards asymmetry.   In order to obtain a better
picture for small energy losses (TKEL* ≤ 160 MeV) in Fig. 4 the element
distributions for the same bins in energy loss are compared for 8.2 and
12.1 MeV/u incident energies.   Whereas at 8.2 MeV/u Gaussian distributions are observed, at 12.1 MeV/u the distributions are asymmetric,
falling off more steeply towards heavier elements than towards lighter
ones.

Summarizing, at the high bombarding energy of 12.1 MeV/u we observe
a favored production of light elements which at small energy losses
(TKEL* ≤ 160 MeV) leads to a skewness in the element distributions and
with increasing energy loss develops into a shift of the maximum towards lighter elements.   This is in striking contrast to the results at
lower bombarding energies [3].

In the following several possible explanations for these differences

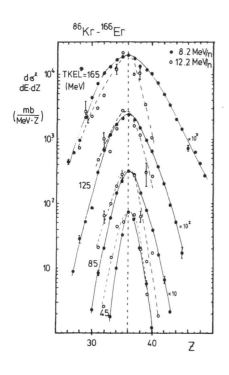

$$\frac{d\sigma^2}{dE \cdot dZ}$$

$$\left(\frac{mb}{MeV \cdot Z}\right)$$

Figure 4
Comparison of element distributions for given bins of TKEL* taken at bombarding energies of 8.2 and 12.1 MeV/u. The relative yield of the distributions is normalized at Z = 36.

shall be discussed.

- Contributions from target-like fragments are for kinematical reasons restricted to very small energies and cannot affect the distributions for TKE > 300 MeV. This has also been verified by comparing the results of three different targets: ^{89}Y, ^{166}Er, and ^{238}U, where in all cases the same distributions for projectile-like fragments have been found.

- Fusion-fission reactions of Kr with light target contaminants, e.g. C or O, cannot be the origin of the light elements as has been shown by the irradiation of a ^{12}C target. This might have been an explanation for the results obtained for ^{166}Er and ^{238}U which had a carbon backing; the ^{89}Y target, however, was self-supporting.

- A reaction mechanism which is considerably different from the one observed at lower energies is rather unlikely, since the widths of the element distributions as a function of energy loss are in reasonable agreement with a parametrization derived from lower energies [6].

- A drift during the primary nuclear exchange can also be excluded since the measurement of the nearly symmetric system $^{86}Kr-^{89}Y$ resulted in the same mass drift towards asymmetry. This is shown in Fig. 3. Since for a symmetric system the element distribution must be symmetric unless a third charged particles is emitted, we must consequently conclude that either the primary interaction is no longer binary or that

the outgoing projectile-like products undergo some sort of fragmentation (evaporation of light charged particles or some sort of fission).

In order to check the latter assumption we have measured the yield
of protons and α particles in coincidence with the projectile-like fragments. A preliminary evaluation of the data shows that the multiplicity
of α particles is about one for energy losses $\gtrsim 300$ MeV and rapidly decreases at smaller energy losses [7]. The overall multiplicity of protons is even smaller. The emission of light charged particles therefore cannot account for the drift in the element distribution of projectile-like fragments and we must assume that the projectiles undergo
a fragmentation into heavier products. In the following we will call
this "splitting" in order to distinguish it on the one hand from the
slow process of sequential fission and on the other from the very fast
process of fragmentation or breakup in the nuclear field.

In order to obtain a direct experimental proof for this splitting
we have tried to measure both splitting fragments in coincidence. This
is rather difficult since a splitting of a fast projectile-like nucleus
will in general lead to two fragments being both emitted under rather
small angles. For apparative reasons the minimum angle between two
ionization chambers is limited to $\geq 30^\circ$. We therefore used the setup
schematically shown in Fig. 5b. The ionization chamber was centered at
4°, covering a range between -9° and $+17^\circ$. A Faraday cup in front of

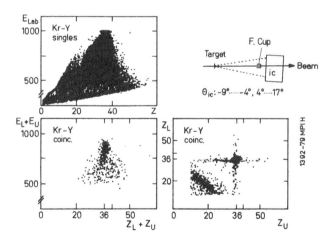

Figure 5   (a) *Scatter plot of laboratory energy versus element number Z for singles
events of the reaction Kr-Y; (b) schematic drawing of the experimental set-
up used to detect coincidences between two fast products; (c) scatter plot
of the summed laboratory energy versus the summed Z of coincidence fragments; (d) scatter plot of the element numbers of coincident events detected
in the upper and lower half of the ionization chamber.*

the ionization chamber shielded angles smaller than 4°. We then used
the independent upper and lower halves [1] of the ionization chamber to
detect coincident fragments at small angles. A scatter plot of the
element numbers of these coincident events is shown in Fig. 5d. Most
of the events are accidental coincidences between either two elastically scattered Kr nuclei ($Z_1 = Z_2 = 36$) or one Kr nucleus together with
some other reaction product ($Z_1 = 36$, $Z_2 \neq 36$). There is, however, a
third component ($Z_1 < 36$, $Z_2 < 36$) which is clearly separated from the
accidental events. It is interesting to note that the latter component
is not peaked around a symmetric mass split but shows a very wide mass
distribution, especially if one considers that in the present preliminary evaluation only events with ($Z_1 > 10$) and ($Z_2 > 10$) have been included in order to avoid coincidences with light target contaminants
like C or O.

In Fig. 5c for coincident events with ($Z_1 < 36$, $Z_2 < 36$) the summed
laboratory energy is displayed versus the summed element numbers. Comparing this distribution with that of singles events shown in Fig. 5a,
a close resemblance can be found, and it is interesting to note that -
trusting the present statistics - the mass drift towards asymmetry seems
to be less pronounced in the summed distribution than for the singles.
The fact that the summed distribution is so similar to that of the singles is direct proof for the supposed splitting of the projectile-like
fragments. It remains to be shown, however, whether this is a slow
process like sequential fission or a fast breakup like the fragmentation
observed at BEVALAC energies.

For this purpose the contour plot of $d^2\sigma/dZ \cdot dv$ is shown in Fig. 6,
where v is the laboratory velocity. In a normal DIC elements far away
from the projectile should have energies close to the Coulomb energies
of two deformed nuclei, or the corresponding velocities as given by the
dashed curve V_{sc} in Fig. 6. Elements heavier than the projectile do
follow this general behavior, in contrast to light elements which show
considerably lower but also much higher velocities (up to 15% above the
incident beam velocity). In case of a breakup or fragmentation process [8,9] the spectator part of the projectile should go on with rather
unchanged velocity. This velocity is indicated by the arrow (a) in
Fig. 6. The complete absence of any enhancement or ridge in the cross
section at this velocity and the presence of very fast light fragments
suggest that the observed splitting is not due to a fragmentation process. An even more stringent proof against nuclear fragmentation comes
from the close agreement between the element distribution as a function
of energy loss for projectile-like fragments and for the reconstructed
splitted events as shown in Fig. 5. This indicates that the splitting

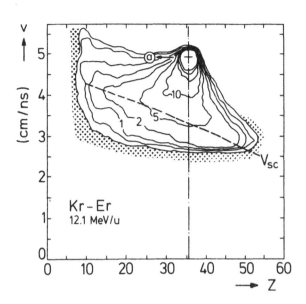

Figure 6   *Contour plot of the double differential cross section $d^2\sigma/(dv \cdot dZ)$ as a function of laboratory velocity and nuclear charge Z integrated over angle in $mb/[(cm/ns) \cdot Z]$. V_{sc} indicates the low velocity limit given by the Coulomb energy of two spheroids (axis ratio 0.6) at a distance between surfaces of 2 fm.*

occurs after the energy dissipation and mass exchange of the deep-inelastic collision.

On the other hand, there are indications that the splitting occurs rather fast before the nuclei have reached statistical equilibrium. One such indication is that for the same energy loss the element distributions at 12.1 MeV/u show a skewness which is not present at 8.2 MeV/u (see Fig. 3). Another proof comes from the probability of splitting shown as a function of energy loss in Fig. 7. Note that the scale at the ordinate gives the probability of observing a splitting event for the detector setup shown in Fig. 5c. An estimate of the total split-

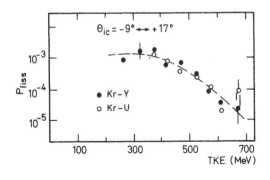

Figure 7
Probability of detecting coincident products as compared to singles events as a function of TKE for the reaction Kr-Y and Kr-U. This probability has not been integrated over angle.

ting probability is ∿10%. The interesting result of Fig. 7 is that there are splitting events already for energy losses as low as 50-100 MeV. Even if one assumes that the excitation energy is not shared proportionally to the masses of projectile and target nuclei at these small excitation energies, an equilibrated Kr nucleus should not undergo sequential fission, since in DIC only relatively modest angular momenta are transferred into the fragments [4]. Moreover, the probability for sequential fission of the Kr projectile which increases rapidly with angular momentum should be higher in the reaction Kr-Y than in Kr-U, whereas nearly the same splitting probability is observed in both cases (see Fig. 7).

In conclusion, the new feature observed in the reaction of 12.1 MeV/u Kr projectiles with several targets is the enhanced production of elements lighter than the projectile, which at small energy losses shows up as a skewness in the element distribution and at larger energy losses develops into a drift towards asymmetry. Coincidences between two light products have been observed, which after adding up the element numbers and laboratory energies show the same distribution as the projectile-like fragments. This suggests that the splitting occurs after the primary deep-inelastic interaction. On the other hand, this splitting is observed already for energy losses as low as 50-100 MeV, which is not sufficient for an equilibrated Kr nucleus to undergo fission. This suggests that in deep-inelastic collisions between heavy nuclei at high incident energies a considerable fraction of the dissipated energy is at first concentrated in a few (collective) degrees of freedom rather than in a small local region (hot spot) and that in the moment of scission this energy is not fully thermalized but leads to the splitting of the outgoing fragment.

We thank G. Augustinski, H. Daues, H.J. Beeskow and M. Ludwig for their assistance in preparing the detection system, H. Folger for the target preparation, and the operators of the UNILAC accelerator for the production of an excellent ^{86}Kr beam of 12.1 MeV/u.

These results have already been presented at the Symposium on Heavy Ion Physics from 10 to 200 MeV/u, Brookhaven National Laboratory, July 16-20, 1979.

1. Sann, H., Damjantschitsch, H., Hebbard, D., Junge, J., Pelte, D., Povh, B., Schwalm, D., Tran Thoai, D.B.: Nucl. Instr. Meth. 124, 509 (1975)
2. Lynen, U., Stelzer, H., Gobbi, A., Sann, H., Olmi, A.: Nucl. Instr. Meth. 162, 657 (1979)
3. Rudolf, G., Gobbi, A., Stelzer, H., Lynen, U., Olmi, A., Sann, H., Stokstad, R.G., Pelte, D.: in print, Nucl. Phys. A119 (1979)
4. Olmi, A., Sann, H., Pelte, D., Eyal, Y., Gobbi, A., Kohl, W., Lynen,

U., Rudolf, G., Stelzer, H., Bock, R.: Phys. Rev. Lett. <u>41</u>, 688 (1978)
5. Eyal, Y., Gavron, A., Tserruya, I., Fraenkel, Z., Eisen, Y., Wald, S., Bass, R., Gould, G.R., Kreyling, G., Renfordt, R., Stelzer, K., Zitzmann, R., Gobbi, A., Lynen, U., Stelzer, H., Rode, I., Bock, R.: Phys. Rev. Lett. <u>41</u>, 625 (1978)
6. Dakowski, M., et al.: submitted to Phys. Lett.
7. Doll, P., et al., GSI annual report 1978, No. 79-11, Darmstadt (1979 (1979)
8. Greiner, D.E., Lindstrom, P.J., Heckman, H.H., Cork, B., Bieser, F.S.: Phys. Rev. Lett. <u>35</u>, 152 (1975).

NEUTRON EMISSION IN HEAVY ION REACTIONS

D. Hilscher

Departments of Chemistry and Physics, University of Rochester,
Rochester, New York, 14627, USA

and

Hahn-Meitner-Institut für Kernforschung Berlin GmbH,
D 1000 Berlin 39, West Germany

and

J.R.Birkelund, A.D.Hoover, W.U.Schröder, W.W.Wilcke, and J.R. Huizenga

Departments of Chemistry and Physics, University of Rochester,
Rochester, New York, 14627, USA

and

A.C. Mignerey and K.L.Wolf

Argonne National Laboratory, Argonne, Illinois, 60439, USA

and

H.F. Breuer and V.E. Viola, Jr.

Department of Chemistry, University of Maryland,
College Park, Maryland, 20742, USA.

1. Introduction

In damped interactions between heavy ions, a large fraction of the kinetic energy of relativ motion in the entrance channel can be converted into intrinsic-excitation energy within short interaction times (10^{-22} - 10^{-21} sec) of the ions. The highly excited fragments produced by this mechanism lose their excitation energy by neutron emission. The relatively small balance of excitation energy is accounted for by proton, alpha and γ-ray emission. This contribution from charged particle emission is calculated with an evaporation program. The experimental observables associated with neutron emission in heavy ion reactions are the following:

1) Neutron energy spectrum and its correlations to the velocity vectors of the heavy fragments:

$$(\frac{d\,\pmb{6}_i}{d\,E}) \; (\; \theta \pmb{;} \; \theta_1, \; \theta_2)$$

2) Neutron multiplicity:

$$\overline{M}_i = \iint M_i \cdot P(M_1, M_2) \cdot d M_1 \cdot d M_2 ,$$

3) Variance of the neutron multiplicity:

$$\sigma^2 (M_i) = \iint (M_i - \overline{M}_i)^2 \cdot P(M_1, M_2) \cdot dM_1 \cdot dM_2$$

4) Cross correlation of neutron multiplicity:

$$\mu (M_1, M_2) = \iint (M_1 - \overline{M}_1) \cdot (M_2 - \overline{M}_2) \cdot P(M_1, M_2) \cdot dM_1 \cdot dM_2$$

where $P(M_1, M_2)$ is the bivariate probability distribution that M_1, M_2
neutrons are emitted by fragment 1 and 2, respectively.

From the neutron multiplicities and the neutron energy spectra in the
rest frames of both fragments we can determine the excitation energy and
temperature of both fragments. That is, it is possible to answer the
question whether the system has reached thermal equilibrium at the time
when the communication between the fragments is switched off. It should
be pointed out, however, that even if thermal equilibrium has been reached at the time of scission, the deformation energy at this point can
cause the excitation energies or temperatures of the two final fragments
to deviate from thermal equilibrium. This latter effect has been observed
[1] for instance in highly energetic proton-induced fission of ^{209}Bi and
^{238}U and possibly also in the reaction ^{197}Au + ^{40}Ar at 240 MeV by Dakows-
ky et al. [2],

Furthermore, we can deduce from the neutron energy spectrum and its correlation with the fragment velocity vectors, the contribution of neutrons
from preequilibrium processes and the time scale for neutron emission.
That is, whether the neutrons are emitted from the dinuclear complex or
from the fragments after full acceleration within their mutual Coulomb
field. As will be shown below, all experiments up to 5 MeV/u of relative
motion above the Coulomb barrier are consistent with the assumption of
statistical equilibrium of the excitation energy degree of freedom within the interaction time.

Though until now no experiment has been performed measuring the second
moments of the neutron multiplicities in heavy ion reactions, it would
be very interesting to test the one-body dissipation model with such

data. This data would yield information on the width of the excitation
energy distribution and on any cross correlation between the excitation
energies of the two fragments. For instance, it would be interesting to
see whether the width of the fragment excitation energies are correlated
with the number of nucleons exchanged as determined from the width of
the Z- and A-distributions. These experiments are, however, extremely
difficult and have been performed until now only in low energy fission
[3].

Many of the basic ideas and techniques used in the analysis of neutron
emission in heavy ion reactions have been used many years ago in the extensive studies of neutron emission in fission [4].

2.   Preequilibrium light particle emission.

Are the neutrons a sensitive probe for preequilibrium light particle
emission? A priori one would give a negative answer to this question.
The simple reason for this is the fact that the neutron multiplicity is
much larger than the charged-particle multiplicity due to the Coulomb
barrier. The charged particles are emitted with a higher probability at
high excitation energies near the beginning of the decay cascade. That
is, charged particles are much more sensitive probes for preequilibrium
particle emission since they are not masked by a large number of evaporated particles. For the investigation of the threshold for preequilibrium
particle emission, charged particles are probably more appropriate than
neutrons. However, depending on the kind of physical processes that are
producing the preequilibrium particles the above conclusion might be drastically changed. For example, in the model where Fermi-jets or promptly
emitted particles (PEP's) are produced the yield of PEP-neutrons is considerably enhanced compared to PEP-protons [4]. On the other hand, for
a hot spot or a localised emission area on the nuclear surface, charged
particles are focussed by the Coulomb field and will be much more easily
identified by this characteristic feature. However, charged particle spectra depend strongly on the unknown deformation of the dinuclear complex
at the time of particle emission.

3.   Techniques

Assuming a binary heavy ion reaction there are in principle three neutron
sources: the composite nucleus and the two final fragments. This assumes
a reaction where sequential fission of the target-like fragment is negligible. If the evaporation residues which have survived fission are identified, the problem is very easy. Then, only one neutron emitter exists,
moving with the average center-of-mass velocity at $\sim 0°$ with respect to
the beam direction. Furthermore, this method preferentially selects small
impact parameters and allows the study of the question of preequilibrium
neutron emission in central collisions, as has been done, for example, by
Westerberg et al. [6]. However, even in this case it is conceivable that
the neutron is emitted from the projectile during the initial stages of
dissipation of the relative kinetic energy, and before the system fuses.
The neutron in this example is emitted from a system that moves faster
than the compound nucleus.

In a strongly damped or fusion-fission binary reaction three possible
sources of neutrons are assumed, namely the composite system and the two
fragments. The emitting fragment is determined from the correlation between the velocity vectors (magnitude and direction) of the fragments and
the neutron velocity vector. In the first iteration step of the analysis
it is assumed that all neutrons that are detected in the direction of the
moving fragment are emitted solely from that respective fragment. Whereas
in the next step the contribution from the respective other fragment is
calculated. To do this certain assumption have to be made: i) the neutrons
are emitted isotropically in the respective restframes and ii) the neutrons
are emitted only from the two final fully accelerated fragments. In figure
1 calculations for ^{165}Ho + ^{56}Fe and ^{165}Ho + ^{136}Xe are shown. It can be
seen for instance that the seperation for more symmetric systems is considerable better than for such asymmetric systems like ^{165}Ho + ^{56}Fe. With
the above described procedure the neutron energy spectra, in the center
of mass of each fragment, can be obtained at any other neutron detection
angle in and out of the reaction plane. These spectra can then be compared with the results at the heavy fragment recoil angles. Alternatively
one can calculate the lab. neutron energy or velocity spectra at any other
angle by using the results at the fragment recoil angles, as has been done
by Eyal et al. [7] and Tamain et al. [8]. Any deviation from the standard
represented by assumption i) and ii) can then be detected, such as noniso-
tropic emission, or emission from the dinuclear complex, or highly energetic preequilibrium neutron emission. That is, the final results or conclusion are independent of the assumptions i) and ii).

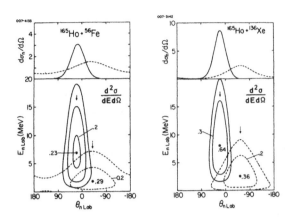

Fig. 1: Calculated neutron cross sections of the projectile like (solid lines) and targets like fragment (dashed lines) for the reactions $^{165}Ho + {}^{56}Fe$ and $^{165}Ho + {}^{136}Xe$ at 8.5 MeV/u using assumptions i) and ii).

However, the sensitivity of this method to highly energetic preequilibrium neutrons depends somewhat on the emission direction of these neutrons. For instance, easy detection is possible for highly energetic neutrons which are emitted in the direction of the slow moving heavy fragment and are produced for instance by a process suggested by Gross and Wylczinsky /9/. Whereas, neutrons coming from a non equilibrium area on the near side of the interaction zone and emitted into the direction of the fast moving projectile like fragment, can be easily masked by evaporation neutrons emitted from this fragment.

The above described method was applied by measurement of the neutron energy or time-of-flight in coincidence with the velocity vectors and masses of both fragments. Eyal et al. /7/ have measured both velocities, whereas Tamain et al. /8/ have measured only the velocity of the projectile like fragment and used two body kinematics to calculate the velocity of the target like fragment. For the experiment described here we have measured the Z and energy of the projectile like fragment and calculated the velocity of this fragment by using A/Z ratios along the line of beta stability. Similar to Tamain et al. /8/, the velocity of the target like fragment was also calculated.

4.   Results and Discussion.

Several groups have measured exclusive neutron spectra in strongly damped heavy ion reactions (see table 1). Eyal et al. [7] have studied the system ^{166}Er + ^{86}Kr at 1.25, 2.5, and 3.35 MeV/u of relative motion above the Coulomb barrier. Tamain et al. [8] have studied ^{197}Au + ^{63}Cu at 0.8 and 1.3 MeV/u and finally we have measured [10] ^{165}Ho + ^{56}Fe at 4.0 MeV/u above the Coulomb barrier. In figure 2 the neutron spectra in the fragment center of mass frames are shown for all three systems. We see that in all systems the exponential slope is the same for both fragments. This exponential slope can be identified with an effective nuclear temperature. Thus, both fragments have the same temperature, which implies that the excitation energy degree of freedom has reached statistical equilibrium during the interaction time. Furthermore, there is no clear indication of preequilibrium highly energetic neutrons. In the case of ^{165}Ho + ^{56}Fe, the contribution of highly energetic neutrons is smaller than 5 %.

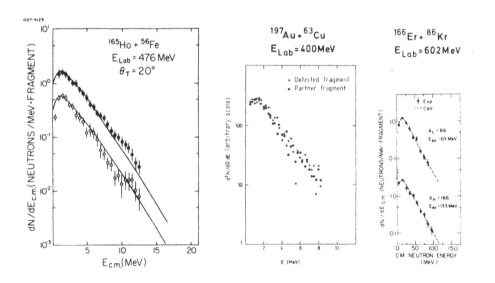

Fig. 2:   Center of mass neutron energy spectra from the reactions ^{165}Ho + ^{56}Fe [10], ^{166}Er + ^{86}Kr [7], and ^{197}Au + ^{63}Cu [8].

Table 1:   Emission of preequilibrium neutrons in exclusive
             measurements

Reaction	E_{lab} [MeV]	$(E_{cm}-V_{CB})/\mu$ [MeV/u]	Measurement a)	Preequi-libr. b)	Ref.
$^{12}C \longrightarrow {}^{158}Gd$	152	8.3	CF	+	[6]
$^{16}O \rightarrow {}^{93}Nb$	204	9.5	SDC/CF	(+) e)	[13]
$^{16}O \rightarrow {}^{154}Sm$	112-169	2.6-6.3	CF	- to +	[14]
$^{20}Ne \longrightarrow {}^{150}Nd$	175	4.6	CF	-	[6]
$^{56}Fe \longrightarrow {}^{165}Ho$	476	4.0	SDC/FF	-	[10]
$^{86}Kr \longrightarrow {}^{166}Er$	496-675	1.25-3.33	SDC	-	[7]
$^{132}Xe \rightarrow {}^{197}Au$	990	2.3	SDC	(+) c)	[15]
$^{63}Cu \longrightarrow {}^{197}Au$	365,400	0.8,1.3	SDC/FF	-	[8]
$^{40}Ar \longrightarrow {}^{197}Au$	240	1.3	SDC/FF	- d)	[2]

a)   CF: complete fusion; FF: Fusion fission; SDC: Strongly damped
      collision.

b)   +: evidence, -: no evidence for pre-equilibrium neutrons

c)   see also reference [7]
      The results of Broek [16], Kumpf et al. [17] and Madey et al.
      [18] are not included since these were inclusive measurements.

d)   evidence for possible shape deformations of fragments.

e)   The lab.energy spectra show highly energetic neutrons but
      since no event by event analysis has been performed yet it
      is not clear whether these neutrons can be associated with
      preequilibrium neutrons.

Equal temperature ($T_1 = T_2$) of both fragments 1 and 2 would require that the excitation energy divides like the mass ratio, if the level density parameter is given by $a = A/c$ and c is independent of A:

$$E_1* \; / \; E_2* \; = \; \frac{a_1 \cdot T_1^2}{a_2 \cdot T_2^2} \; = \; A_1 \, / \, A_2$$

Since most of the excitation energy is carried away by neutrons, the neutron multiplicities measure the excitation energy of each fragment and one would expect, for the case of statistical equilibrium, the ratio of the neutron multiplicities to be equal to the mass ratio. This is shown in the upper part of figure 3. To be more exact, however, it's necessary to calculate the neutron multiplicities with an evaporation code that takes into account charged particle emission and angular momentum effects. The filled triangles in the lower part of figure 3 are the result of such a calculation using the code MBII [11]. Similar agreement between the results of evaporation calculations and the measured neutron multiplicities has been obtained by Eyal et al. [7].

The in to out of plane anisotropy can be determined by the measured neutron yield out of the reaction plane compared with the expectation according to assumption i), using the neutron yield at the fragment recoil angles. In the case of the reaction ^{166}Er + ^{86}Kr, Eyal et al. [7] report no out of plane anisotropy whereas Tamain et al. [8] in the case of ^{197}Au + ^{63}Cu and this experiment [10] in the case of ^{165}Ho + ^{56}Fe find a 20 % anisotropy for the neutrons emitted by the heavy fragment. This agrees qualitatively with model calculations for the spin alignment. However, in neither case is the accuracy sufficient for a proof of spin alignment.

Finally, the neutron multiplicity is also sensitive to the N/Z ratio of the primary fragments. This effect is shown in fig. 4. The measured ratio of neutron multiplicities of the heavy and light fragment at a TKE-loss of 100 MeV is consistent with an N/Z ratio larger than 1.26. In an independent experiment [12] N/Z has been determined to be 1.28 at this TKE-loss. The equilibrium N/Z according to liquid drop potential surface calculations [10] varies between 1.30 to 1.49, depending on different assumptions used in these calculations, whereas the N/Z of the composite system is 1.38. Thus, the neutron-to-proton ratio is almost equilibrated [10] at interaction times as short as (4 or 8) $\cdot$ 10^{-22} sec.

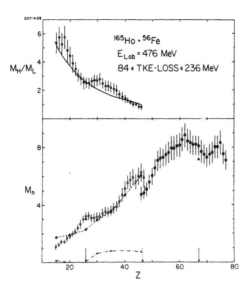

Fig. 3: Neutron multiplicities of the light fragment (open circles) and heavy fragment (filled squares) as a function of Z. The filled triangles are evaporation calculations. In the upper part the ratio M_H/M_L is plotted and compared to the mass ratio m_H/m_L (solid line).

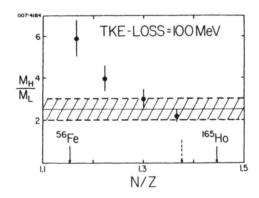

Fig. 4: The calculated ratio M_H/M_L is shown as a function of the N/Z of the projectile like fragment. The hatched area represents the experimental value of M_H/M_L.

The findings for all three investigated systems indicate that the energy dissipation reaches thermal equilibrium very fast, that is within $(1 - 5) \cdot 10^{-22}$ sec, and consequently no highly energetic preequilibrium neutrons have been observed.

Until now only data up to 4 MeV/u of relative motion above the Coulomb barrier has been discussed. In table 1 a summary of published exclusive data is shown. It is seen that, for energies above (5 - 6) MeV/u, several groups have reported a clear indication of preequilibrium neutrons. In the reported cases, the experimental selection of the evaporation residues via characteristic γ-rays of these compound nuclei, selects the impact parameters of mainly central collisions, in which case Fermi-jet neutrons are especially enhanced [5]. However, for the reaction of ^{16}O + ^{93}Nb at 9.5 MeV/u above the Coulomb barrier Petitt et al. [13] have possibly also observed preequilibrium neutrons for more peripheral strongly damped collisions.

Within the exciton model one would expect preequilibrium emission to occur when the energy per excitation is comparable to the particle effective binding energy. In the case of very asymmetric systems one can identify the initial exciton number with the number of nucleons in the projectile [19]. Thus we would expect the occurence of preequilibrium neutron emission at (7 - 8) MeV/u.

A similar conclusion can be reached by comparing, as a function of the temperature, the relevant time constants involved. These are the particle evaporation time [20] τ_p, the nuclear relaxation time [21] τ_R, and the interaction time [22] τ_{INT}. In figure 5 this has been done. The interaction time is given for ^{165}Ho + ^{56}Fe at 4 MeV/u above the Coulomb barrier. It can be seen that for nuclear temperatures which can be reached at this bombarding energy, if a hot spot (T ≈ 5 MeV) were formed on contact of the two ions, it would decay within a typical nuclear relaxation time, τ_R, which is small compared to the evaporation time τ_p. Thus it is expected in this case that the neutrons are emitted from the fully equilibrated and accelerated fragments.

For incident energies of (7 - 8) MeV/u above the Coulomb barrier temperatures of ~ 8 MeV can be reached momentarily and then the condition $\tau_p \gg \tau_R$ is not fulfilled and preequilibrium particle emission is expected to occur.

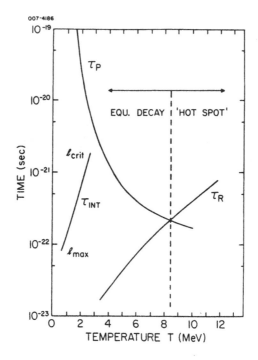

Fig. 5:   Relevant time scales (see text).

[1]   E. Cheifetz, Z. Fraenkel, J. Galin, M. Lefort, J. Péter,   and
       X. Tarrago, Phys.Rev. C2 (1970) 256

[2]   M. Dakowski, R. Chechik, H. Fuchs, F. Hanappe, B. Lucas, C. Mazur,
       M. Morjean, J. Péter, M. Ribrag, C. Signarbieux, B. Tamain, to be
       published in Phys.Rev.Lett.

[3]   C. Signarbieux, R. Babinat, H. Nifenecker, J. Poitou, Proceeding
       of the Symposium on Physics and Chemistry of Fission in Rochester,
       1973, p. 179

[4]   H. Nifenecker, C. Signarbieux, R. Bobinet, J. Poitou, Proceedings
       of the Symposium on Physics and Chemistry of Fission in Rochester,
       1973, p. 117

[5]   J.P. Bondorf, J.N. De, G. Fai, A.O.T. Karvinen, B. Jakobsen, and
       J. Randrup, NORDITA - 79/27
       J.P. Bondorf, J.N. De, A.O.T. Karvinen, G. Fai, and P. Jakobsen,
       Phys.Lett. 84B, 162 (1979)

[6]   L. Westerberg, D.G. Sarantites, D.C. Hensley, R.A. Dayras, M.L.
       Halbert, and J.H. Barker, Phys.Rev. C18 (1978) 796

[7]   Y. Eyal, A. Gavron, I. Tserruya, Z. Fraenkel, Y. Eisen, S. Wald,
       R. Bass, C.R. Gould, G. Kreyling, R. Renfordt, K. Stelzer, R. Zitz-
       mann, A. Gobbi, U. Lynen, H. Stelzer, I. Rode, and R. Bock,
       Phys.Rev. Lett. 41 (1978) 625, and WIS-79/18-Ph

[8]   B. Tamain, R. Chechik, H. Fuchs, F. Hanappe, M. Morjean, C. Ngô,
       J. Péter, M. Dakowski, B. Lucas, C. Mazur, IPNO-RC-79-02

[9]   D.H.E.Gross and J. Wilczynski, Phys.Lett. 67B (1977) 1

[10]   D. Hilscher, J.R. Birkelund, A.D.Hoover, W.U.Schröder, W.W.Wilcke,
       J.R. Huizenga, A.C. Mignerey, K.L. Wolf, H.F.Bruer, and V.E.Viola,
       Phys.Rev. C20 (1979) 576

[11]   M. Beckermann and M. Blann, UR-NSRL-135 (1977)

[12]   H. Breuer, B.G.Glagola, V.E.Viola, K.L.Wolf, A.C. Mignerey, J.R.
       Birkelund, D. Hilscher, A.D.Hoover, J.R. Huizenga, W.U.Schröder,
       and W.W. Wilcke, Phys.Rev.Lett. 43 (1979) 191

[13]   G.A.Pettit, R.L.Ferguson, D.C. Hensley, F.E.Obenshain, F. Plasil,
       A.H.Snell, G.R.Young, J.E.Gaiser, K.A.Geoffrey, and D.G.Sarantites,
       Bull.of the Am.Phys.Soc. 24 (1979) 630

[14] K.A. Geoffrey, D.G. Sarantites, L. Westerberg, J.H. Barker, D.C. Hensley, R.A. Dayras, and M.L. Halbert, Bull.of Am.Phys.Soc. 23 (1978) 950, and 24 (1978) 695

[15] C.R.Gould, R.Bass, J.v.Czarnecki, U. Hartmann, K. Stelzer, R. Zitzmann, and Y. Eyal, Z.Physik, A284 (1978) 353

[16] H.W. Broek, Phys.Rev. 124 (1961) 233

[17] H.Kumpf, L.Kumpf, and Shih Shuang-hui, Sov.Journ. of Nucl.Phys. 1 (1965) 186

[18] R. Madey, B.D. Anderson, A.R.Baldwin, R. Cecil, W.Schimmerling, J. Kort, Bull.of Am.Phys.Soc. 23 (1978) 515

[19] M. Blann, Nucl. Phys. A235 (1974) 211

[20] R.G. Stokstad,
      Proc.Topical Conf. on Heavy-Ion Collisions, Falls Creek, Falls State Park, Tennessee, CONF. 770602
      (U.S.Department of Commerce) (1977)

[21] A. Kind and Patergnani, Nuovo Cimenta 10 (1953) 1375

[22] W.U. Schröder and J.R. Huizenga, Ann.Rev. of Nucl.Sci. 27 (1977) 465

FUSION-FISSION TYPE COLLISIONS

H. Oeschler

DPh-N/BE, CEN Saclay, BP 2, 91190 Gif-sur-Yvette, France

and

H. Freiesleben[*]

Gesellschaft für Schwerionenforschung, 6100 Darmstadt, Germany

ABSTRACT

Three examples of fusion-fission type collisions on medium-mass nuclei are investigated whether the fragment properties are consistent with fission from equilibrated compound nuclei. Only in a very narrow band of angular momenta the data fulfill the necessary criteria for this process. Continuous evolutions of this mechanism into fusion fission and into a deep-inelastic process and particle emission prior to fusion have been observed. Based on the widths of the fragment-mass distributions of a great variety of data, a further criterion for the compound-nucleus-fission process is tentatively proposed.

I. INTRODUCTION

In heavy-ion reactions compound nuclei can be formed at very high angular momenta. Due to the large centrifugal forces which greatly reduce the fission barriers, fission may occur for all nuclei across the periodic table. As the rotating liquid-drop model (RLDM) [1] predicts the limits of stability against fission (fission barrier height, $B_f \approx$ neutron binding energy, B_n) much effort was devoted to exploit this region experimentally. The RLDM further predicts the stability limits of rotating nuclei (vanishing fission barrier, $B_f = 0$). The impact of the latter on the reaction dynamics is less known. The intension of this contribution is to investigate, whether these limits establish themselves in the fission properties of the decaying nuclei.

Fission from fully equilibrated nuclei which have been formed in the complete amalgamation of target and projectile is called "compound nucleus fission" (CNF). Necessary conditions for this classification are given in sect. 2.1. This type of reaction is a subgroup of the more general class "fusion fission" (FF) exhibiting the typical fission properties, too, however the statistical equilibrium of the decaying nucleus is not required.

The new data in the mass region 90-190 suggest that the considered systems undergo fusion-fission type collisions. Close inspection of the data reveals, however, tnat there exists not only a transition between CNF and FF, but also paths towards incomplete fusion followed by fission and towards deep inelastic collisions (DIC). In sect. 2.1 results of the reaction ^{132}Xe + ^{30}Si forming ^{162}Er [2] are presented. The similarities and differences between this reaction and the ^{132}Xe + ^{56}Fe system [3] are discussed in sect. 2.2. Deviations from symmetric fission are observed in ^{32}S induced reactions on light target nuclei (^{59}Co - ^{89}Y) [4] as shown in sect.2.3.

Finally in sect. 3 we summarize experimental widths of fission fragment mass distributions of medium-mass nuclei. It is discussed, whether the trend of the width can serve as a further criterion for classification.

2. FUSION-FISSION TYPE COLLISIONS ON MEDIUM-MASS NUCLEI

2.1. The compound nucleus ^{162}Er

The compound nucleus ^{162}Er has attracted much attention as it can be formed by several entrance channels. Five reactions have been studied so far : ^{16}O + ^{146}Nd [5], ^{32}S + ^{130}Te [5,6], ^{40}Ar + ^{122}Sn [7,8], ^{86}Kr + ^{76}Ge [7,8] and ^{132}Xe + ^{30}Si [2] applying the γ-multiplicity technique [5,7,8] and the time-of-flight method [2,6]. In ref. [5] the measured fusion-evaporation cross sections were used to determine the maximum angular momenta leading to fusion, ℓ_{crit}. Between these values and the measured γ-multiplicities a linear dependence was found. This justifies the use of γ-multiplicity data in order to determine ℓ_{crit}. Its dependence on excitation energy for the reaction ^{40}Ar + ^{122}Sn [7] is shown in Fig.1 taken from ref. [7]. At low energies the experimental values are nicely reproduced by a fusion model (dashed-dotted line). At about 65 $\hbar$ ℓ_{max} saturates. It is concluded, that compound nuclei at higher angular momenta decay by fission as it is suggested by the RLDM [1] which predicts the fission barrier, B_f to be lower than the neutron binding energy, B_n for $\ell \gtrsim 69 \hbar$.

In order to study the compound nucleus ^{162}Er at high angular momenta we investigated its fission properties using the asymmetric entrance channel ^{132}Xe + ^{30}Si. With Xe-beam energies from 5.4 MeV/u to 8.2 MeV/u one covers the region from the expected onset of fission at 69 $\hbar$ up to and beyond the critical value of 95 $\hbar$, where the RLDM predicts the fission barrier to vanish. At the highest energy up to 130 $\hbar$ are available in the entrance channel.

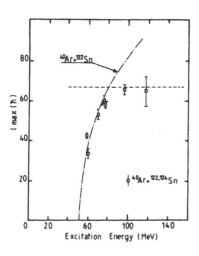

Fig. 1 - The maximum angular momentum of the fusion-evaporation process determined from γ-multiplicity measurements vs the excitation energy of the compound nucleus ^{162}Er formed in the reaction ^{40}Ar + ^{122}Sn. (from ref. [7]).

The experiments were carried out using a time-of-flight telescope in combination with an ionization chamber [2]. Fig. 2 shows raw data from the latter. At all energies fully relaxed fragments concentrated about Z = 34, being half of the compound atomic number, are observed which is typical for symmetric fission. At the highest energy of 8.2 MeV/u deep inelastic reaction products can clearly be seen around the projectile. It is obvious from Fig. 2 that fission fragment properties can undisturbedly be studied over the whole energy range, as they are well separated from deep inelastic processes. Due to the high center-of-mass velocity of this inverted reaction we also observe at forward angles fission

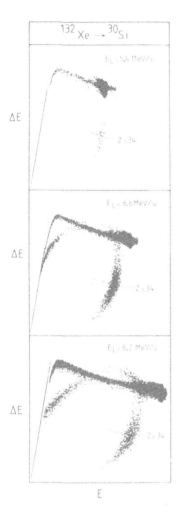

Fig. 2 - Scatter plots of the energy loss in the ionization chamber vs the laboratory energy.

Fig. 3 - Double differential cross sections $d^2\sigma/d\theta dZ$ of the fully relaxed reaction products vs atomic number. The lines are to guide the eye.

fragments at low laboratory energies being emitted towards backward angles in the c.m. system. They cause the funny horse-shoe-like shape of the kinematic curve determined by the Coulomb repulsion between the reaction products.

From the raw data Z-spectra of the relaxed reaction products have been obtained. They are shown in Fig. 3 and quantitatively demonstrate the separation between fission fragments and deep inelastic reaction products. These spectra have been measured at angles close to the grazing angle (for details see table 1). As the fission fragment distributions are rather narrow, long interaction times can be assumed resulting in $1/\sin\theta$ angular distributions. This has not been verified in the present experiment but has been proven in many similar reactions [9]. In the following, two aspects will be discussed : firstly, the fission excitation function and secondly, the properties of the fragment distributions.

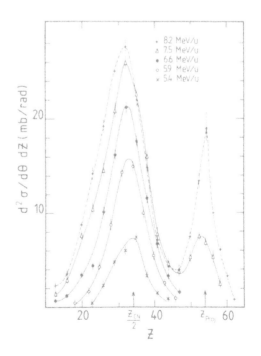

Table I

Survey of results and quantities characterizing the reaction

E_{lab}/A (MeV/u)	E_{cm} (MeV)	E_{cm}/V_{Coul}	θ_{Lab}	θ_{cm}^{sym}	E_X^{CN} (MeV)	θ_{saddle}	σ_{ER} a) (mb)	$\sigma_{fission}$ b) (mb)	σ_{reac} c) (mb)	$t_{T=1/2}$ c) (ℏ)	ℓ_{crit} (exp) (ℏ)
5.4	132	1.43	9°	29°	85	1.51	865	130	1403	82	70
5.9	144	1.56	8°	27°	97	1.68	790	306	1656	93	77
6.6	161	1.75	7°	24°	114	1.89	703	466	1938	107	84
7.5	183	1.99	5.5°	20°	136	2.14	621	653	2222	122	93
8.2	200	2.17	5°	19°	153	2.26	569	713	2397	133	98

a) assuming $\sigma_{ER} = \pi \lambdabar^2 (65)^2$.

b) from optical-model fit to the measured elastic scattering.

c) the total errors are estimated to be of the order of $\pm$ 10 %.

Fig. 4 displays the fission excitation function ; the result of the very similar entrance channel $^{32}S + ^{130}Te$ [6] is given, too. A threshold behaviour is observed at an energy where the compound nucleus has reached approximately 65 ℏ, in accordance

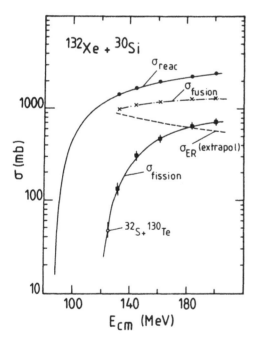

Fig. 4 - Measured fission excitation function (■), the solid line is to guide the eye. The dashed line represents the evaporation-residue cross section which has been assumed to be due to partial waves ≤ 65 ℏ. σ_{fusion} represents the sum of $\sigma_{fission}$ and σ_{ER}. The reaction cross-section (●) result from fits to the elastic scattering data and the connecting solid line shows the predictions to other energies.

with the limiting angular momentum found for fusion evaporation [7,8]. Assuming that at all bombarding energies the compound nuclei with spins between 0 and 65 ℏ decay by particle evaporation only, the evaporation-residue cross section has been calculated. This estimate is shown as dashed line in Fig. 4. Summing σ_{ER} and the measured $\sigma_{fission}$ to a fusion cross-section (dashed-dotted line) the maximum angular momenta leading to fusion can be extracted (see table 1). At the highest energy a maximum angular momentum of 98 ℏ has been deduced and at 95 ℏ the RLDM predicts the fission barrier to vanish. This seems, however, to be an accidental coincidence, as the fission cross section shows no saturation and a short-run experiment at higher energies indicate that it will still increase.

From the limiting angular momentum found for the fusion-evaporation process and the fission excitation function one can conclude that the limit at 65 ħ marks the frontier between particle evaporation and fission, in good agreement with the RLDM. The total fusion cross section seems to rise up to ℓ-values of nearly 100 ħ.

In order to study whether fully equilibrated compound nuclei undergoing fission have been formed we discuss the properties of the fission fragments. Necessary conditions for CNF are :

1. angular distributions proportional to 1/sinθ,

2. full relaxation in energy : i.e. fragment kinetic energies given by the Coulomb repulsion,

3. symmetric mass- and Z-distributions primarily centered at half the combined target and projectile mass and Z, respectively, (for fissility parameters $x > x_{BG}$; (see sect. 2.3)).

Condition 1 and 2 are fulfilled in the data in this article. The relaxation in energy is a fast process ; hence these criteria are rather weak and are often satisfied by deep inelastic reactions, too. Full relaxation of the mass-(charge-) asymmetry degree of freedom seems to be a more restrictive condition.

In order to verify condition 3 one has to take into account, that the observed distributions differ from the primary ones due to particle emission from the excited fragments. In the reaction $^{132}Xe + ^{30}Si$ their spins are low ($\lesssim 20$ ħ) and statistical model calculations [10] predict only neutron evaporation. The Z-distribution should hence be centered at $Z_{CN}/2$, while the centroids of the mass distributions are shifted below $A_{CN}/2$. This is observed at low excitation energies, as displayed in Fig. 5.

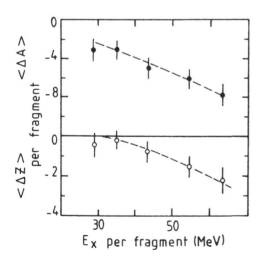

Fig. 5 – Shifts of the centroids of the observed mass and Z-distributions with respect to $A_{CN}/2 = 81$ and $Z_{CN}/2 = 34$ vs excitation energy per fragment.

At high energies a shift of up to two charge units per fragment is observed, as already can be seen from Fig. 3. As mentioned, it is excluded that these charged particles originate from the fission fragments. It is furthermore unlikely that charged particles are emitted from the compound nucleus prior to fission. Statistical model calculations [10] predict even at 100 ħ a predominance of neutron over charged particle emission.

The observed loss of charge is probably due to emission prior to fusion a process not unlikely at energies of 4 MeV/u above the barrier (see contributions by J. Bondorf, M. Halbert, J. Wilczyinski).

It seems plausible that the width of the fission-fragment mass distribution is an essential quantity concerning the question whether CNF has been observed. However, it is difficult at present to state a fourth condition in clear terms. In the case of fission from an equilibrated system, the width should be mainly governed by the temperature. This dependence is shown in Fig. 6 for the ^{132}Xe + ^{30}Si system. The temperature at the saddle point has been chosen as abscissa in order to allow comparison with Nix's calculations [11] which are included as dashed-dotted line. This calculation is based on decaying equilibrated nuclei having spin zero, whereas the measured fission fragments originate from nuclei with spins around 65 ℏ to 100 ℏ. The absolute values agree surprisingly well ; the slope of the experimental points is steeper. The narrow width in agreement with Nix's prediction suggests fission to originate from equilibrated nuclei. As discussed before, incomplete fusion occurs at higher bombarding energies leading to the formation of various fissioning nuclei. The observed increase of the mass width (Fig. 6) may thus find a simple explanation.

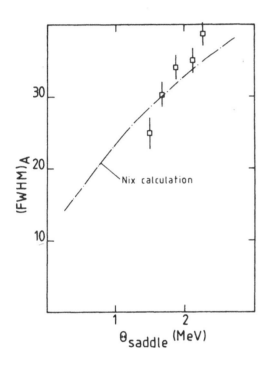

Fig. 6 - FWHM of the observed mass distributions vs the temperature at the saddle point. The dashed-dotted line represents the prediction of ref. [11].

From the reaction ^{132}Xe + ^{30}Si and the other entrance channels one concludes that fission sets in at about 65 ℏ as expected from the RLDM. At the onset the fission fragments exhibit all properties of CNF. At higher energies pre-equilibrium particles are emitted. The observed narrow widths of the mass distributions indicate that after incomplete fusion the nuclei being lighter than the compound nucleus are equilibrated when they fission. The pre-equilibrium emission occurs already at angular momenta lower than the predicted stability limit of the compound nucleus.

2.2. The reaction ^{132}Xe on ^{56}Fe

The narrow mass distributions around symmetry observed in the reaction ^{132}Xe on ^{30}Si contrast the broad distribution found in the reaction ^{132}Xe on ^{56}Fe studied at

5.7 MeV/u [3] (Fig. 7). The hatched areas represent the quasi-elastic reaction products, the broad distribution the fully relaxed components. This distribution peaks at Z = 40 ($Z_{CN}/2$). The width is twice as large as predicted by Nix [11]. In spite of a rather low bombarding energy, high angular momenta are available in the entrance channel. The highest ℓ-value contributing to the relaxed reaction products is 120 ℏ. But already around 84 ℏ the fission barrier has vanished according to the RLDM. Therefore two or perhaps three components may be mixed undistinguishable and a speculative decomposition is given as dashed line in Fig. 7. It marks a fraction due to CNF with ℓ-value between 33 ℏ ($B_f = B_n$) and 84 ℏ ($B_f = 0$) and with a width as predicted by Nix [11]. The higher ℓ-value then correspond to fusion-fission processes probably not passing through an equilibrated compound nucleus and resulting then in a broader distribution. Furthermore it is possible that deep inelastic collisions contribute to this distribution.

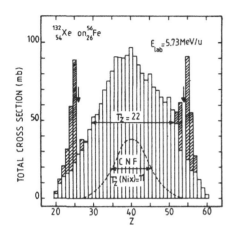

Fig. 7 - Element distribution of quasi-elastic reaction products (hatched area) and of fully relaxed fragments (open area). The dotted lines hypotetically indicates the fraction due to fission from compound nuclei.

This decomposition is on one hand speculative and on the other hand bears an academic character, as no limits between the several components are visible and no criteria exist how to separate them. It is rather a continuous evolution between the various processes and the question concerning the stability limit of the compound nucleus has no apparent answer.

2.3. Heavy-ion reactions forming nuclei with A = 90-120

According to the RLDM the nucleus ^{162}Er has very similar stability limits as ^{121}Cs. The latter has been formed in the reaction ^{32}S on ^{89}Y [4]. At a bombarding energy of 4.8 MeV/u the angular momenta are just sufficient to allow fission. The laboratory-energy-versus-mass scatter plot obtained at 35° is shown in the upper part of Fig. 8. The isolated area of fission events peaking around $A_{CN}/2$ and having energies corresponding to the Coulomb repulsion between two fragments reminds of the reaction ^{132}Xe on ^{30}Si around 5.4 MeV/u. The width of the mass distribution is even smaller than predicted by Nix [11]. The fission excitation function (Fig. 9) shows the same steep rise as seen in the reaction ^{132}Xe + ^{30}Si. Consequently, all properties indicate that fission originates in this reaction from fully equilibrated nuclei.

This picture, however changes in the reactions between ^{32}S and lighter target nuclei, down to ^{59}Co [4]. The angular momenta are always high enough to allow fission as shown in table 2, and indeed in all reactions events are observed along the line given by the Coulomb repulsion between the fragments. The mass distribution however, vary drastically for the different reactions as can be seem in Fig. 8. Whereas in the reaction ^{32}S on ^{85}Rb still a peak around $A_{CN}/2$ is seen, this typical behavior of symmetric fission is lost in the reactions ^{32}S on ^{74}Ge and ^{65}Cu.

Table 2
Summary of the conditions for compound nucleus formation and its decay by fission

Reaction	Compound nucleus	Fissility parameter x	E_L (MeV)	E_x^{CN} (MeV)	(ref.12) ℓ_{crit} ($\hbar$)	$\ell_{B_f = B_n}$ ($\hbar$)	$\ell_{B_f = 0}$ ($\hbar$)
^{32}S + ^{59}Co	^{91}Tc	.40	160 170	91.5 98.0	60.5 63.4	54	76
^{32}S + ^{65}Cu	^{97}Rh	.41	160 170	96.5 103.2	62.5 65.6	59	82
^{32}S + ^{74}Ge	^{106}Cd	.43	160 170	99.4 106.4	64.7 68.1	53	91
^{32}S + ^{79}Br	^{111}Sb	.47	170	99.8	68.0	61.5	88
^{32}S + ^{85}Rb	^{117}I	.48	170	96.2	69.0	63.5	93
^{32}S + ^{89}Y	^{121}Cs	.50	153 184	76.0 98.8	61.9 74.1	62	97

In Fig. 10, the mass distributions of the relaxed reaction products of all six reactions are summarized. The double differential cross section $d^2\sigma/d\theta dA$ is given as function of the asymmetry in the exit channel. The dashed line marks symmetric mass plits, the arrows the projectile and target mass. This comparison shows a narrow symmetric fission mass distribution for the reaction ^{32}S on ^{89}Y. It broadens when forming lighter compound or composite nuclei until in the reaction ^{32}S on ^{65}Cu a minimum develops. Finally, the ^{32}S + ^{59}Co system exhibits a maximum around the target mass. A similar behaviour has been observed in the reaction ^{32}S + ^{50}Ti [13] where the binary nature of the process was verified. For the presented data, measured cross sections at various angles are compatible with $1/\sin\theta$ angular distributions.

The continuous evolution of the mass distribution in Fig. 10 might not surprise as the fissility parameters of the compound nuclei range from 0.5 to 0.4 (see table 2). In this region one expects the Businaro - Gallone limit [14]. Only nuclei having fissility parameters higher than this limit are expected to fission symmetrically with increasing widths when approaching this limit.

If the Businaro-Gallone limit is an explanation, than the mass distributions should

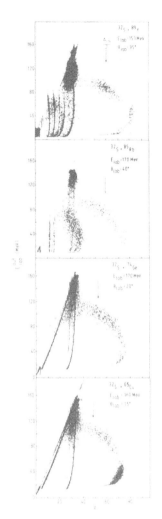

Fig. 8 - Scatter plots of laboratory energy vs mass of four reactions demonstrating a change in the mass distributions of the relaxed fragments.

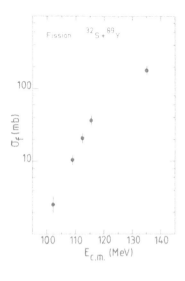

Fig. 9 - Measured fission excitation function for the reaction ^{32}S on ^{89}Y.

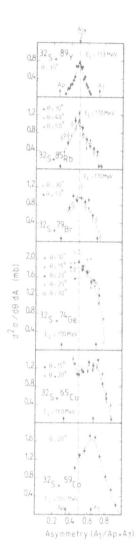

Fig. 10 - Mass spectra of the relaxed reaction products vs asymmetry varying from a narrow distribution to broad ones peaked at target (and projectile) mass for the lightest systems.

be independed of the entrance channel. (Bohr's definition of the compound nucleus [15]). Unfortunately, there are not many data to compare with. The compound nucleus ^{97}Rh formed with ^{32}S on ^{65}Cu has also been studied by the reaction ^{35}Cl on ^{62}Ni [16] and the reported Z distributions agree well with our data when interpreted in mass units. Due to the similarity this in not a stringent test of Bohr's criterion. Part of the mass range studied here is covered by the more asymmetric entrance channels ^{12}C on ^{89}Y, ^{98}Mo and Ag [17]. In order to produce compound nuclei with spins favouring fission, high incident energies are needed. Therefore a large yield from deep inelastic collisions is observed. In Fig. 11 (taken from ref. [17]) the Z-distribution from ^{12}C on ^{89}Y at 197 MeV is shown. It suggests a decomposition (added as dashed and dotted lines) into a deep inelastic collision component and one centered at $Z_{CN}/2$ attributed to fission. In all mentioned ^{12}C-induced reactions the widths of the fission distributions correspond to 20-24 mass units. They agree only with the widths observed in the ^{32}S-induced reactions on the two heavy nuclei ^{85}Rb and ^{89}Y. For the lighter systems, ^{32}S on ^{74}Ge and ^{65}Cu which are best suited to be compared with the ^{12}C + ^{89}Y-system, much broader and dissimilar distributions are observed. These pronounced discrepancies establish a violation of Bohr's criterion [15] and hence the CNF picture seems not applicable for ^{32}S induced reactions on nuclei with A ≤ 80. The strong change in the mass distributions shown in Fig. 10 is probably not due to the proximity of the Businaro-Gallone limit.

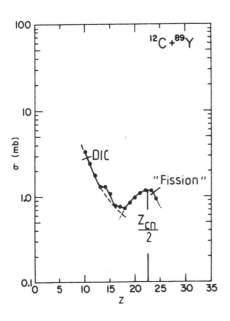

Fig. 11 - Element distribution obtained in the reaction ^{12}C on ^{89}Y at 197 MeV (from ref. [17]). A decomposition into deep inelastic processes (dashed line) and fission fragments (dotted line) has been added.

Assuming that deep inelastic processes are present in the light systems one might be tempted to explain the change in the mass distributions shown in Fig. 10 by coexistence of two processes, CNF and DIC, with changing dominance. This interpretation encounters several difficulties : the ratio of the center-of-mass energy over the barrier is in all reactions about 1.7. Therefore it is difficult to understand why deep inelastic processes should be strong in the lighter systems and negligible in the heavier ones, in particular when one calculates the mass-transport coefficients in a diffusion model [18] for the various reactions. They turn out to be rather similar and very small. Furthermore the change in the mass spectra begins with a broadening of the fission peak and not by an appearance of additional side peaks as can be seen from Fig. 10 for the reactions

^{32}S on ^{85}Rb and ^{79}Br. Therefore the interpretation of the observed change as a competition between CNF and DIC in terms of a simple diffusion mechanism is not satisfying.

Instead of a competition between the two processes an evolution from one mechanism to the other may occur. This seems plausible as compound-nuclei lifetimes and relaxation times of light nuclei are comparable [19], and hence the memory of the entrance channel might be preserved. Then the observed transition would reflect different stages of equilibration of a fusion-fission process. It is interesting to note that this transition occurs for ℓ-values, where the fission barrier has not yet vanished. No conclusion on the validity of the RLDM in this mass region can be reached at present.

3. WIDTHS OF THE MASS DISTRIBUTIONS

In this chapter we discuss the widths of fission mass distributions for various reactions. The intention is to compare with Nix's calculations [11] and to derive a further criterion for CNF based on the width of the mass distribution.

In Nix's calculations the widths is given as function of the fissility parameter x and the temperature at the saddle point θ. The theoretical predictions are shown in Fig. 12. The width is strongly increasing when approaching the Businaro-Gallone limit (x_{BG}) at x = 0.396 and slowly increasing with temperature. In order to compare various data with the theory the curve at θ = 2 MeV is chosen as reference. This choice was guided by the fact that most heavy-ion data have been obtained at this temperature. The FWHM at different temperatures have then been scaled according to the theoretical dependence. This is mostly a minor correction. The theory [11] considers only decaying nuclei with spin zero, however in heavy-ion reactions, fission arises from high-spin nuclei. Not much is known about the influence of angular momentum on the width of the distributions. Recently it has been stated [20] that this influence is negligible if the fission barrier has not vanished. Therefore we have chosen for our comparison only data for which the fission barrier does not vanish.

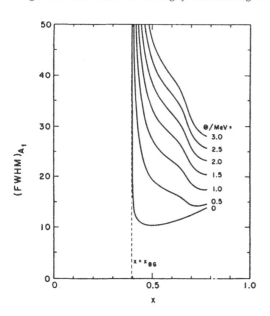

Fig. 12 – Predictions of the FWHM of the mass distribution vs the fissility parameter x for various values of the saddle-point temperature θ. (from ref. [11]).

The experimental results are given in Fig. 13 in analogy to Fig. 12. The full black points present the ^{32}S-induced reactions discussed in sect. 2.3 and the reaction ^{132}Xe on ^{30}Si discussed in sect. 2.1. The other data are taken from the literature. In spite of the scattering, a trend can be stated : at masses around 180 the measured widths exceed the predictions by 20-50 % which has already be mentioned in Nix's paper [11]. Around mass 150-160, the measured widths seem to agree with the prediction and for lighter masses the experimental values are far below the theoretical values. The general trend might be characterized by the hatched area, reflecting a functional depence $\Gamma_A/A = 0.2$, thus being a continuation of the trend found for heavier systems [23]. This trend, however is in strong contrast to the existence of a Businaro - Gallone point around x = 0.4 as has been pointed out in ref. [21]. The two black symbols from ^{32}S-induced reactions (put in brackets) leaving the general trend at x ≈ 0.43 indicate, that they do not belong to the class of CNF, as argued already in sect. 2.3.

It is to early to deny on the basis of present data the existence of the Businaro-Gallone point. One is tempted, however, to add a further criterion of the kind $\Gamma_A/A \approx 0.2$ to the list of necessary conditions for CNF given in sect. 2.1.

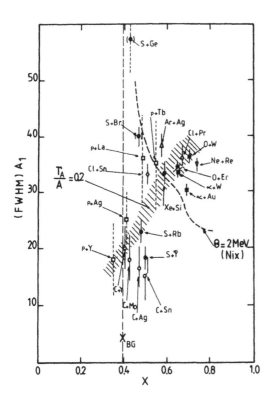

Fig. 13 - Same graph as Fig. 12. The experimental points (● in chapter 2 and ref. [2,4], * ref. [11], ◇ ref. [16], o ref. [17], ∇ ref. [20], ◻ ref. [21], Δ ref. [22]) scaled in order to correspond to θ = 2 MeV are compared to Nix's calculation. The hatched area represents a ratio $\Gamma_A/A = 0.2$.

4. CONCLUSION

The presented data show that at high angular momenta various reaction mechanisms proceeding on different time scales may contribute undistinguishably, evidenced in the ^{132}Xe + ^{56}Fe reaction. In medium-mass nuclei (e.g. ^{121}Cs, ^{162}Er) the domain of CNF appears to be limited to a rather small ℓ-window above $\ell_{B_f=B_n}$. At higher angular momentum and/or higher bombarding energies pre-equilibrium emission has been observed. In ^{32}S-induced reactions on nuclei with A < 80 the CNF process seems to evolve towards a faster deep inelastic process. The transitions occur at ℓ-values for which the fission barrier has not vanished. Therefore it is difficult, may be hardly possible to explore the stability limits of compound nuclei in heavy-ion-induced reactions, a situation perhaps different from that in fusion between light heavy ions [24]. The data emphasize the need for a dynamic model describing the various mechanisms.

Comparing various results of fusion-fission type collisions of medium-mass nuclei one can conclude that the change from symmetric to asymmetric mass distributions observed in ^{32}S-induced reactions on nuclei around A ≈ 80 cannot be explained by the proximity of the Businaro-Gallone limit. Its existence has to be questioned. A further criterion for the CNF process based on the widths of the mass distributions ($\Gamma_A/A \simeq 0.2$) is tentatively proposed.

The authors would like to acknowledge helpful discussions with K.D. Hildenbrand.

* Present address : Abteilung für Physik und Astronomie, Ruhr-Universität Bochum, 4630 Bochum, Germany.

REFERENCES

[1] S. Cohen, F. Plasil and W.J. Swiatecki, Ann. of Phys. 82 (1974) 557.

[2] H. Oeschler, H. Freiesleben, K.D. Hildenbrand, P. Engelstein, J.P. Coffin, B. Heusch and P. Wagner, to be published.

[3] B. Heusch, C. Volant, H. Freiesleben, R.P. Chestnut, K.D. Hildenbrand, F. Pühlhofer, W.F.W. Schneider, B. Kohlmeyer and W. Pfeffer, Z. Physik A288 (1978) 391.

[4] H. Oeschler, P. Wagner, J.P. Coffin, P. Engelstein and B. Heusch, Phys. Lett. 87B (1979) 193.

[5] B. Herskind, Proc. of the Symposium on macroscopic features of heavy ion collisions, Argonne, Illinois (1976), ANL Report ANL/PHY-76-2, vol.I, p. 385.
B. Andrews, I. Beene, C. Broude, J. Ferguson, O. Häusser, B. Herskind, M. Lone and D. Ward, to be published.

[6] H. Oeschler, Selected topics in nuclear structure, Bielsko, Poland, vol.I (1978) 276 and annual report 1976, Max Planck Institut für Kernphysik, Heidelberg, Germany.

[7] H.C. Britt, B.H. Erkkila, P.D. Goldstone, R.H. Stokes, B.B. Back, F. Folkmann,
O. Christensen, B. Fernandez, J.D. Garrett, G.B. Hagemann, B. Herskind, D.L.
Hillis, F. Plasil, R.L. Ferguson, M. Blann and H.H. Gutbrod, Phys. Rev. Lett.
39 (1977) 1458.

[8] D.L. Hillis, J.D. Garrett, O. Christensen, B. Fernandez, G.B. Hagemann,
B. Herskind, B.B. Back and F. Folkmann, Nucl. Phys. A325 (1979) 216.

[9] A.M. Zebelman, L. Kowalski, J.M. Miller, K. Beg, Y. Eyal, L. Yaffe, A. Kendil,
D. Logan, Phys. Rev. C10 (1974) 200 ;
H.C. Britt, B.H. Erkkila, R.H. Stokes, H.H. Gutbrod, F. Plasil, R.L. Ferguson,
M. Blann, Phys. Rev. C13 (1976) 1483 ;
W. Scobel, H.H. Gutbrod, M. Blann and A. Mignerey, Phys. Rev. C14 (1976) 1808 ;
M.L. Halbert, R.A. Dayras, R.L. Ferguson, F. Plasil and D.G. Sarantites, Phys.
Rev. C17 (1878) 155.

[10] Code CASCADE, F. Pühlhofer, Nucl. Phys. A280 (1977) 267 ;
Code GROGI2, J.R. Grover and J. Gilat, BNL-report 50246.

[11] J.R. Nix, Nucl. Phys. A130 (1969) 241.

[12] C. Ngô, Thesis, Université de Paris Sud (1975).

[13] C.K. Gelbke, P. Braun-Munzinger, J. Barrette, B. Zeidman, M.J. LeVine, A. Gamp,
H.L. Harney and Th. Walcher, Nucl. Phys. A269 (1976) 460.

[14] U.L. Businaro and S. Gallone, Nuovo Cimento 1 (1955) 629 and 1277.

[15] N. Bohr, Nature 137 (1936) 344.

[16] J. Bisplinghoff, P. David, M. Blann, W. Scobel, T. Mayer-Kuckuk, J. Ernst and
A. Mignerey, Phys. Rev. C17 (1978) 177.

[17] J.B. Natowitz, M.N. Namboodiri and E.T. Chulick, Phys. Rev. C13 (1976) 171.

[18] W. Nörenberg, Phys. Lett. 52B (1974) 289 and J. Phys. 37 (1976) C5-141.

[19] J. Bisplinghoff, A. Mignerey, M. Blann, P. David and W. Scobel, Phys. Rev. C16
(1977) 1058.

[20] C. Lebrun, R. Chechik, F. Hanappe, J.F. Lecolley, F. Lefebvres, C. Ngô, J. Péter
and B. Tamain, Nucl. Phys. A321 (1979) 207.

[21] G. Andersson, M. Areskoug, H.-Å. Gustafsson, G. Hylton, B. Schrøder and E. Hagebø
Phys. Lett. 71B (1977) 279 and Proc. Int. Symposium on physics and chemistry of
fission, Jülich, Germany (1979).

[22] H.C. Britt, B.H. Erkkila, R.H. Stokes, H.H. Gutbrod, F. Plasil, R.L. Ferguson
and M. Blann, Phys. Rev. C13 (1976) 1483.

[23] M. Lefort, these proceedings.

[24] S. Harar, these proceedings.

DEEP INELASTIC COLLISIONS AT ENERGIES CLOSE TO THE COULOMB BARRIER

K.E. Rehm, H. Essel, P. Sperr, K. Hartel, P. Kienle,
H.J. Körner, R.E. Segel[+], and W. Wagner

Physik Department, Technische Universität München,
8046 Garching, W-Germany

I. Introduction

The study of deep inelastic collisions has revealed several interesting quantities for the interaction of two bulks of nuclear matter.
Especially drift and diffusion coefficients for mass, charge, angular
momentum and energy exchange could be determined from the experimental data [1-3]. Most of these experiments were performed at high incident energies which results in short collision times. Thus it is
not possible to study the approach phase during which a partial
equilibrium of the two interacting nuclei is established. The study
of these reactions at energies close to the Coulomb barrier is experimentally quite difficult due to the low energies of the ejectiles
emitted at backward angles, which makes precise mass and charge
determinations impossible. Most of these difficulties can be overcome by use of an inverse reaction kinematics, i.e. bombarding a
lighter target with a heavy projectile. Under this condition the
target-like ejectiles from central collisions are emitted with high
energies at forward angles which strongly simplifies their identification. In this contribution we report on some results obtained with
a ^{208}Pb beam from the UNILAC accelerator at GSI.

II. Experimental Details

The experiments were performed with a Si-detector ΔE-E-time of
flight telescope consisting of a 9.2 µm thick ΔE-detector and a
175 µm E-detector, separated by a flight path of 25 cm. With a time
resolution of ～ 200 psec [FWHM] we obtained a mass resolution of
about 4 %. The M and Z calibration was performed with recoil particles from elastic scattering in the Z-range Z = 28 - 92. The Z resolution was 5 % mainly limited by inhomogenities in the ΔE-detector.
The energy resolution of 5 % was determined by the aperture of the
telescope. With this experimental setup[4] we measured the four-fold
differential cross section $d^4\sigma/(d\Omega\ dMdZdE)$ in the angular range

Θ_L = 10°-60°. The experimental data were transformed to the c.m. system by an event by event analysis including corrections for neutron evaporation.

The systems investigated in these experiments and bombarding energies are summarized in Table I. Also included are the ratios of the c.m. energy to the Coulomb barrier calculated with the strong absorption radii as derived from the relevant quarter-point angles of the elastic angular distributions. Compared to the well studied reaction[5] ^{136}Xe + ^{209}Bi (last line in table I) the systems in our case were studied at energies much closer to the Coulomb barrier. Similarly the grazing angular momenta are smaller in our case by a factor of about 2 - 3 as compared to the ^{136}Xe + ^{209}Bi reaction.

System	E_{Lab} [MeV]	$E_{c.m.}/E_{coul}$	l_{gr} [h]
^{208}Pb+^{94}Zr	1280	1.21	211
^{208}Pb+^{110}Pd	1180	1.09	155
^{208}Pb+^{110}Pd	1280	1.16	208
^{208}Pb+^{148}Sm	1280	1.08	194
^{208}Pb+^{170}Er	1180	1.03	140
^{208}Pb+^{170}Er	1280	1.08	209
^{208}Pb+^{181}Ta	1280	1.08	219
^{136}Xe+^{209}Bi	1130	1.59	480

Table I: Targets and bombarding energies used in this work for ^{208}Pb-induced reactions. Also included are the ratios between the c.m. energies and the corresponding Coulomb barriers and the grazing angular momenta.

III. Experimental Results

Fig. 1 shows two Wilczynski-plots for the systems ^{208}Pb + ^{110}Pd and ^{170}Er measured at 1180 MeV bombarding energy. We observe that the corresponding deflection functions are somewhere between Coulomb-like (^{208}Pb + ^{170}Er) and strongly focussed (^{208}Pb + ^{110}Pd). The observed total kinetic energy losses (TKEL) are usually below 200 MeV, with a maximum at around 100 MeV. Due to the simultaneous measurement of

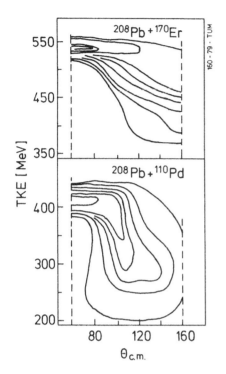

160 - 79 - TUM

Fig. 1 Wilczynski diagrams for the systems ^{208}Pb+^{170}Er, ^{110}Pd measured at 1180 MeV incident energy.

mass and charge we are able to determine the evolution of the mass and charge exchange between the two interacting nuclei as function of TKEL which will be used as a clock. Fig. 2 shows two views of the M-Z-plane for different values of the energy loss. Looking along the valley of stability, shown as thick line in Fig. 2 (see right part in Fig. 2), we observe that the M-Z-distribution stays well inside the valley of stability, which has a parabolic shape as function of (N-Z). On the other hand we observe a strong broadening of the distributions along the valley of stability with increasing TKEL, characteristic for a diffusion-like process. The shapes of these distributions can be approximated by Gaussians with variances σ^2 which increase as function of TKEL. In the following we shall discuss two aspects of the results in greater detail, namely the neutron to proton ratio

for the nucleons transferred between the two interacting nuclei and the time dependence of the variances for charge and mass exchange.

A) Neutron to proton ratio of the transferred nucleons

Fig. 3 shows the measured variances of the charge distributions as function of TKEL for the systems investigated in this work. In addition we have plotted the results for Kr and Xe induced reactions from Huizenga et al.[6] (dotted curve) and the data obtained in U + U collisions[7] (dashed curve). Two striking facts are seen in Fig. 3:

i)   We observe a strong system dependence of the measured variances σ_Z^2. The system ^{208}Pb + ^{110}Pd studied at 1280 MeV bombarding energy shows the largest widths observed in these experiments.

ii) We observe a strong energy dependence of σ_Z^2. Doubling the energy available above the Coulomb barrier for the system ^{208}Pb + ^{110}Pd (open and solid squares in Fig. 3) increases the widths for a

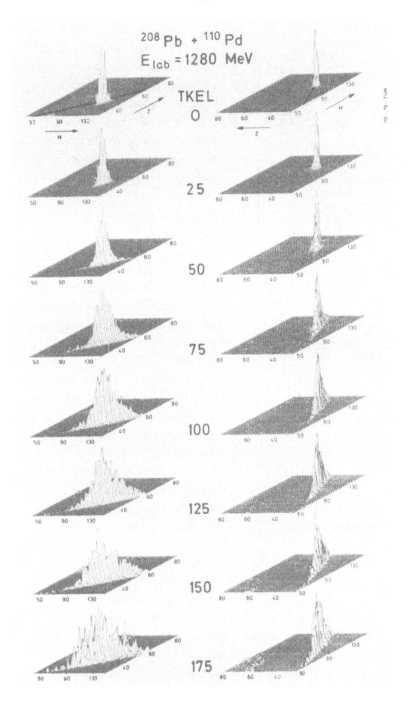

Fig. 2 M-Z distributions for the reaction ^{208}Pb + ^{110}Pd at 1280 MeV incident energy for different values of the total kinetic energy loss (TKEL). The solid line indicates the valley of stability.

given energy loss by about a factor of 2.

These effects might be due to a large extent to the strong influence
of the Coulomb barrier on the exchange of charged particles at these
low incident energies.

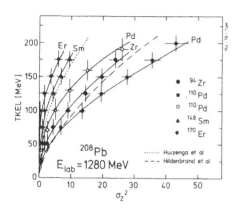

Fig. 3 Variances σ^2 for the
charge distributions for
^{208}Pb induced reactions on
different target nuclei. The
open squares correspond to
the results at 1180 MeV
incident energy. The dotted
line is taken from Ref. 6,
the dashed curve from Ref.7.

The corresponding results of the
widths of the mass distribution
are shown in Fig. 4 Again we
observe a strong system dependence
of the variances σ_A^2 with ^{110}Pd
showing the largest width for a
given energy loss. If we consider
the energy dependence of σ_A^2, however, we observe no change in
going from 1180 to 1280 MeV bombarding energy (open and solid squares in Fig. 4). Together with the
results of σ_Z^2 (see Fig. 3) we can
thus conclude, that at 1180 MeV
incident energy more neutrons must
have been exchanged between the
interacting nuclei if compared to
1280 MeV. This neutron rich nucleon
flow[8] is not specific to the
system ^{208}Pb + ^{110}Pd. In Fig. 5 we
show the ratios of σ_Z^2/σ_A^2 as function of TKEL for the systems

studied in this work. According to their behaviour they may be put into two groups. The systems studied at the highest energy above the
Coulomb barrier (^{208}Pb + ^{94}Zr, ^{110}Pd at 1280 MeV incident energy)
show an independent ratio σ_Z^2/σ_A^2 which is consistent with the value
$(Z/A)^2$ expected from a correlated exchange of neutrons and protons.
The other systems studied at energies closer to the Coulomb barrier
(^{208}Pb + ^{148}Sm, ^{170}Er at 1280 MeV, ^{208}Pb + ^{110}Pd at 1180 MeV) show
an increase of σ_Z^2/σ_A^2 with increasing TKEL, which again points at a
preferential exchange of neutrons in the early reaction stages for
these systems. This effect is different from the results observed at
higher incident energies[9] and indicates again the strong influence
of the Coulomb barrier at low bombarding energies.

B) The time dependence of the variances σ^2

In order to extract diffusion and drift coefficients for charge and

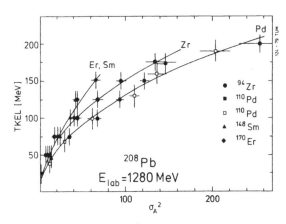

Fig. 4 Variances σ^2 for the mass distributions for ^{208}Pb induced reactions on different target nuclei. The open squares are the result at 1180 MeV incident energy.

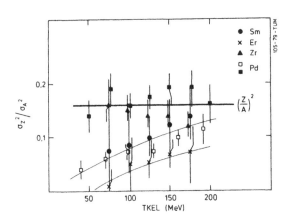

Fig. 5 Ratio of the variances σ_Z^2/σ_A^2 as function of the total kinetic energy loss (TKEL). The open squares are the results obtained at 1180 MeV incident energy.

mass exchange from the measured M-Z-distributions a time scale for the interaction time between the two colliding nuclei has to be introduced. So far there are two models which calculate nuclear interaction times from experimental quantities[10,11]. Schröder et al.[10] generate a deflection function from the experimental Wilczynski plot. The interaction time t is then obtained from the rotation angle $\Delta\Theta$ of the composite system. A more sophisticated treatment has been given by Riedel et al.[11]. In this model the nuclear part of the deflection function is parameterized with parameters adjusted to describe the experimental angular distribution. The interaction time is then calculated analytically taking relaxation of radial and tangential motion and deformation effects into account. Due to their model character both methods are subject to uncertainties in the absolute value of the interaction time. This fact will be transferred into the same uncertainty of the diffusion coefficients described below. It does, however, not effect the general conclusions drawn. In our case the time scales obtained from the two methods agree within 20 %. For systems measured at higher incident energies[10] the variances for the charge distributions show a linear time dependence which can be explained within simple diffusion models. Our results for the system $^{208}Pb + ^{94}Zr$ are shown in Fig. 6. We observe a nonlinear dependence of σ_A^2 and σ_Z^2 as function of the interaction time t. This effect can be explained if inertia effects are included in the Fokker-Planck equations. This has not been done in the treatment of Ref. 12. As shown in Ref. 13 the inertia effects cause an approach phase during which a local equilibrium between the two nuclei is established. The duration of this phase is governed by a relaxation time $\tau = \mu/\gamma$, where μ and γ are the inertia and friction coefficients, respectively, which are connected to the relevant degree of freedom. The analytical time dependence of $\sigma^2(t)$ is given by[13]:

$$\sigma^2 = 2Dt + (v_o\tau)^2 \ [1- \exp(-t/\tau)]^2 -$$

$$-D\tau \ [1- \exp(-t/\tau)][3- \exp(-t/\tau)].$$

(1)

v_o is the average initial exchange rate of mass or charge, which describes the time dependence at short interaction times, while D is a diffusion coefficient responsible for the linear time dependence of σ^2 for long interaction times. The solid lines for σ_A^2 and σ_Z^2 in Fig.6 are least squares fits to the data using eq. (1). Similar results are obtained for the other systems investigated in this work. The results for the parameters v_{OA}, v_{OZ}, D_A, D_Z and τ_A, τ_Z are summarized in Fig.7. Several interesting facts may be noticed from Fig. 7.

i) The relaxation time τ, which represents the duration of the approach phase is of the order of 10^{-21} sec. Increasing the energy available above the Coulomb barrier by a factor of 2 leads to a decrease of the relaxation time by the same factor. The relaxation times for the charge and mass degree of freedom are identical within the error bars.

ii) The diffusion coefficients D which control the time dependence of σ^2 for long interaction times are system and energy independent in our case within the error bars. The ratio of the diffusion coefficients D_A/D_Z is consistent with the value $(A/Z)^2$ expected from a correlated transfer of neutrons and protons. The strong system and energy dependence of the variances as function of TKEL seen in Fig. 3 - 4 is due to the initial exchange rates (dA/dt) and (dZ/dt), respectively.

From the fact that the duration of the approach phase decreases with increasing energy, we thus expect according to eq (1), that for high incident energies the time dependence of σ^2 is given by

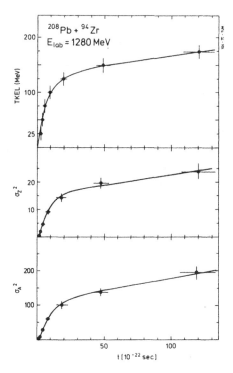

Fig. 6 Total kinetic energy loss and variances σ^2 for the charge and mass distributions as function of the interaction time t calculated according to Ref. 11. The solid lines for the variances are the results of least squares fits to the data.

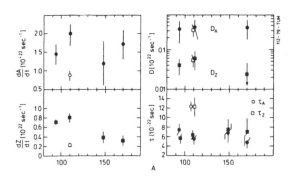

Fig. 7 Parameters obtained from least squares fits to the data at 1280 MeV incident energy with eq. (1) (see text). The open symbols are the results at 1180 MeV.

$$\sigma^2 = 2Dt$$

which was observed experimentally[10]. The detailed investigation of the approach phase in our case was therefore possible due to the relatively long relaxation time of $\sim 10^{-21}$ sec at these low incident energies.

To summarize we have observed that deep inelastic reactions at energies close to the Coulomb barrier show some interesting features which were not observed in previous investigations at higher bombarding energies. Due to the small overlap of the interacting nuclei at the low energies we have a small average friction coefficient which in turn results in long relaxation times of $\sim 10^{-21}$ sec. Therefore effects of the approach phase during which a local equilibrium between the interacting nuclei is established can favourably be studied at these energies. Furthermore we observe an enhancement of the neutron exchange as compared to proton exchange at energies very close to the Coulomb barrier, which might be an important effect for the production of very heavy neutron rich nuclei.

This work was supported by the BMFT.

+ A. v. Humboldt Awardee, present address Northwestern University, Evanston, Ill. USA.

/1/   W.U. Schröder and J.R. Huizenga, Ann. Rev. Nucl. Sci. 27, 465 (1977)
/2/   M. Lefort and Ch. Ngô, Ann. Phys. 3, 5 (1978)
/3/   H.A. Weidenmüller, to be publ. in Progr. Part. and Nucl. Phys.
/4/   H. Essel, P. Sperr, K. Hartel, P. Kienle, H.J. Körner, K.E. Rehm, W. Wagner, sub. to Nucl. Instr. Meth.
/5/   W.U. Schröder, J.R. Birkelund, J.R. Huizenga, K.L. Wolf, V.E. Viola, Jr., Phys. Lett. C 45, 301 (1978)
/6/   J.R. Huizenga, J.R. Birkelund, W.U. Schröder, K.L. Wolf, and V.E. Viola, Jr. Phys. Rev. Lett. 37, 885 (1976)
/7/   K.D. Hildenbrand, H. Freiesleben, F. Pühlhofer, W.F.W. Schneider, R. Bock, D.v. Harrach, and H.J. Specht, Phys. Rev. Lett. 39, 1065 (1977)
/8/   H. Essel, K. Hartel, P. Kienle, H.J. Körner, K.E. Rehm, P. Sperr, and W. Wagner, Phys. Lett. 81B, 161 (1979)
/9/   J.R. Huizenga, Contribution to this Conference
/10/   W.U. Schröder, J.R. Birkelund, J.R. Huizenga, K.L. Wolf, and V.E. Viola, Jr., Phys. Rev. C 16, 623 (1977)
/11)   C. Riedel, G. Wolschin, and W. Nörenberg, Z. Phys. A 290, 47 (1979)
/12/   W. Nörenberg, Phys. Lett. 52 B, 289 (1974)
/13/   K.E. Rehm, H. Essel, K. Hartel, P. Kienle, H.J. Körner, P. Sperr, and W. Wagner, Phys. Lett. 86 B, 256 (1979).

Promptly Emitted Particles in Nuclear Collisions

J. P. Bondorf

The Niels Bohr Institute, University of Copenhagen,

DK-2100 Copenhagen Ø, Denmark

Abstract

Discussion of a mechanism in which nucleons transferred in nuclear collisions are emitted promptly into the continuum via the coupling of internal nucleon motion to the relative motion of the nuclei.

Introduction

Collisions between nuclei will usually give rise to emission of nucleons or other light particles. Only under very special circumstances is it possible to detect directly the time of emission (measured with respect to the time of impact). This can for example be done by channeling techniques which allows time intervals down to $\sim 10^{-17}$ sec. This time is already so long that most nuclear systems will be equilibrized when it has passed. For all shorter times only indirect determination of the time from the spectra is possible. It is, however, of great value for the understanding of heavy ion reaction mechanisms to be able to extract such information from the spectra of emitted light particles. This would give some additional and independent checks on the various reaction mechanisms which are presently accepted for heavy ion collisions, especially the early stages of the collisions. In this paper we think of reactions in which the bombarding energy is typically from a few MeV per nucleon above the Coulomb barrier to maybe the Fermi energy per nucleon.

As a first rough classification one can divide emitted light particles (with mass number ≤ 4, say) into two different kinds:

i) They can be emitted from the dinuclear complex when the two nuclei are in close contact in the beginning of their mutual interaction.

ii) They can be emitted at a later time from excited bigger reaction products such as the two products in deep inelastic reactions or they can be emitted at a later time from a fused system.

This division is schematic and somewhat arbitrary, but by that we

want to distinguish between a promptly emitted particle (PEP) and a
delayed emitted particle (DEP) which comes from an equilibrized
system.   One can of course also imagine intermediate situations, but
we stick to the extremes here.   In fig. 1 a schematic space-time
diagram with only one spatial variable (along the beam axis) shows
the two categories.

SPACE - TIME :

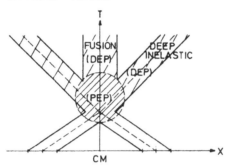

CM

SPACE :

Fig. 1.   Schematic space time diagram for a collision between two
identical nuclei in the CM system.   The space is represented by one
variable only, the travel distance along the beam axis (deflection
angle zero).   The hatched areas represent regions in which promptly
(PEP) and delayed (DEP) particles are emitted.

In this lecture I shall call your attention to a possible
mechanism for nucleon PEPs in heavy ion collisions.   The mechanism
is able to account for a major part of the hard component of observed
neutron spectra in fusion reactions.   The basis of the model and detailed calculations are presented elsewhere [1] .   I shall therefore
stick to a relatively short presentation of the model and devote most
of the talk to comments about the basic assumptions and about conse-

quences of this PEP mechanism.

Quasi Free PEPs

In the nuclear ground state the nucleons can be considered as
almost freely moving in a constant potential and their velocities
$\vec{v}_D$ have a Fermi distribution. In close collisions the two nuclei
make contact via a common interface area, the window, through which
nucleons can move. Seen from the receiving nucleus R the velocity
of a transferred nucleon is then

$$\vec{v}_R = \vec{v}_D + \vec{v}_{rel} \tag{1}$$

where $\vec{v}_{rel}$ is the velocity of the donor nucleus D seen from R
at the time of transfer. The Fermi distributions for the two nuclei
seen from R are shown in fig. 2. It is seen that a large part of

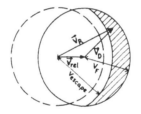

Fig. 2. Distribution of velocities $\vec{v}_R$ in velocity space for the
center of mass of the recipient R. The hatched area represents the
condition that (2) is positive. It is particles from this volume
which can eventually become PEPs.

the transferred nucleons can flow freely through R , and if they
have sufficient energy for overcoming the nuclear binding, they can
escape. We call such nucleons quasifree PEPs. Not all of them
succeed the escape, because there is a probability of absorption on
the way through R . This PEP flow continues as long time as the
window is open and $\vec{v}_{rel}$ has not been damped completely or has
changed its direction so that the two nuclei no longer approach each

other.

The nucleons escape from R opposite the window by the energy

$$\frac{1}{2}m_N v_u^2 = \frac{1}{2}m_N v_R^2 + U \qquad (2)$$

where U is the negative nucleon potential. Besides the requirement
that (2) should be positive there are two more escape criteria,
1) that the nucleons conserve angular momentum with respect to R
and 2) that the nucleons overcome the nucleon barrier in R (zero
for neutron PEPs). Probabilities of PEPs can now be calculated
assuming a nucleon bombarding rate on the window which is proportional
to

$$\rho_0 \cdot v_D \cdot A \cdot \cos\alpha$$

where ρ_0 is the nucleon density, A the area of the time dependent
open window and α the angle between $\vec{v}_D$ and the window normal. By
assuming an absorption of nucleons in R as given by the optical
model for nucleon scattering and by using some classical model for
the ion ion relative motion, one can now calculate the PEP spectra as
function of the heavy ion impact parameter. In the present calculation we assume that the ion momentum is frozen during a PEP passage.
It should be remembered that there are two kinds of PEPs, the above-
mentioned, and PEPs which have R as donor and D as recipient. One
can associate calculated PEPs with either deep inelastic or fusion
reactions depending on how the ion-ion motion proceeds in the reaction.

Fig. 3 shows a comparison between calculated quasifree neutron
PEP cross sections and the measured hard component of neutron spectra
in the heavy ion fusion reaction $^{12}C + ^{158}Gd$ at $\varepsilon_{LAB} = 12.7$ MeV per
nucleon [2]. The hard component has a temperature of 6 MeV and in
contrast to the soft component with a compound temperature of 2 MeV
it can therefore be interpreted as emitted during the early stages of
the reaction when the energy was concentrated on a few degrees of
freedom. In the calculation a more sophisticated window geometry
than the above-mentioned has been used by introduction of proximity
geometry [1]. The overall absolute magnitude of the cross section
is very sensitive to parameters, such as radii and absorption and the
fit is therefore surprising, maybe a little bit too good.

The present classical model does not take into account certain
important effects. First of all, the average PEP angular momentum is
low, in the example 2-3$\hbar$ and the angular distribution should therefore be smeared accordingly, Quantum PEPs have been seen in TDHF
calculations [3] and they have qualitatively the same character as
the PEPs of our classical model. When a PEP is emitted, it drains a

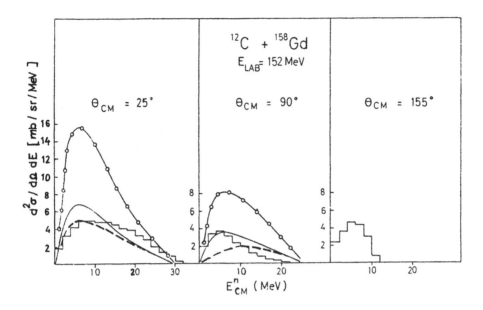

Fig. 3. The histograms show the calculated energy distribution of neutron PEPs at three angles in the CM system for the reaction $^{12}C + ^{158}Gd$ at E_{LAB} = 152 MeV. The experimental PEP spectrum is represented by one of the three continuous curves which are extracted from the data [2] under three different assumptions (see [1]).

rather big energy from the relative motion. This is especially important if two or more PEPs are emitted in the same ion-ion event, but since the probability of that is low, we have not included the recoil losses, but only changed $\vec{v}_{rel}$ by damping from proximity friction.

One finds experimentally an enhanced forward PEP probability whenever the separation energy of the least bound nucleon of the projectile is lowered [4]. It is worth noting that since the separation energy of the least bound nucleon enters in eq. (2) and in the cross section, the shell effects and the isospin of the donor nucleus affect the calculated cross sections and thresholds of the PEPs dramatically and with the same trend as in the reported experiment. The energy dependence of the PEP cross section is rather steep. In fig. 4 there is shown an energy dependence for the above-mentioned reaction. The curve marked "2-body PEPs" is the calculated cross section for the emission of nucleons which are results of one nucleon-

nucleon collision between a transferred nucleon and a nucleon in
R whenever a transferred nucleon is absorbed.

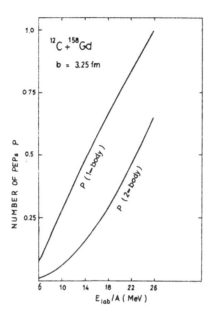

Fig. 4. Calculated number of PEPs, P , emitted in the reaction
^{12}C + ^{158}Gd at various E_{LAB} energies for a particular impact parameter
b = 3.25 fm.

Finally we shall make some comments about the PEP model and the
preequilibrium theory of Griffin (and Blann) [5] . In the Griffin
theory, which is designed for light particle induced reactions, there
is in the beginning of the reaction created a number of initial
excitons usually of the order of the number of nucleons in the projectile. In a heavy ion reaction one cannot in a similar way define
a number of initial excitons based upon the number of particles in
the projectile. This is because the initial phases of the heavy ion
reaction may be largely dominated by the excitation of collective degrees of freedom [6] . On the other hand it may still be possible
to use the Griffin theory in heavy ion collisions, provided that the
problem of initial excitons is solved. The early transferred nucleons
which are treated in this paper have exactly the character of initial
excitons and it is tempting to define an initial number of excitons
on the basis of the transferred nucleons. This is, however, not the
purpose of the present paper.

Conclusions

The model for quasifree PEPs is still too primitive to be able to make precise quantitative predictions, but the inclusion of quantum effects and shell and isospin effects gives a hope for prediction of more specific spectra which can test the assumptions. The PEPs associated with complete fusion reactions are of particular interest because they can be separated from compound spectra which are symmetric around 90° while the PEP spectra are not symmetric whenever the target and projectile are different.

In the introduction we discussed the importance of obtaining information on the time of emission of light particles in a heavy ion reaction. The PEP mechanism which is discussed in this talk is based upon very definite assumptions on how and when the nucleon emission takes place. The most important condition for emission is that it can only be initiated while v_{rel} is still big, i.e. before the damping has had too much effect. The emission is therefore indeed very early in the reaction.

Finally we shall point out that there are probably several other mechanisms for PEP production of nucleons and composite particles than the simple mechanism outlined in this paper.

Acknowledgements

The author acknowledges M. Halbert and A. Karvinen for stimulating comments concerning this work.

References:
[1] J. P. Bondorf, Journal de Physique C5(1976)195.
    J. P. Bondorf, J. N. De, A. O. T. Karvinen, G. Fái and
    B. Jakobsson, Phys. Lett. 84B(1979)162.
    Same authors + J. Randrup, NORDITA-79/27, Nuclear Physics
    (in press).
[2] L. Westberg et al., Phys. Rev. C18 (1978)796.
[3] H. S. Köhler, Proc. of the Orsay Conf. on TDHF Saclay, May 1979,
    ed. P. Bonche et al.
[4] M. Halbert, contribution to this conference, and private
    communication.
[5] J. J. Griffin, Phys. Rev. Lett. 17(1966)478.
    M. Blann, Phys. Rev. Lett. 21(1968)1357.
[6] R. A. Broglia, C. H. Dasso, A. Winther, Phys. Lett. 53B(1974)301.

LIGHT PARTICLE EMISSION IN HEAVY ION REACTIONS AT 10 AND 20 MeV/Nucleon

H. HO

Max-Planck-Institut für Kernphysik, Heidelberg, W.-Germany

and

Texas A&M University, College Station, Texas

Abstract. Energy and angular correlations between light particles and heavy ions were measured for the reaction ^{40}Ar+^{93}Nb at 10 MeV / nucleon and for the reaction ^{16}O+Ti at 20 MeV/nucleon. The correlation data reveal several pre-equilibrium components: one component is characterized by "beam velocity" α particles and shows up in coincidence with both fusion-like fragments as well as quasi-elastic and deep-inelastic scattered fragments. The other pre-equilibrium component is observed only in coincidence with deep-inelastic scattered fragments and is characterized by two maxima separated by a pronounced minimum in the direction of the detected projectile-like fragment. It is proposed that the first pre-equilibrium component stems from the very early stages of heavy ion collision before the system has had a chance to rotate whereas the second component originates from the later stages of heavy ion collision and the first instants of separation.

1. Introduction

In the beginning of my talk I would like to remind you of an experiment which we performed three years ago at the Max-Planck-Institut für Kernphysik in Heidelberg [1]. At that time we bombarded ^{58}Ni targets with 6 MeV/nucleon ^{16}O and measured the light particle emission in coincidence with deep-inelastic scattered fragments. The result was that a considerable fraction of light particles showed the characteristic features of pre-equilibrium emission. To explain the data we proposed the formation of a "hot spot" during the heavy-ion collision. This hot spot would reveal itself by a higher yield of light particles and a higher temperature of the spectra into specific directions which are determined by the dynamics of the heavy-ion collision [1,2]. However, one experiment is not conclusive enough and today I want to talk about two

other experiments. In the first one we studied the light particle emission in the system ^{40}Ar+^{93}Nb at 10 MeV/nucleon [3,4]. The results of this
experiment will be free of any α cluster structure effect of the ^{16}O-
projectile. In the second experiment we used a much higher bombarding
energy - 20 MeV/nucleon ^{16}O+Ti[5]-which should in principle favor the
formation of a hot spot in heavy ion collisions.

2. The ^{40}Ar+^{93}Nb-Reaction.

Let me demonstrate the basic ideas of the experiments in the case of the
^{40}Ar+^{93}Nb-reaction. Fig. 1 shows a velocity diagram of deep-inelastic
scattered ^{40}Ar with ^{93}Nb. The projectile-like fragments were detected
at a scattering angle of approximately 37° and are indicated by the
velocity vector $\vec{v}_{Ar}$. $\vec{v}_{BEAM}$ and $\vec{v}_{CM}$ are the velocity vectors of the beam
and of the total center of mass, respectively. Let us first assume that
we deal with a binary reaction and subsequent α emission from the fragments. Then the recoiling target-like fragment would fly onto the opposite side of the beam with the velocity vector $\vec{v}_{Nb}$. The coincident α
particles would have most probable velocities as indicated by the two
circles centered around the velocities of the fragments $\vec{v}_{Ar}$ and $\vec{v}_{Nb}$.
The radii of the two circles correspond to the Coulomb barrier between
α particles and the Ar-like and the Nb-like fragment. It is obvious
from the velocity diagram that on the side of the beam opposite to the
detected heavy ion we have only α particles from the Nb-like fragments
whereas on the same side of the beam we get contributions from both
fragments. Fig. 2 shows two α velocity spectra taken at 120° and -48°
as indicated by the dashed vectors in Fig. 1 (α detection angles on the
opposite side of the beam are defined to be positive whereas angles on
the same side are defined to be negative). From the velocity scale in
Fig. 1 we verify that the spectrum at 120° peaks close to the expected
value of 2 cm/ns. The spectrum at -48° has two components at 2 cm/ns
and 4 cm/ns which we ascribe to α emission from the Nb-like and the
Ar-like fragment, respectively. The second kinematical solution for
α particles from the Ar-like fragment is not present because it is below our experimental threshold. For equilibrium α emission from fragments with average spins perpendicular to the reaction plane one expects
isotropic distribution in the reaction plane.[1] However comparing the two
spectra in Fig. 2 it is obvious that the coincident yield for α partic-

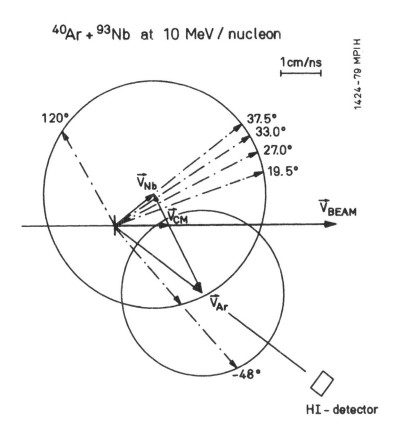

^{40}Ar + ^{93}Nb at 10 MeV / nucleon

1cm/ns

1424-79 MPIH

120°

37.5°
33.0°
27.0°
19.5°

$\vec{V}_{Nb}$

$\vec{V}_{CM}$

$\vec{V}_{BEAM}$

$\vec{V}_{Ar}$

-48°

HI - detector

Fig. 1:   Velocity diagram for deep-inelastic scattering of
             20 MeV/nucleon ^{40}Ar with ^{93}Nb. The Ar-like fragment
             was detected at approximately 37°. The two circles
             represent the most probable velocities for equilibrium
             α emission from the projectile-like fragment and the
             target-like fragment.

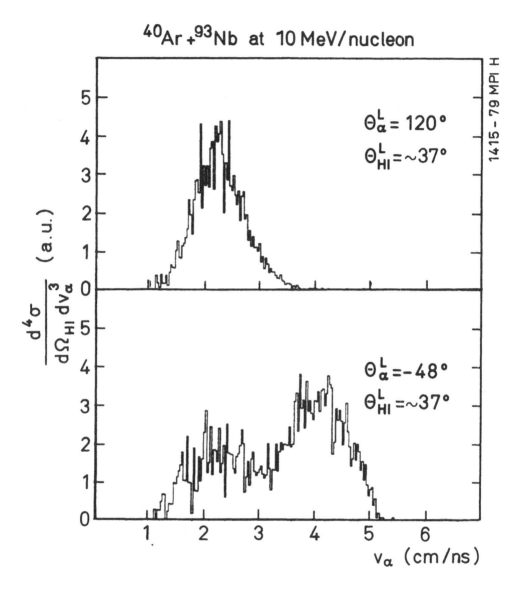

Fig. 2:   Two velocity spectra for α particles in coincidence
with Ar-like fragments at 37°. The α detection angles
were 120° (top) and -48° (bottom). Plotted is the
Galilei-invariant cross section $d^4\sigma/d\Omega_{HI}dv_\alpha^3$ versus the
α velocity. The two spectra are normalized to the single
rate in the heavy ion detector.

les from the Nb-like fragment is different at those angles. At this
point I would like to stress the fact that we can compare the yield in
the two spectra directly because we have plotted the Galilei-invariant
cross section $d^4\sigma/d\Omega_{HI}dv_\alpha^3$. The finding that the coincidence rate at
$\theta_\alpha^L = -48^\circ$ is smaller than the one at $\theta_\alpha^L = 120^\circ$ seems to indicate some
kind of "shadowing" in the direction of the detected projectile-like
fragment.

Fig. 3 shows four α velocity spectra measured at $\theta_\alpha^L = 37.5^\circ$, 33.0°,
27.5°, and 19.5° as indicated in the velocity diagram (Fig. 1). The
dashed line in Fig. 3 marks the most probable velocity of 3.5 cm/ns as
we observe it at 37.5°. Again we find this most probable velocity close
to the expected one from the velocity diagram. However, in the velocity
spectrum at 33.0° a shoulder shows up in the high velocity tail and in
the spectrum at 27.0° the maximum has obviously shifted. Eventually in
the spectrum at 19.5° we observe the maximum at 4.4 cm/ns. This is exactly the velocity of the beam and indicates a process where a fast α
particle is emitted at the first instant of the heavy-ion collision.

In Fig. 4 the in-plane angular correlation is plotted for θ_α^L from
0° to 180° where mainly α particles from the Nb-like fragment contribute, and for some angles on the other side where the contribution of
α particles from the Ar-like fragment could be separated. In order to
study deviations from equilibrium emission from the Nb-like fragment
we have transformed the angular correlation into the restframe of the
Nb-like fragment. We have plotted the differential α multiplicity i.e.
the number of α particles per unit solid angle and per deep-inelastic
scattered fragment as a function of the in-plane angle θ_α^{Rec}. $\theta_\alpha^{Rec} = 0^\circ$
corresponds to the direction of the recoiling target-like fragment. At
backward angles the differential multiplicity levels at 0.07 sr^{-1}
whereas towards forward angles it increases strongly. For angles close
to the detection angles of the projectile-like fragment the differential
α multiplicity is even lower than the level of 0.07 sr^{-1} at backward
angles. We ascribe the constant differential multiplicity at backward
angles to equilibrium emission of α particles from the Nb-like fragment
whereas at forward angles the strong increase is due to pre-equilibrium
emission. Admittedly a clear separation between α particles from the
Nb-like and the Ar-like fragment is not possible at the most forward
angles and also the direct component with beam velocity contributes to
the differential multiplicity at these forward angles. However, already
at $\theta_\alpha^{Rec} \lesssim 60^\circ$ we observe an increase of the differential multiplicity
where we are quite safe from contributions from both processes.

Integrating over the whole sphere we obtain multiplicities for the

148

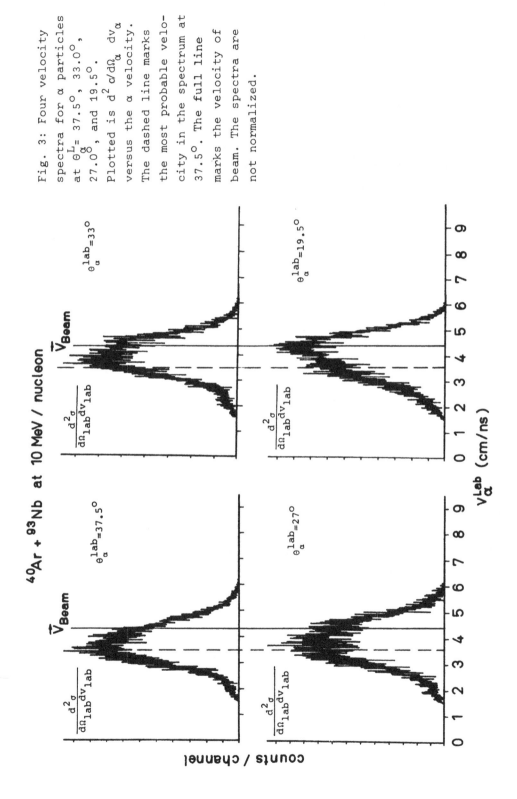

Fig. 3: Four velocity spectra for α particles at θ_α^L = 37.5°, 33.0°, 27.0°, and 19.5°. Plotted is $d^2\sigma/d\Omega_\alpha\, dv_\alpha$ versus the α velocity. The dashed line marks the most probable velocity in the spectrum at 37.5°. The full line marks the velocity of beam. The spectra are not normalized.

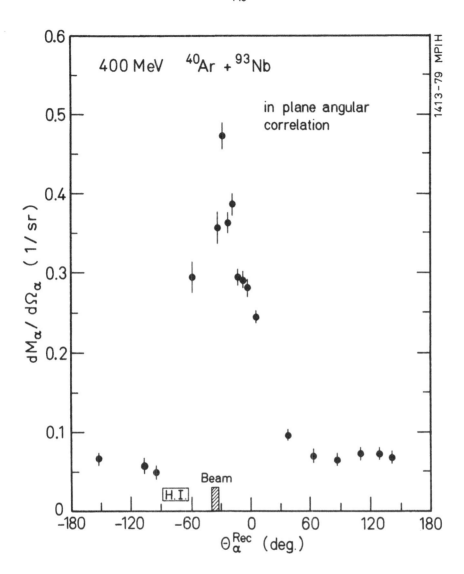

Fig. 4:   In-plane angular correlation for α particles in the
           restframe of the Nb-like fragment. Plotted is the dif-
           ferential α multiplicity versus the in-plane angle
           Θ_α^{Rec} . Θ_α^{Rec} = 0 corresponds to the momentum transfer
           axis.

150

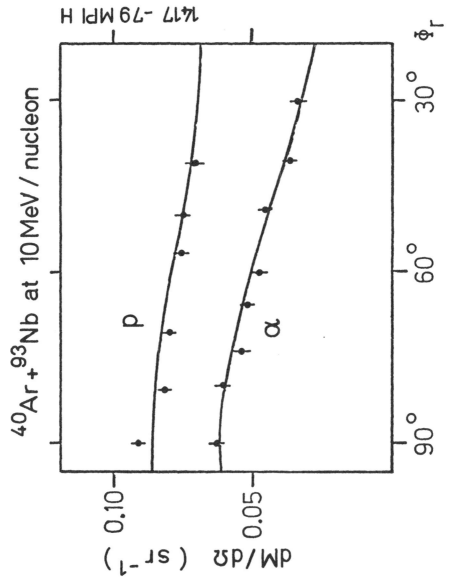

Fig. 5: Out-of-plane angular correlation for α particles (bottom curve) and protons (top curve) in the restframe of the Nb-like fragment. Plotted is the differential multiplicity versus the polar angle Φ_r. $\Phi_r = 90°$ corresponds to the in-plane direction at $\theta_{p,\alpha}^{Rec} = 110°$.

equilibrium emission of 0.6±0.1 and 1.0±0.2 for α particles and protons, respectively. For the nonequilibrium component we estimate M_α^{non}=0.3±0.1.

Let me mention that we can extract information on the spin distribution of the Nb-like fragment from the equilibrium component at backward angles. Fig. 5 shows the out-of-plane angular correlation for α particles (bottom curve) and protons (top curve) measured at those angles. The finding that the differential multiplicity is constant over an angular range of 150° in the reaction plane and that there is an out-of-plane anisotropy for light particles is indicative of fragment spins which are in the average perpendicular to the reaction plane. With a Gaussian distribution for the fragment spins we derive an average spin of ⟨I⟩= 29±5 ℏ, which is close to the value in the sticking limit I_{st}= 33 ℏ, and an average alignment of ⟨P_{zz}⟩= 0.75±0.25 .

3. The ^{16}O+Ti-Reaction.

Let me come to presentation of the ^{16}O+Ti data[5]. Fig. 6 shows a calculated velocity plot for α emission from equilibrated deep-inelastic scattered fragments. The projectile-like fragment e.g. carbon, was detected at 20° as indicated by $\vec{v}_C$. $\vec{v}_{BEAM}$, $\vec{v}_{CM}$, and $\vec{v}_{Ti}$ are the velocity vectors of the beam, the total center of mass, and the recoiling target-like fragment. Instead of two simple circles centered around $\vec{v}_C$ and $\vec{v}_{Ti}$ as we had them in Fig. 1 we have now two circular ridges represented by contour lines of constant cross section for the α velocity vectors. Again, I want to stress the fact that we have plotted $d^4\sigma/d\Omega_{HI}dv_\alpha^3$ which is Galilei-invariant. The two circular ridges give rise to a maximum in the overlapping region. We could improve this plot by taking into account the energy and angular distribution of the projectile-like fragment which would distort the circle around $\vec{v}_C$ to an ellipse and which would enhance the coincident yield in that part of the circle which is closer to the beam axis.

Let us compare the prediction with the experiment. Fig. 7 shows an experimental velocity plot for α particles in coincidence with projectile-like fragments at 20°. To improve the statistics we summed over all atomic numbers Z from 4 to 10 but one should point out that there are no significant differences between α velocity plots for deep-inelastic scattered carbon and oxygen. For quasi-elastic scattered carbon we can identify the α break-up of the oxygen. Having still in mind the obser-

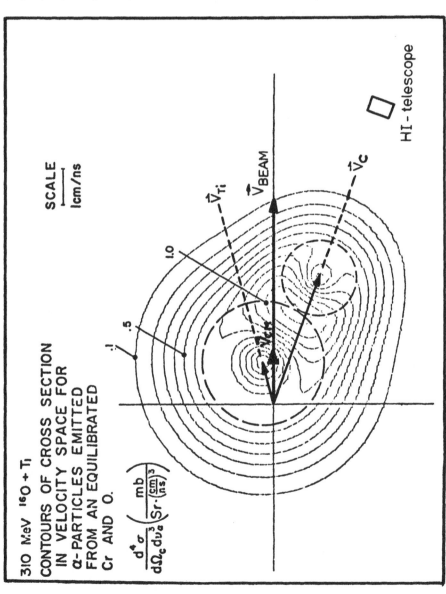

Fig. 6: Calculated velocity plot for deep-inelastic scattering of 20 MeV/nucleon ^{16}O with Ti. Deep-inelastic scattered carbon is detected at 20^0. The two circular ridges represent subsequent emission of α particles from both fragments. The closed curves represent contour lines of constant cross section. In this simple picture the overlapping circular ridges give rise to a local maximum.

310 MeV $^{16}O + Ti$

CONTOURS OF CROSS SECTION IN VELOCITY SPACE FOR α-PARTICLES EMITTED FROM AN EQUILIBRATED Cr AND O.

$$\frac{d^4\sigma}{d\Omega_c \, dv_a^3} \left(\frac{mb}{Sr \cdot \left(\frac{cm}{ns}\right)^3}\right)$$

SCALE
1cm/ns

.1
.5
1.0

$\vec{V}_{Ti}$
$\vec{V}_{BEAM}$
$\vec{V}_{CN}$
$\vec{V}_c$

HI-telescope

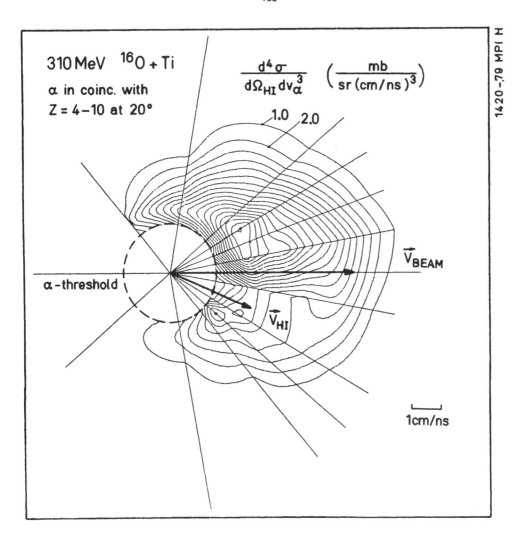

Fig. 7 :   Experimental velocity plot for α particles in coincidence
with projectile-like fragments (Z=4-10). Plotted is the
Galilei-invariant cross section $d^4 c/d\Omega_{HI} dv_\alpha^3$. Step between
neighbouring contour lines is 1.0 mb/sr·$(cm/ns)^3$. The
straight lines correspond to measured angles (except the
one in beam direction). The dashed circle corresponds to
the experimental α threshold.

vation of the direct component in the $^{40}Ar+^{93}Nb$-reaction we rediscover the beam velocity component as a broad shoulder in the high velocity tail at 20° on the opposite side of the beam. However, in addition we see two maxima at lower velocities on both sides of the beam axis separated by a pronounced minimum in the direction of the detected projectile-like fragment. This is not at all what one expects from the prediction in Fig. 7. Let me demonstrate some features of these different components. For this purpose we have plotted in-plane angular correlations with windows on v_{α}^{Rec} i.e. the α velocity in the restfrAme of the target-like fragment. Fig. 8 shows these in-plane angular correlations between α particles and projectile-like fragments detected at 20° (left side) and 40° (right side). Again we have summed over all Z from 4 to 10. Five angular correlations are shown for each heavy ion detection angle and the numbers at these correlations refer to the average velocity in the velocity bin chosen. Let us begin with the angular correlation with the highest velocity v_{α}^{Rec} = 6 cm/ns. The angular correlations are strongly peaked on that side of the beam opposite to the detected heavy ion and they are insensitive on the detection angle of the heavy ion as one can see by comparing the uppermost angular correlations. This is exactly what one expects for a process where a fast α particle is emitted in the first stages of heavy-ion collision. The α multiplicity for this direct component is estimated to M_{α}^{dir}= 0.7±0.2 .

Let me draw your attention to the angular correlation with the lowest velocity v_{α}^{Rec}= 2 cm/ns . Here we see in the case of θ_{HI}= 20° (left side) that the angular correlation shows two maxima and a minimum at -10° which was evident already in the velocity plot of Fig. 7. However, in contrast to the beam velocity component this low velocity component is sensitive on the detection angle of the heavy ion: the corresponding angular correlation for θ_{HI}= 40° (right side) shows the minimum at -50°. For the maxima on both sides of the minimum we observe a similar shift in the same direction. How can one explain this experimental finding? Is it possible to explain it by equilibrium emission from the two fragments at times when the fragments are still close to each other? We have studied this question and have carried out three-body-Coulomb-trajectory calculations. The conclusion from these calculations is the following: Coulomb effects and reabsorption of the α particle play an important role for emission times shorter than $4 \cdot 10^{-22}$ sec (after separation of the two fragments). However, in order to explain the observed coincidence cross section distribution we have to assume in addition that the α particles are emitted from the nuclear surfaces which face each other i.e. that they are emitted from the contact zone at the very end of the heavy

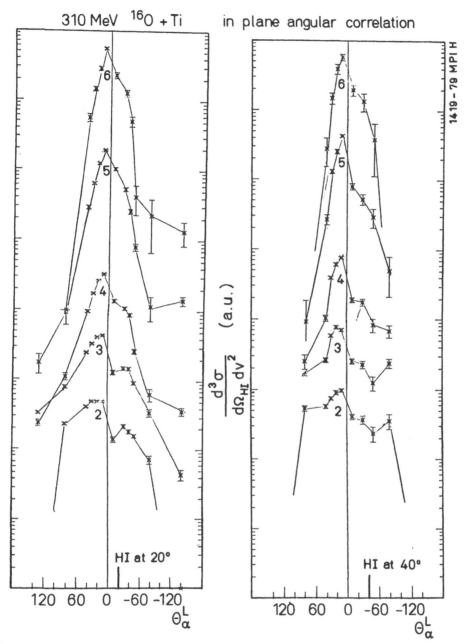

Fig. 8: In-plane angular correlation for α particles in coincidence with projectile-like fragments (Z=4-10) detected at 20° (left side) and 40° (right side). For further details see text. Note the logarithmic scale!

ion collision. The asymmetry in cross section for the two maxima can be
easily produced by an asymmetric shape of the emitting surfaces or by
rotation of the fragments after separation. This rises the interesting
question if we have a positive or a negative deflection function for
this system. The α multiplicity integrated over both maxima is M_α^{pre} =
1.5+0.5 .

   Let me come to my final point. We have also measured light particles in coincidence with fusion-like fragments. Fig. 9 shows a velocity
plot for α particles in coincidence with fusion-like fragments ($Z \gtrless 16$)
detected at 20° as indicated by $\vec{v}_{FR}$. $\vec{v}_{BEAM}$ and $\vec{v}_{CM}$ are the velocity
vectors of the beam and the total center of mass, respectively. We ob-
serve the major part of coincident α particles on the opposite side of
the beam. This is merely a consequence of momentum conservation: only
if many light particles are emitted onto the opposite side of the beam
a heavy nucleus can be detected at 20°. The loci of the most probable
α velocity vectors lie on a circle whose center is close to the velocity
vectors $\vec{v}_{CM}$ and $\vec{v}_{FR}$. The radius of this circle if converted to energy
corresponds to the Coulomb barrier between α particles and the fusion-
like fragments. Statistical model calculations   have been performed
and show reasonable agreement with the data. However, again we find
beam velocity α particles at 20° on the opposite side of the beam. This
suggests a process where a fast α particle is emitted prior to the fusion of the rest of the projectile with the target. It is very likely
that we observe here the typical features of the "incomplete fusion" or
"massive transfer" reaction [6]. The velocity plot changes considerably
if one detects the fusion-like fragments at 40° (Fig. 10). The circular
ridge has moved towards the beam axis and has revealed the beam velocity
component in its full extension. It is remarkable that this component
extends to rather large angles while the average velocity degrades gradually. Comparing Fig. 10 with Fig. 9 we also see that there might be a
considerable fraction of beam velocity α particles hidden under the
equilibrium part. In the case of fusion-like fragments we derive M_α^{dir} =
0.8+0.2 .

4. Concluding Remarks.

Comparing the $^{40}Ar+^{93}Nb$ data with the $^{16}O+Ti$ data it is obvious that
pre-equilibrium emission of light particles increases with higher bombarding energy. Two pre-equilibrium components have been observed in

157

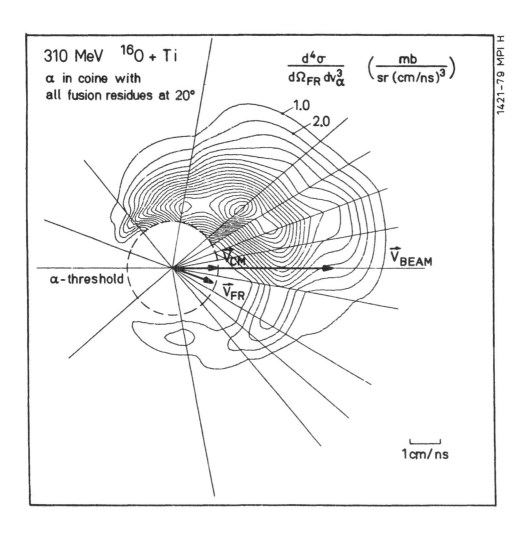

Fig. 9:   Velocity plot for α particles in coincidence with fusion-
like fragments (Z>16) detected at 20°. For further details
see figure caption of Fig.7

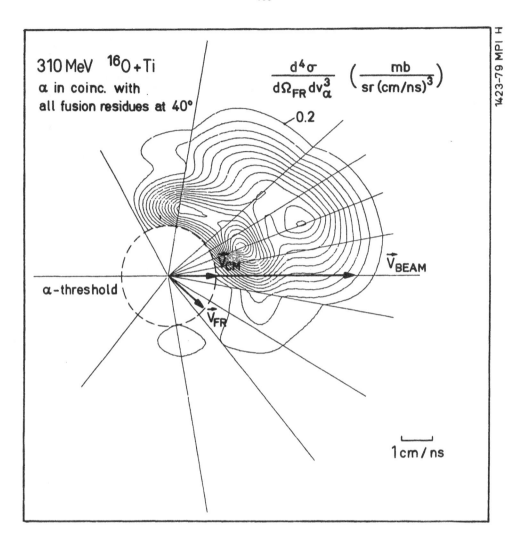

Fig. 10:   Velocity plot for α particles in coincidence with fusion-like fragments (Z>16) detected at 40°. For further details see figure caption in Fig. 7.

both reactions: one component is characterized by beam velocity and
shows up in coincidence with both fusion-like fragments as well as projectile-like fragments. It is suggested that these particles are emitted at the first instant of heavy-ion collision prior to the fusion or
the deep-inelastic collision of the rest of the projectile with the
target. The fact that we observe this beam velocity component in coincidence with projectile-like fragments indicates that these light particles are generated more or less in peripheral collisions. This is also
supported by the experimental finding that in incomplete fusion reactions the "compound" nuclei are produced in high spin states [6].

The second pre-equilibrium component is observed only in coincidence with deep-inelastic scattered fragments and is characterized by
two maxima separated by a pronounced minimum in the direction of the
detected heavy ion. This component can be explained by pre-equilibrium
α emission at the final stages of heavy ion collision or the first instants of separation (an emission time shorter than $4 \cdot 10^{-22}$ sec is estimated). It is in addition necessary to restrict the α emission to the
nuclear surfaces which experience or have experienced strong interaction
at these times. The features are similar to those observed in terniary
fission. However, in these special fission processes the angular correlation is symmetric with respect to the scission axis which is not the
case in our data: the maximum on the opposite side of the beam is approximately three times higher than the one on the same side.

We would like to resume the question of a hot spot formation in
heavy ion collisions: the experimental data are consistent with the existence of a hot spot but it is evident that due to the strong Coulomb
effects and due to the presence of the beam velocity component a determination of a hot spot temperature is not possible. However, it is expected that the correlation data with other light particles like protons
should give us additional information to explore this question. If we
- on the other hand - assume that we have produced a hot spot in these
reactions then it follows from the estimate of the emission time that
the equilibration time in nuclear matter cannot be considerably smaller
than $4 \cdot 10^{-22}$ sec. This in turn would imply that the equilibration time
would be comparable with the collision time of heavy ions thus invalida-
ting the premises in transport theories of heavy ion reactions [7].

The author wants to express his sincere thanks to all his coworkers
at the Max-Planck-Institut für Kernphysik in Heidelberg and at the Texas
A&M University in College Station. He thanks especially P. Gonthier,

W. Kühn, and J. Slemmer whose dissertations treat the light particle emission in 20 MeV/nucleon ^{16}O+Ti-reactions and 10 MeV/nucleon ^{40}Ar+^{93}Nb reactions. He wants to thank for fruitful discussions with N. Namboodiri, J. B. Natowitz, and J. P. Wurm. He acknowledges the hospitality of the Gesellschaft für Schwerionenforschung in Darmstadt and of the Cyclotron Institute in College Station where the two experiments were performed. He also acknowledges the support from the Deutscher Akademischer Austauschdienst by granting a NATO-scholarship for his stay in College Station.

References

1. Ho, H., Albrecht, R., Dünnweber, W., Graw, G., Steadman, S.G., Wurm, J.P., Disdier, D., Rauch, V., Scheibling, F.: Z. Physik A283, 235 (1977),
    Ho, H., Albrecht, R., Demond, F.-J., Wurm, J.P., Dünnweber, W., Graw, G., Disdier, D., Rauch, V., Scheibling, F.: Proc. Int. Conf. Nuclear Structure, Tokyo, 1977.

2. Weiner, R., Weström, M.: Phys. Rev. Lett. 34, 1523 (1975) and Nucl. Phys. A286, 282 (1977),
    Gottschalk, P.A., Weström, M.: Nucl. Phys. A314, 232 (1979).

3. Slemmer, J.: Ph. D. Thesis, Heidelberg,
    Slemmer, J., Albrecht, R., Damjantschitsch, H., Ho, H., Kühn, W., Wurm, J.P., Rode, I., Scheibling, F., Ronningen, R.M.: to be publ.

4. Kühn, W.: Ph. D. Thesis, Heidelberg,
    Kühn, W., Albrecht, R., Damjantschitsch, H., Dössing, T., Ho, H., Slemmer, J., Wurm, J.P., Rode, I., Scheibling, F., Ronningen, R.M.: to be publ.

5. Gonthier, P., Ho, H., Namboodiri, M.N., Natowitz, J.B., Adler, L., Hartin, O., Kasiraj, P., Khodai, A., Simon, S., Hagel, K.: Proc. Int. Symp. Continuum Spectra of Heavy Ion Reactions, San Antonio, 1979.

6. Inamura, T., Ishihara, M., Fukuda, T., Shimoda, T.: Phys. Lett. 68B 51 (1977),
    Zolnowski, D.R., Yamada, H., Cala, S.E., Kahler, A.C., Sugihara, T.T.: Phys. Rev. Lett. 41, 92 (1978)
    Siwek-Wilczynska, K., du Marchie van Voortshuysen, E.H., van Popta, J., Siemssen, R.H., Wilczynski, J.: Phys. Rev. Lett. 42, 1599 (1979)

7. Weidenmüller, H.A.: MPI-report MPI H-1978-V29 and to be published in "Progress in Particle and Nuclear Physics", ed. by D. Wilkinson, Pergamon Press, and references therein.

Geometry and Dynamics in the Hot Spot Model*

N.Stelte[+] and R.Weiner

Physics Department, Philipps University

D-355o Marburg, Fed. Rep. of Germany

Abstract

An attempt is made to separate geometrical aspects from dynamical ones in the hot spot phenomenon. We discuss the influence
of the finite size of the excitation region and the temperature
dependence of transport and thermodynamical coefficients on the
solution of the diffusion equation. The hot spot model is also
generalized and applied to inclusive reactions. The results
of a comparison between theory and experiment in the 1oo-8oo MeV
range are found to be satisfactory.

+ Invited talk presented by N.Stelte at the Symposium on Deep-
Inelastic and Fusion Reactions with Heavy Ions
Hahn-Meitner Institut Berlin - October 1979

*Work supported in part by the GSI-Darmstadt

1. Introduction

The main purpose of new heavy ion accelarators is the study of collective phenomena. The hot spot is such a phenomenon and there are several experiments [1] which indicate appearance of hot spots in heavy ion reactions.

By "hot spot" we understand a local concentration of energy. The name suggests a thermodynamical treatment and in all applications done so far this has been the case, albeit it is nonequilibrium thermodynamics we are dealing with. Whether thermodynamics and in some generalised versions of the hot spot model hydrodynamics applies to nuclear matter is a subject in itself, connected with the relation between the mean free path in excited nuclear matter and the dimensions of the system. While the creation of the hot spot depends on the reaction mechanism, the dissipation depends on the intrinsic properties of nuclear matter only, the equation of state, heat conductivity and viscosity. Thus the problem is determined on the whole by the initial and boundary conditions (geometry) and a set of hydrodynamical equations, which express the conservation of mass, momentum and energy (dynamics).

$$\frac{\partial}{\partial t} \rho + div (\rho \vec{v}) = 0$$

$$\frac{\partial}{\partial t} (\rho v_i) \sum_{k=1}^{3} \frac{\partial}{\partial x_k} (p \delta_{ik} + \rho v_i v_k - \sigma'_{ik}) = 0 \tag{1}$$

$$\frac{\partial}{\partial t} (\tfrac{1}{2} \rho v^2 + \rho \bar{\varepsilon}) + div [\rho \vec{v} (\tfrac{1}{2} v^2 + \omega) - \vec{v} \sigma' - \kappa \, grad \, T] = 0$$

Here ρ, p, $\vec{v}$, σ', $\bar{\varepsilon}$, ω, κ are density, pressure, velocity, viscosity stress tensor, internal energy, enthalpy, heat conductivity and temperature respectively. If the velocity and temperature distributions are known, the emission of light fragments from the nuclear surface can be calculated with the formula of Blatt and Weißkopf

$$\frac{d\sigma}{d\varepsilon \, d\Omega} \propto \varepsilon \cdot \iint exp \left(- \frac{\varepsilon + \vec{p} \cdot \vec{v} + s}{T} \right) df \, dt \tag{2}$$

where ε and $\vec{p}$ are kinetic energy and momentum and s is the separation energy of the secondary. To carry out this calculation in its generality implies an extensive numerical effort. While work in this direction is in progress it is useful to consider approximations of the model for selected processes so that the physics involved should not be hidden by numerics; where data exists these approximations appear to be confirmed by experiment.

If the effect of momentum transferred to the hot spot can be ne-glected, we get the static hot spot picture, for which the dissipation of the thermal energy is treated by the classical diffusion equation only.

$$\rho c_p \frac{\partial T}{\partial t} = \text{div} \left(\kappa \ \text{grad} \ T \right)$$

(3)

Here c_p is the heat capacity at fixed pressure. For T-independent coefficients ρ, c_p, κ the problem is <u>linear</u> and can be solved analy-tically [2] . If one considers heat propagation in a sphere and axial symmetry with respect to the initial excitation the solution of Eq. (3) reads

$$T(\cos \theta, r, t) = \sum_{m=0}^{\infty} \sum_{n=0}^{\infty} c_{nm} e^{-\alpha_{nm}^2 \chi t} r^{-\frac{1}{2}} J_{n+1/2} (\alpha_{nm} r) P_m (\cos \theta)$$

(4)

Here $\chi = \kappa/(\rho c_p)$ is the thermal conductivity; the coefficients c_{nm} are determined by the initial condition and α_{nm} are determined from the boundary condition $\partial T/\partial r|_{r=R} = 0$, which expresses energy conservation until the first emission takes place.

2. Heavy Ion Reactions

a) Geometry

Projectile energies of a few MeV above the Coulomb barrier lead to grazing collisions with more or less pointlike excitations. For this purpose a delta-function was used as an initial condition [2] . It was applied to the reaction 96 MeV $^{16}O+^{58}Ni$. [3]

On the other hand higher energies well above the Coulomb barrier mean less peripheral reactions with a spatially extended excitation, which might resemble a spherical cap [4] (see fig.1). This geometry should be preferred with increasing mass number of the projectile if we look at hot spots in the target. As a matter of fact the terms target and projectile are in this context interchangeable since the hot spot can be produced in both nuclei.

As an example we show in fig.2 the temperature distribution at the nuclear surface for different times in the linear approximation. To get an impression of the importance of an "extended" initial condition several values of the cap parameter r_{in} are compared: large differencies in the maximum of the temperature at the early stage of preequilibrium are obtained. The application of the delta-function solution to nonperipheral collisions would thus lead to an overestimation of the observed asymmetry effect and an underestimation of the width of the temperature distribution as a function of the polar angle θ.

b) Dynamics

We expect the thermodynamical coefficients c_p, ρ and κ to be T-dependent and regard T-independent coefficients only as a linear approximation for regimes with small temperature gradients. In particular two cases of physical interest are considered in the following. A low energy regime corresponding to a Fermi gas, where

$$c_p \approx c_v = \frac{\pi^2}{2} \frac{k^2}{m \, \varepsilon_F} T$$

$$\kappa = \frac{7}{48\sqrt{2}\,\pi} \frac{\varepsilon_F^{3/2}}{\sqrt{m}\,Q} \frac{1}{T} \tag{5}$$

Here ε_F, k , m, Q are the Fermi energy, Boltzmann constant, nucleon mass, and the effective nucleon-nucleon cross section.

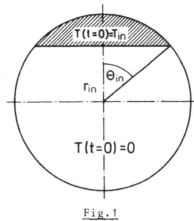

Fig.1

Spherical cap with temperature T_{in} as initial condition on a sphere. The parameter $r_{in} = R \cos \Theta_{in}$, determines the size of the cap.

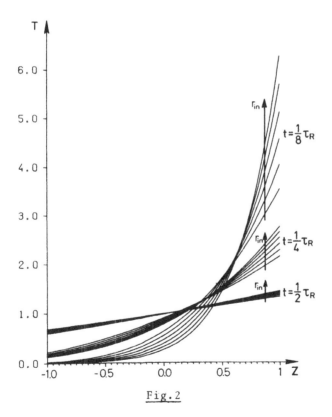

Fig.2

Temperature distribution at the nuclear surface as function of $Z = \cos \Theta$ (Θ polar angle) for three times and cap parameter $r_{in}=0.5$, 0.6, 0.7, 0.8, 0.9, 1. R (fig.1). The excitation energies are the same for all cap parameters.

For an intermediate energy regime (excitation energy up to few hundreds MeV/nucleon) corresponding to a Boltzmann gas one has

$$c_p = \frac{5}{2} \frac{k}{m}$$

$$\kappa = \frac{1}{\pi} \frac{k}{m}^{3/2} \frac{1}{Q} \sqrt{T} \tag{6}$$

In fig.3a and b we compare the results of a numerical integration of the diffusion equation (3) for these two cases for a cap parameter r_{in}=0.7R. The t and z=cosθ dependence of T shows that for a Fermi gas high temperatures are to be found in the system for a rather long time (t=10^{-22}s). At the very end of the diffusion process the temperature falls down quickly, so that there should be essentially only one high temperature observed in the experiment besides the equilibrium temperature. This means that in a first approximation for a Fermi gas regime a hot spot picture with fixed geometry can be used without considering heat diffusion.

The situation is quite different for a Boltzmann type gas where the temperature shows a smooth behaviour in space and time which is similar to that of T-independent coefficients. In this case the whole continuum of temperatures is relevant for the description of the process.

These results have important implications for the investigation of the equation of state via the hot spot effect.

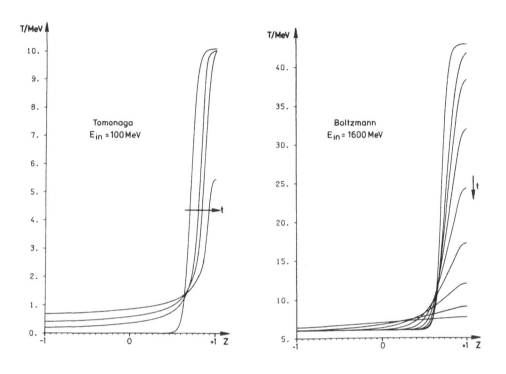

Fig.3a   Fig.3b

Temperature distribution at the nuclear surface as function of $Z = \cos \Theta$ at different times $t = 2^n \cdot 10^{-23}$ s, $n = 0,1,2,\ldots$ E_{in} is the initial excitation energy.

3. Proton — nucleus inclusive reactions

While the ideal experiment for the investigation of the hot spot phenomenon is a coincidence measurement [2] there are several experiments of inclusive nature of the form

$$a + A \longrightarrow b + x$$

where a is usually a proton, b a proton or a light nucleus, A a target nucleus and x stands for anything (not measured). The question arises what, if anything, can be learned from these experiments and in particular whether the hot spot mechanism is consistent with these data although it is clear that the amount of information available from such an experiment is rather limited.

In proton-nucleus collisions hot spots should be created all over the nucleus through the local loss of energy (and momentum) by the particle. We use the delta-function initial condition and the linear expansion but consider now for the first time the hydrodynamical aspect of the problem which leads to the picture of the moving hot spot [5] . The general formula reads

$$\frac{d\sigma}{d\varepsilon d\Omega} \propto \int_0^R 2\pi b\, db \int_0^{\ell(b)} \ell^{-\Delta/\lambda}\, d\Delta \int_0^{2\pi} d\varphi \int_0^{\pi} \cos\Psi(\varphi, \vartheta, \Theta)\, d\vartheta \int_{t_0}^{t_R} \varepsilon\, \ell^{-\frac{\varepsilon - \vec{p}\vec{v}(b, \Delta, \varphi, \vartheta, t) + S}{T(b, \Delta, \varphi, \vartheta, t)}}\, dt \quad (7)$$

The first two integrals describe the formation of hot spots (HS) in the volume of the nucleus (λ is the mean free path) while the next two integrals describe the emission of secondary particles from the nuclear surface. As in the previous section an effort was made to evaluate this integral so that the physics of the problem (i.e. the dynamical part) should be separated from the geometrical aspects. If we assume that the diffusion of energy and momentum are processes of similar nature and can be treated by the same diffusion equation (3), the movement of the hot spot (characterised by a drift parameter d) can be separated from the diffusion process. It is useful to introduce now the temperature weight function on the nuclear surface [4]

$$\rho(T, u) = 2\pi R^2 \int_0^{\frac{1}{2}\vartheta'} d\vartheta' \cos\vartheta' \int_0^{t_R} dt\, \delta\left(T(t, \vartheta', u(b, \Delta)) - T\right) \quad (8)$$

in a coordinate frame (ϑ', φ') where T is symmetric in the azimutal angle φ'. For $\varepsilon \gg$ T only $\vartheta' < \Delta \vartheta' \ll \pi/2$ contribute and (7) becomes

$$\frac{d\sigma}{d\varepsilon d\Omega} \propto \int_0^R 2\pi b \, db \int_0^{\ell(b)} e^{-\Delta/\lambda} d\Delta \, \cos \Psi(\varphi(u), \vartheta(u), \Theta) \int_0^{T_{max}} g(T, u) \varepsilon \cdot e^{-\frac{\varepsilon + \vec{p} \cdot \vec{v}(T) + S}{T}} dT \tag{9}$$

The crucial quantity $g(T, u)$ $\left(u = | \vec{r}_{HS}(b, \Delta) - \vec{r}_{emission}(b, \Delta)| \right)$ which obeys a transformation law in the variable u is calculated on the computer and can be fitted by a simple formula. Together with a related expression for the velocity it contains the dynamical part.

Simplifications for the geometry (first four integrals) can be made for the emission in different hemispheres (see fig.4). The most important fact connected with the physics of the problem is that the hot spots look quite different in the forward and in the backward hemisphere. Due to the momentum transferred to the hot spots they drift for a certain time essentially in the direction of the incident proton. For the backward hemisphere that means that they drift away from the surface while for the forward hemisphere they drift towards   the surface. They can reach the forward surface at different stages of their lifetime; some of them are "young", very hot and quickly moving. At the central backward surface they are never young; they have shared their initial momentum with many other nucleons and only the heat diffusion brings back some temperature to the surface. These hot spots can be treated as being static, but shifted by a mean drift length from their point of creation to the forward direction. Important conclusions for secondary spectra follow from these qualitative considerations; they are summarized in table 1.

Fig.5 shows a fit to the reaction loo MeV p + ^{58}Ni $\longrightarrow$ p + X [6] for $\Theta < 9o°$. The data are consistent with the thermodynamic coefficients of the Fermi gas model, with a relaxation time $\tau_R = R^2/\chi = 2o \, \tau_o$ where τ_o is of the order of the nucleon interaction time. The same value was found in previous fits to heavy ion coincidence experiments [3] . The drift length is 2-3 fm [5] .

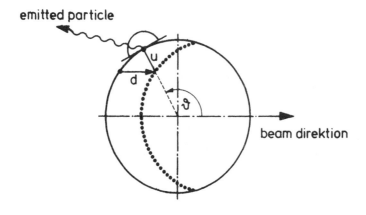

Fig.4

Distribution of hot spots in a nucleus (represented by the dotted line) assumed to be created at the backward surface and drifted by a mean path d in beam direction. The shortest distance to the surface u depends on the polar angle.

Table 1

Comparison of forward and backward secondary spectra in proton-nucleus reactions within the "moving hot spot" model.

Backward ($\Theta \gg 90^\circ$)	Forward ($\Theta < 90^\circ$)
low T, low v: large slopes	high T, high v: small slopes
small variation of spectrum with angle	strong variation of spectrum with angle
$\dfrac{d\sigma}{d\varepsilon\,d\Omega}$ ($\varepsilon = 0$) $\propto A^{2/3}$	$\dfrac{d\sigma}{d\varepsilon\,d\Omega}$ ($\varepsilon = 0$) $\propto A^{1/3}$

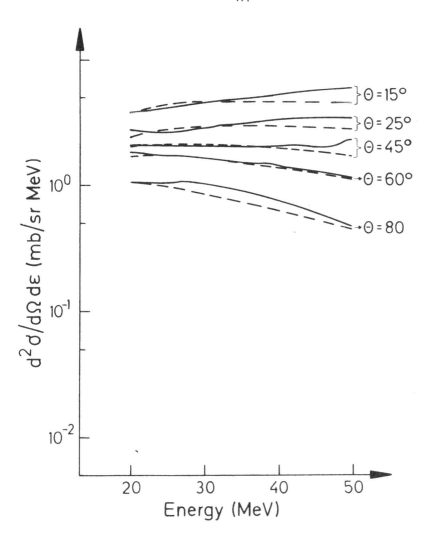

Fig.5

Comparison of experimental (continuous lines) and theoretical (hot spot model) differential cross sections for the reaction p + A ⟶ p + X at E = 1oo MeV. The data are from Ref.[6].

For the backward hemisphere one gets eventually the formula

$$\frac{d\sigma}{d\epsilon \, d\Omega} \propto d \, R^2 \frac{E}{\varsigma \, c_p} \, e^{-\frac{d^3 \, (\epsilon + s)}{0.161 \, E/(\varsigma c_p)}} \quad , \tag{1o}$$

which we expect to work for A $\gtrsim$ 5o, we have compared the results
of the model with the data of ref. [7] for proton production at
$\Theta = 18o^\circ$ in reactions induced by 6oo MeV and 8oo MeV protons. The
results of this comparison are summarized in table 2. The dependence of the slope of the spectra can be used to relate the drift
d to the incident energy E. Comparison with similar data [6, 8]
at other energies suggest that the E-dependence of the intercept
is experimentally not yet settled so that the $E^{4/3}$-dependence
of our model can be regarded as a prediction. The results of this
section suggest that the hot spot model in its generalized (moving)
version is consistent with inclusive reactions data in the energy
range 1oo - 8oo MeV.

Table 2

Comparison between theory and experiment [7] for backward proton
production in the reaction:6oo- 8oo MeV p + A $\longrightarrow$ p + X .

	Experiment	Theory
$\dfrac{d\sigma}{d\epsilon \, d\Omega}$	$Be^{-\beta\epsilon}$	$Be^{-\beta\epsilon}$
slope $1/\beta$	$A^{\approx 0}$	A^{0}
slope $1/\beta$	$E^{\approx 0}$	E^{0} , if $d \propto E^{1/3}$
factor B	$A^{\approx 2/3}$	$A^{2/3}$
factor B	$E^{\approx 5}$	$E^{4/3}$

4. Conclusion

The separation of geometry from dynamics in the investigation
of the hot spot mechanism may ultimately lead to a method for the
investigation of the temperature dependence of thermodynamic observables and transport coefficients. In this way important information about the equation of state of nuclear matter might be
obtained since it appears that more or less the same mechanism
works in different energy regimes.

List of References

1) H.Ho et al. Zs.f.Physik A283, 235 (1977)
   T.Nomura et al., Phys.Rev.Lett. 4o, 694 (1978)
   L.Westerberg et al., Phys.Rev. C18, 796 (1978)
2) R.Weiner, M.Weström, Nucl.Phys. A286, 282 (1977)
3) P.Gottschalk, M.Weström, Nucl.Phys. A314, 232 (1979)
4) N.Stelte, to be published
5) N.Stelte,M.Weström,R.Weiner, to be published
6) J.R.Wu, C.C.Chang,H.D.Holmgren, Phys.Rev. C19, 659 (1979)
7) S.Frankel et al., Phys.Rev.Lett. 36, 642 (1976)
8) A.M.Baldin et al., Communication of the Joint Institute for
   Nuclear Research, Report No. Dubna-113o2, 1978 (unpublished)

SYMPOSIUM ON DEEP-INELASTIC AND FUSION REACTIONS WITH HEAVY IONS

HAHN-MEITNER INSTITUT FUR KERNFORSCHUNG, BERLIN

October 23-25, 1979

AN EXPERIMENTAL APPROACH OF THE FRICTION PHENOMENON
IN DEEP INELASTIC COLLISIONS BASED ON SECONDARY LIGHT-PARTICLE EMISSION STUDIES.

J. ALEXANDER[+], T.H. CHIANG[++], J. GALIN, B. GATTY, D. GUERREAU, X. TARRAGO

Institut de Physique Nucléaire, B.P. n°1, 91406-Orsay Cedex, France

and

R. BABINET, B. CAUVIN, J. GIRARD

DPHN/MF, Centre d'Etudes Nucléaires de Saclay, B.P. n°2, 91190-Gif/Yvette, France

Abstract : It is shown how one can take advantage of evaporated charged particles emitted by fully thermally equilibrated deep inelastic fragments to extract both spin and degree of alignment of these fragments. The investigated system is $^{40}Ar(280$ MeV$) + ^{58}Ni$.

INTRODUCTION

The macroscopic properties of nuclear matter have been extensively investigated during the last few years through[1-5] deep inelastic collisions between heavy nuclei.

The classical concepts of friction and viscosity have been used successfully to describe such collisions. On the one hand, the strong damping of the kinetic energy in the relative motion can be understood in terms of a radial component of a friction force acting between the two colliding nuclei. On the other hand, the transfer of orbital angular momentum into intrinsic spin of the reaction products is interpreted as a manifestation of the tangential component of the friction force.

Thus, energy dissipation and angular momentum transfer are certainly closely related and must be investigated simultaneously. The natural way to get some insight into the reaction mechanism is to look at all the particles (neutrons and

---

[+] Present address : State University of New York, Department of Chemistry, Stony Brook New York 11794, U.S.A.

[++] Present address: Department of Physical Technics, University of Peking, Peking, China.

charged particles) and γ-rays that carry off some excitation energy at any stage of the process. However, depending on the entrance channel characteristics, one has to face quite different situations. For heavy systems (typically Ar + Au or Cu + Au) particle emission is most likely restricted to neutrons that carry away most of the excitation energy but very little angular momentum. Thus, charged particles may be disregarded since neutrons and γ-rays play a prominent role in taking away most of the energy for the first ones, and most of the angular momentum for the latters. In such cases the energy balance is established by measuring the neutrons[6-12] and the angular momentum sharing is obtained by γ-ray multiplicity measurements[13].

Moreover, when the target like nucleus is heavy enough to undergo fission, the spin of this nucleus can be reached classically by measuring the out of plane angular[19-22] distribution of the corresponding fission fragments.

For light or medium mass systems (typically Ar + Ni) neutron and charged particles and γ-rays emission play an important role in the deexcitation process. One cannot neglect any more both energy and angular momentum removed by the charged particles[23]. In principle it would be necessary to investigate these three channels in order to control the energy dissipation and angular momentum transfer. However, as it will be shown in the following, most of the information can be reached through proton and α-particle investigation only.

Due to the large amount of angular momentum they can carry off, as compared to protons and neutrons, α-particles are essentially emitted in the first deexcitation step of a thermally equilibrated nucleus with large spin. Thus, they can reveal, better than any other particle the main characteristics of the emitting nucleus (i.e. temperature and spin) as it has been left after the deep inelastic interaction. Once it has been established that the observed α-particles are issued from thermally equilibrated nuclei, then, one can apply the classical statistical theory in order to carry out the temperature of the emitting nucleus from the α-particle energy spectra, and its spin from their out of plane angular distribution.

It is in this spirit that we have undertaken a detailed study of the light charged particles (mainly protons and alpha particles) in coincidence with the main fragments from DIC in the reaction 280 MeV ^{40}Ar + ^{58}Ni.

The aim of this experiment was to answer the following questions which will be discussed in three different sections.

1/ Are all the charged particles observed in coincidence with the deep inelastic fragments understood in terms of statistical evaporation from fully thermalized and fully accelerated fragments ?

Is there any evidence for rapid emission from hot spots[24] or promptly emitted particles (PEP's Fermi jets) such as the ones suggested by Bondorf[25] or Gross-Wilczynski[26] ?

2/ What is the associated spin of the emitting fragment and what information do we then get on the tangential friction ?

3/ What are the advantage of using such a method as compared with standard γ-ray multiplicity measurements ? How can we get information on the spin alignment ?

I - THE ORIGIN OF THE PROTON AND α-PARTICLES OBSERVED IN COINCIDENCE WITH THE FRAGMENTS.

The choice of the ^{40}Ar(280 MeV) + ^{58}Ni system was essentially motivated by the large amount of information which is already available on this reaction[27-31].

In particular, fragment-fragment coincidence data[31] and γ-ray multiplicity data[29] already gave us some clues about the important role played by charged particles in the deexcitation process. Detailed information can be found in references[27-31].

For a good understanding of the present paper two important results must be stressed. First, it was previously measured that DIC were taking place in collisions with angular momenta in a narrow range (74 ħ < ℓ_{DIC} < 98 ħ) which will allow us in the following to consider an average value as well representative of the phenomenon. Second, in the present investigation the D.I. fragments are detected at 30° which is well above the grazing angle. It was shown that the corresponding collisions are completely damped and correspond to long reaction times with trajectories crossing at least once the beam direction.

Let us now consider the in-plane angular distributions. As light particles may originate from different emission sources it is very convenient to present the data in terms of invariant cross sections, $\sigma_I = \frac{1}{P} \frac{d^2\sigma}{dEd\Omega}$, as a function of parallel and transverse momenta ($P_{//}$ and $P_{\perp}$). This quantity is proportional to $\frac{d^3\sigma}{d^3p}$, so that it is Galilean invariant.

For a single source emitting particles isotropically, iso-invariant cross sections would show up as circles centered around the tip of the velocity vector characterizing the source.

Figure 1 shows the invariant cross section plots for α-particles in coincidence with deep inelastic fragments of charge Z = 23 and 16 corresponding respectively to symmetric (or nearly symmetric) and asymmetric mass splitting in the exit channel. Each raw of dots corresponds to a single angular measurement and the size of the

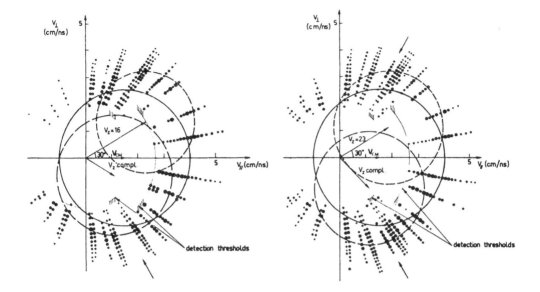

Figure 1 : Invariant cross-sections plots for α-particle in coincidence
with fragments of charge Z = 16 and Z = 23. The size of the dots is
an increasing function of σ_I(. < .02, .02 < . < .05, .05 < $\bullet$ < .1,
$\bullet$ > .1 μb/sr^{2}MeV^2). The thin lines represent the experimental velocity thresholds. The dashed circles are centered at the tip of
velocity vectors for the fragments, with their radius corresponding
to the most probable α-particle velocity. The full line circle indicates what would be the expected most probable velocity for
α-emission for the composite system.

dots is an increasing function of the invariant cross section. Experimental velocity
thresholds are indicated as well as the average velocity vectors of the two main fragments (the heavy fragment average velocity is computed assuming two-body kinematics).

Two α-particle emission sources appear very clearly in these diagrams, they
are the two fragments. The maxima in the invariant cross sections are stressed in
figure 1 by two dashed circles centered at the tip of the fragment average velocity
vectors. At forward angles, in the region of velocity space where the two velocity
circles overlap, there is a clear pile up of the cross section and a very strong asymmetry of the α-particle energy spectra with respect to the beam axis. This is best
seen in the diagram corresponding to Z = 16. The most probable velocities, measured
at φ_α = +10° and -10° are completely different, and in good agreement with a preferential emission by the light fragment Z = 16 at +10° and by the heavy one at -10°.

In order to check more quantitatively the hypothesis of statistical emission
by fully accelerated fragments the cross section in the rest frame of the emitted

fragment should be found as isotropic. One example is given in figure 2 for the case
of symmetric splitting. As suggested by figure 1, the cross sections (or particle

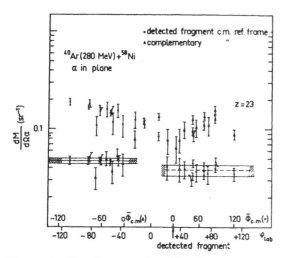

multiplicities) can be expressed in
the rest frame of the detected fragment for $+30° < \varphi_{lab} < +120°$ and in
the rest frame of the complementary
fragment for $-30° < \varphi_{lab} < -120°$.
In both angular ranges the distributions are flat. Moreover, within
the experimental uncertainties, the
same α-particle multiplicity is observed for both fragments which
again suggests a complete statistical equilibrium with similar excitation energies for both similar fragments.

Figure 2 : In-plane angular distribution for
the symmetric splitting case $(Z_{detected}=23)$
plotted in the rest frames of both the detected fragment (•) and the complementary
fragment (▲).

Other clues of the statistical evaporation origin of the detected particles may be found in the
analysis of their energy spectra.

Some of them are shown in figure 3 in the rest frame of the emitting fragments
(Z = 23,30 and 36) and are compared with classical spectra of the shifted Maxwell
type :

$$P(E) \ dE = \frac{E - B_S}{T^2} \ exp \left[- (E - B_S)/T \right]$$

where T is the nuclear temperature and B_S is a threshold energy.

The most probable energy is carried out from these spectra and compared
with the expected ones given by : $T + B_S$ (figure 4), where T is the nuclear temperature computed assuming the fragments to be in complete statistical equilibrium and
B_S has been fitted on available experimental data on particle emission by compound
nuclei. The agreement is quite reasonable, at least for the high energy tail of the
spectra which is most sensitive to the temperature. The maxima are also rather well
reproduced. In contrast, the threshold energies are not very well accounted for due
to the neglect of penetrability effects in such a crude representation.

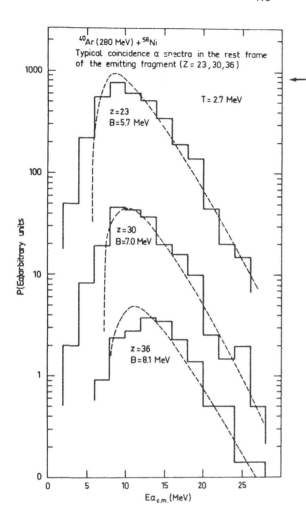

Figure 3 : Typical experimental α-energy spectra for three emitters Z = 23, 30 and 36 plotted in the rest frame of these emitters (histogramms) compared the calculation (dashed curve).

Figure 4 : Most probable α-energies as a function of Z (the charge of the emitter) compared with the simple expression B_α + T.

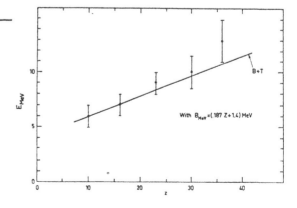

At last, an interesting result is given in figure 5 where it is shown that the same temperature fits the spectra of α-particles issued from the two complementary fragments.

This result is understood if a thermodynamical equilibrium has been reached in the composite system before scission. Similar conclusions were obtained on different systems when looking at the neutrons[6-12].

Finally, we conclude from the above analysis that most of the light particles seem to be evaporated by the fully accelerated fragments and that, within our experimental uncertainties, there is no need to introduce other processes to explain the data. The onset for an additional contribution observed by Ho et al.[32] on the Ar + Nb system can be related to the increase of the bombarding energy. (7 MeV/A in the Ar + Ni case ; 10 MeV/A in the Ar + Nb one).

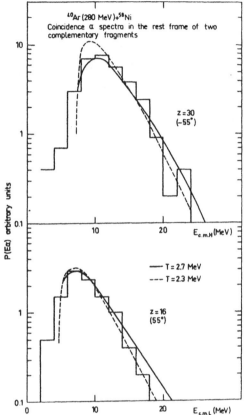

Figure 5 : α-spectra of two complementary fragments (Z = 16 and 30) in the rest frame of their emitter. The same temperature fits the spectra.

II - FRAGMENT SPIN DETERMINATION FROM OUT OF PLANE ANGULAR DISTRIBUTIONS. COMPARISON WITH CLASSICAL MODELS

Due to the transfer of part of the orbital angular momentum into spin of the fragments, the latters will tend to have their spin perpendicular to the reaction plane. This is the reason for which the in-plane angular distributions were found isotropic in the reference frame of the emitters, whereas the out-of-plane distribution should show an anisotropy that characterizes both the spin value and degree of alignment.

In the preceding section we could see that, by an appropriate choice of the in-plane angle (say ± 60°) where to make an out-of-plane angular distribution, one is only sensitive to the α-emission by a single fragment. This is a great advantage of light particle measurements over the γ-ray multiplicity technique for which such a clear cut cannot be achieved.

The results of some out-of-plane α-particle distribution plotted in the center of mass of the corresponding emitting fragment are shown in figure 6. They

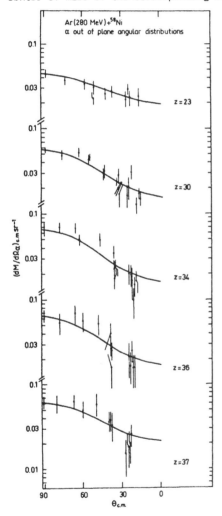

were analyzed following the classical development of Ericson-Strutinsky[33]. The probability for a particle to be emitted at an angle θ with respect to the spin I of the emitter is given by :

$$W_{I\ell\epsilon}(\theta) \quad \alpha \quad J_0 \left(\frac{\hbar^2(I+1/2)(\ell+1/2)}{\mathcal{J}T} \sin\theta\right)$$

where J_0 is the zeroth order associated Bessel function, ℓ the orbital angular momentum of the evaporated particle, $\mathcal{J}$ and T the moment of inertia and temperature of the residual nucleus.

Following Dossing[34] or Catchen et al.[35] the integration over all the possible ℓ values and energies ϵ leads to :

$$W_I(\theta) \quad \alpha \quad Exp \left(\frac{h^2(I+1/2)^2}{2\mathcal{J}T} \frac{mR^2}{mR^2+\mathcal{J}} \sin^2\theta\right)$$

where mR^2 represents the relative moment of inertia of the particle at the surface of the nucleus.

The spin I of the emitting fragment is obtained by fitting the experimental out-of-plane distributions using for T the temperature carried out from the energy spectra analysis. The other parameters entering the formula are taken as the rigid body value $\mathcal{J} = 2/5 \, mR^2$ with $R = 1.2 \, A^{1/3}$ fm and mR^2 evaluated following McMahan and Alexander[36] as:

$$mR^2 = m(r_0 A^{1/3} + R_\alpha)^2$$

with $r_0 = 1.42$ fm, $R_\alpha = 2.53$ fm

Figure 6 : Typical α out-of-plane angular distributions plotted in the rest frame of their corresponding emitter indicated in the figure.

Note that the above relations assume total spin alignment. A possible disalignment would wash out the out-of-plane anisotropy and thus lead to a lower estimate of the spin values.

As shown in figure 7, the spin value is a rapid increasing function of the Z (or mass) of the emitter. It is also obvious that the ratio of the spins of the two complementary fragments reflects more a sticking configuration (where $I_1/I_2 = \mathcal{J}_1/\mathcal{J}_2 = (A_1/A_2)^{5/3}$) than a rolling configuration (where $I_1/I_2 = R_1/R_2 = (A_1/A_2)^{1/3}$).

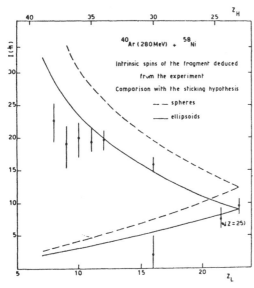

Figure 7 : Experimental intrinsic spins of the individual fragments compared with the results of calculations assuming a sticking configuration of rigid body.

For comparison, the sticking limits for both spherical and deformed nuclei are plotted in figure 7, for an average ℓ in the entrance channel. The deformation of the nuclei has been obtained taking into account the measured kinetic energy of the fragments and the centrifugal energy corresponding to the average orbital angular momentum obtained by difference between the entrance channel average one and the measured spins of the fragments. From the deduced Coulomb energy the deformation is easily reached.

Once this deformation has been taken into account, the overall agreement with the experimental data is rather fair. However, it can be noticed that the experimental anisotropies for heavy emitters (Z ≥ 35) seem to be systematically lower than the predicted ones. Several hypothesis can be invoked to explain this possible discrepancy. First, for very asymmetric configurations the shape of the composite system as described by the model may not be quite realistic. Then, in these model calculations only an average ℓ value was considered as if all the ℓ waves in the deep inelastic window were equally contributing to the different reaction products in the exit channel. In fact, the large mass asymmetries are most likely to result from long interaction times and from the smallest ℓ waves involved within the ℓ window for DIC. Therefore slightly lower spin values may be expected for such asymmetric exit channels. Finally, as already pointed out previously, the low spin value may simply originate from some disalignment of the emitter. This possibility will be discussed in some details in the next section.

III - SPIN ALIGNMENT. CONCLUDING REMARKS

It is possible to further check the degree of spin alignment of the emitters by determining the spin absolute value and by comparing with the value deduced from out-of-plane measurements. Indeed the relative proton and α-particle emission probabilities are only sensitive to the absolute value of the spin.

In principle, these probabilities could be evaluated as a function of excitation energy and spin through standard evaporation computer codes. However, for a particular case, we can directly compare our data with experimental ones from Reedy et al.[37]. For a ^{75}Br compound nucleus the multiplicity ratio M_α/M_p was measured as a function of spin[37]. A very similar nucleus Z = 36, with similar excitation energy is formed in the deep inelastic collision of Ar + Ni. In the later case the integrated experimental multiplicities for protons and α-particles are respectively 1.5 and .56 leading to M_α/M_p = .37. From the ^{75}Br experimental data a corresponding value of I = 22 ℏ can be deduced, which is very close to the α-anisotropy experimental value (I = 20 ℏ). Such a result clearly indicates a strong alignment of the heavy fragment spin.

Finally, it is quite interesting to compare the spin distributions issued from α-anisotropy data with those derived from γ-ray multiplicity measurements[29] (figure 8). As expected the α-anisotropy results lie systematically well above the γ-multiplicity data (assuming 100 % of stretched E2 transitions, which must be considered as an upper limit). The difference between the two sets of data corresponds to the part of angular momentum removed by the evaporated particles. From the statistical model[35], one can calculate the average angular momentum carried out by such particle and knowing their multiplicity one can deduce the total amount, which appears to be in fair agreement with the experimentally deduced data.

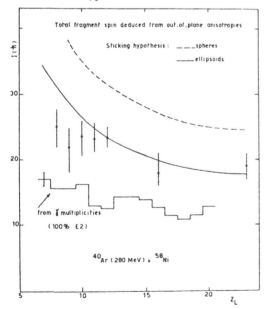

It is clear that, for such light systems, light charged particle measurements give the best picture of the reaction products as they are left after the deep inelastic interaction. Also,

Figure 8 : The total fragment spins of the two complementary fragments deduced from this experiment are compared with γ-multiplicity measurements of ref.[29].

they have already shown at higher bombarding energy[32] that a faster emission mechanism takes place. In conclusion, they appear to be a very nice tool to further investigate the dissipative phenomena in heavy ion induced reactions.

REFERENCES

[1] L. Moretto and R.P. Schmitt, J. de Phys. (Paris) 37, C5 109 (1976).

[2] J. Galin, ref.1), page 83.

[3] W.V. Schroeder and S. Huizenga, Ann. Rev. Nucl. Sci. 27, 465 (1977).

[4] M. Lefort and C. Ngô, Ann. Physique (Paris) 3, 5 (1978).

[5] D.K. Scott, Int. Report Berkeley, LBL 77 27 (1978)

[6] D. Hilcher, this conference.

[7] J. Péter, M. Berlanger, C. Ngô, B. Tamain, B. Lucas, C. Mazur, M. Ribrag, C. Signarbieux, Z. Phys. A283, 413 (1977).

[8] B. Tamain, R. Chechik, H. Fuchs, F. Hanappe, M. Morjean, C. Ngô, J. Péter, M. Dakowski, B. Lucas, C. Mazur, M. Ribrag, C. Signarbieux, Nucl. Phys. in press.

[9] M. Berlanger, R. Chechik, H. Fuchs, F. Hanappe, M. Morjean, J. Péter, B. Tamain, M. Dakowski, B. Lucas, C. Mazur, M. Ribrag, C. Signarbieux, Lac Balaton Meeting, Juin 1979 (Hungary).

[10] Y. Eyal, A. Gavron, I. Tserruya, Z. Fraenkel, Y. Eisen, S. Wald, R. Bass, C.R. Gould, G. Kreyling, R. Renfordt, K. Stelzer, R. Zitzmann, A. Gobi, U. Lynen, H. Stelzer, I. Rode and R. Bock, Phys. Rev. Lett. 41 (1978) 625.

[11] C.R. Gould, R. Bass, J.V. Czarnecki, V. Hartmann, K. Stelzer, R. Zitzmann, Y. Eyal, Z. Phys. A284, 353 (1978).

[12] D. Hilscher, J.R. Birkelund, A.D. Hoover, W.U. Schröder, W.W. Wilcke, J.R. Huizenga, A. Mignerey, K.L. Wolf, H.F. Breuer, V.E. Violo Jr, Internal Report UR-NSRL-189 (1979).

[13] C. Gerschel, M.A. Deleplanque, M. Ishihara, C. Ngô, N. Perrin, J. Péter, B. Tamain, L. Valentin, D. Paya, Y. Sugiyama, M. Berlanger, F. Hanappe, Nucl. Phys. A317, 473 (1979).

[14] A. Olmi, H. Sann, D. Pelte, Y. Eyal, A. Gobbi, W. Kohl, U. Lynen, G. Rudolf, H. Stelzer, R. Bock, Phys. Rev. Lett. 41, 688 (1978).

[15] R. Albrecht, W. Dünweber, G. Graw, H. Ho, S.G. Steddman, J.P. Wurm, Phys. Rev. Lett. 34, 1400 (1975).

[16] K.V. Ribber, R. Ledoux, S.G. Steadman, F. Videback, G. Young, C. Flaum, Phys. Rev. Lett. 38, 334 (1977).

[17] P. Glassel, R.S. Simon, R.M. Diamond, R.C. Jared, I.Y. Lee, L.G. Moretto, J.O. Newton, R. Schmitt, F.S. Stephens, Phys. Rev. Lett. 38, 331 (1977).

[18] M.M. Aleonard, G.J. Wozniak, P. Glassel, M.A. Deleplanque, R.M. Diamond, L.G. Moretto, P.P. Schmitt, F.J. Stephens, Phys. Rev. Lett. 40, 622 (1978).

[19] J.G. Wozniak, R.P. Schmitt, P. Glässel, R.C. Jared, G. Bizard, L.G. Moretto, Phys. Rev. Lett. 40, 1436 (1978).

[20] P. Dyer, J.J. Puigh, R. Vandenbosch, T.D. Thomas, M.S. Zisman and L. Nunnelley, Nucl. Phys. A322, 205 (1979).

[21] M. Rajagopalan, L. Kowalski, D. Logan, M. Kaplan, J.M. Alexander, M.S. Zisman, J.M. Miller, Phys. Rev. C19, 54 (1979).

[22] D.V. Harrach, P. Glässel, Y. Civelokoglu, R. Manner and H.J. Specht, Phys. Rev. Lett. 42, 1728 (1979).

[23] M.N. Namboodiri, J.B. Natowitz, P. Kasiras, R. Eggers, L. Adler, P. Gonthiers, C. Cerruti and S. Simon, Phys. Rev. 20, 982 (1979).

[24] N. Slelte, This conference.

[25] J. Bondorf, This conference.

[26] D.H.E. Gross and J. Wilczynski, Phys. Rev. Lett. 67B, 1 (1977).

[27] B. Gatty, D. Guerreau, M. Lefort, X. Tarrago, J. Galin, B. Cauvin, J. Girard, H. Nifenecker, Nucl. Phys. A253, 511 (1975).

[28] J. Galin, B. Gatty, D. Guerreau, M. Lefort, X. Tarrago, R. Babinet, B. Cauvin, J. Girard, H. Nifenecker, Z. Phys. A278, 347 (1976).

29) A. Albrecht, B.B. Bock, R. Bock, B. Fischer, A. Gobbi, K. Hildenbrand, W. Kohl, U. Lynen, I. Rode, H. Stelzer, G. Auger, J. Galin and J.M. Lagrange, European Conference on Nuclear Physics with Heavy Ions, Caen (1976), com. p. 167.

30) H. Gauvin, D. Guerreau, Y. Le Beyec, M. Lefort, F. Plasil and X. Tarrago, Phys. Lett. 58B, 163 (1975).

31) R. Babinet, B. Cauvin, J. Girard, H. Nifenecker, B. Gatty, D. Guerreau, M. Lefort, X. Tarrago, Nucl. Phys. A296, 160 (1978).

32) H. Ho, This conference.

33) T. Ericson and V. Strutinski, Nucl. Phys. 8, 284 (1958).
    T. Ericson, Adv. of Physics 9, 425 (1960).

34) T. Dossing, preprint 1978.

35) G.L. Catchen, M. Kaplan, J.M. Alexander and M.F. Rivet, preprint 1979 (to be published).

36) M.A. Mc Mahan and J.M. Alexander, preprint 1979.

37) R.C. Reedy, M.J. Fluss, G.F. Herzog, L. Kowalski and J.M. Miller, Phys. Rev. 188, 1771 (1969).

MULTI-PARTICLE PRODUCTION IN ^{32}S-INDUCED REACTIONS

D. Pelte

Physikalisches Institut der Universität Heidelberg and
Max-Planck-Institut für Kernphysik, Heidelberg, Germany

Reactions between heavy ions are usually accompanied by the production of light particles. The interesting questions related to the
light-particle production are: (i) how many and which particles are
produced? and (ii) what is the production mechanism? With respect to
the latter question one can imagine two extreme situations, on the one
hand the statistical emission of light particles from the thermalized
nuclei of the primary reaction. This process is relatively slow as
thermalization of energy and angular momentum requires a sufficient
length of time. The alternative is a fast process by which all particles are produced during the time the two colliding nuclei interact,
see ref. [1].

In an attempt to investigate these problems a kinematical coincidence
spectrometer has been set up. The spectrometer consists of two large-
area ionization chambers [2] that measure the total energy, energy loss,
locus of impact and arrival time of heavy ions. For a sufficiently
accurate measurement of the time of flight, the ionization chambers
have a distance of 1 m from the target. But because of their large
front area (40 x 16 cm^2) they nevertheless cover solid angles of 50 msr
each. A schematical drawing of the experimental arrangement is shown
in fig. 1. The reactions studied with this spectrometer are ^{32}S + ^{27}Al,

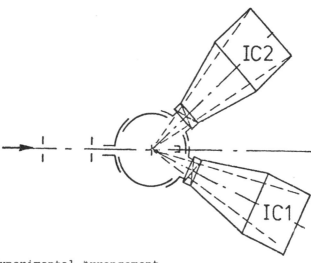

Figure 1   Experimental arrangement

^{28}Si at 135 MeV and ^{32}S + ^{40}Ca at 190 MeV bombarding energies. The pulsed ^{32}S beam was accelerated by the MP postaccelerator combination of the Max-Planck-Institut für Kernphysik, Heidelberg, and had a cycle time of 143 ns. The time resolution was approximately 800 ps at the target position. The targets consisted of isotopically pure material of approximately 500 µg/cm^2 thickness that was evaporated onto thin C backings.

The analysis of the experimental data is based on the conservation laws for charge, mass, linear momentum and energy of the heavy fragments:

$$Z_0 = Z_1 + Z_2 + \Delta Z$$
$$M_0 = M_1 + M_2 + \Delta M$$
$$P_0 = P_1 + P_2 + \Delta p$$
$$E_0 = E_1 + E_2 + \Delta E + Q_T$$

The subscript 0 specifies the entrance channel whereas the subscripts 1,2 refer to the detectors 1 and 2, respectively. Quantities with the symbol Δ correspond to the deficits with respect to a binary reaction. The quantity Q_T characterizes the total reaction Q value. It can only be determined if the additional assumption $E = (\Delta p)^2/2\Delta M$ is made. This assumption is correct for a reaction with three outgoing particles. In all other cases where the number of particles is larger than three the deficits cannot be associated with specific particles. It appears disputable, however, whether this association is necessary and wanted. The complete kinematic reconstruction of the n-particle process is rather complex, and it remains to be shown by future experiments whether or not the determination of the deficits alone suffices to characterize the reaction.

As two heavy fragments are detected, this experimental arrangement is slightly different from the usual setup where one heavy fragment and one light particle are measured. The advantages of the present arrangement are: (i) Because of the large detector areas and because of the reaction kinematic a large fraction of all light particles is registered. One thus gains almost complete knowledge of their kinematic properties. This is demonstrated in fig. 2 which shows the reconstructed pattern of the α production from the reaction ^{32}S + ^{28}Si → ^{32}S + ^{24}Mg + α.
(ii) The emission of light particles into the forward direction can be analyzed without restrictions in the angular and energy ranges. On the other hand, the main disadvantage of the present arrangement is that often the deficits cannot be related to specific light particles. For

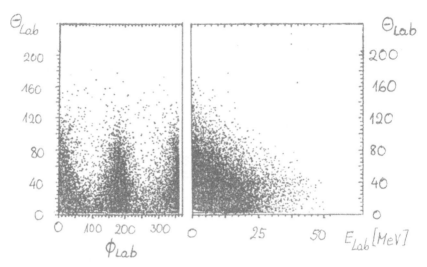

Figure 2 Reconstructed distributions of α particles from the reaction
^{32}S + ^{28}Si → ^{32}S + ^{24}Mg + α

example, $\Delta Z = 2$, $\Delta M = 4$ may indicate the production of an α particle
but it could also mean the production of two protons and two neutrons.
In some cases, however, these ambiguities can be resolved by means of
Q_T.

The experimental data are first ordered according to ΔZ as the
nuclear charge is determined with highest accuracy. Figure 3 displays
the charge for the three reactions studied. It is obvious that even
for the reactions with the smaller bombarding energy (^{32}S + ^{27}Al, ^{28}Si)
the majority of events belongs to the two-fragment inclusive type, i.e.
more than two particles are produced. This is even more the case for

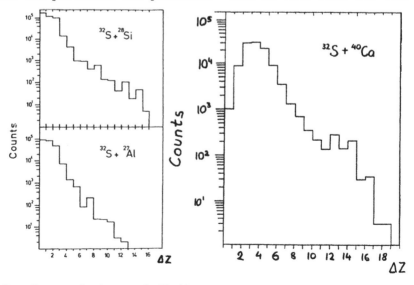

Figure 3 Measured charge deficits

the ^{32}S + ^{40}Ca reactions at 190 MeV bombarding energy where less than 1% of all events are two-particle exclusive. One also observes, in all cases, charge deficits of as large as, e.g., $\Delta Z = 6$ which can either indicate the production of ^{12}C or of three α particles.

The elements produced in different ΔZ bins are shown in fig. 4.

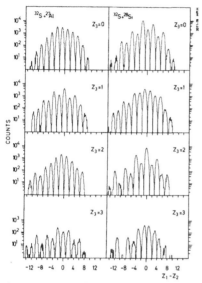

<u>Figure 4</u>   Measured element distributions

On the abscissa of this figure $Z_1 - Z_2$ is plotted, i.e. the difference of the nuclear charges measured by detector 1 and 2, respectively. It is then clear that the $Z_1 - Z_2$ values of the same absolute magnitude but different signs belong to the same fragmentation. The $Z_1 - Z_2$ distributions show, on the average, the Gaussian distributions. On top of these Gaussian distributions, however, one finds strong oscillations. The reactions clearly seem to favor the production of fragments with even Z. This behavior is observed not only for the $\Delta Z = 0$ channels (mainly two-fragment exclusive) but also for the $\Delta Z \neq 0$ channels (two-fragment inclusive). The reason for the staggering phenomenon may lie in the low Q values of α-transfer reactions and α emission. For the understanding of the selectiveness it might be interesting to use larger bombarding energies in order to study whether or not the staggering is still present at larger energy transfers. First results with the ^{32}S + ^{40}Ca reaction at 190 MeV bombarding energy indicate that one still observes the preferential production of fragments with even Z.

Selective phenomena are also observed in the energy-loss spectra of the two-fragment exclusive reactions. As an example, fig. 5 displays

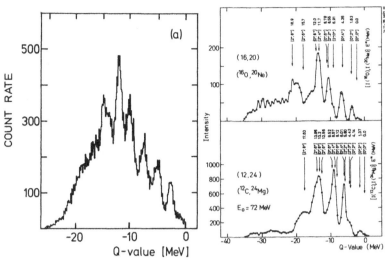

Figure 5   Measured Q-value spectra for $^{32}S + ^{28}Si \rightarrow ^{32}S + ^{28}Si$ (a) and $^{12}C + ^{24}Mg \rightarrow ^{12}C + ^{24}Mg$ (b)

$d\sigma/dQ$ for the inelastic $^{32}S + ^{28}Si$ scattering. On top of a broad distribution between -25 and 0 MeV one finds structures of approximately 3 MeV spacing and 1.5 MeV width. Similar structures are also found in the other two-fragment exclusive channels although they are not as pronounced. We also like to mention that the preferential population of a certain class of states was also observed [3] in the $^{12}C + ^{24}Mg$ reaction (cf. fig. 5). In the latter case these states could be identified as the high-spin states of the combined $^{12}C + ^{24}Mg$ system. Such an identification is not as straightforward in reactions with the ^{32}S beam, since the structure of the nuclei involved is not as well known. It is an interesting conjecture, however, that reactions between light ions selectively populate the high-spin states.

At energy losses larger than 25 MeV the $^{32}S + ^{28}Si$ reactions lead to more than two particles in the exit channel (two-fragment inclusive reactions). The analysis has concentrated on $\Delta Z=2$ events as the upper limit of Q_T indicates that these events predominantly correspond to the production of an α particle. As an example, fig. 6 displays the Q-value spectrum of the primary reaction $^{32}S + ^{28}Si \rightarrow ^{36}Ar + ^{24}Mg$ and the measured disintegration energies E_{rest} of the subsequent α decay of ^{24}Mg. The Q_B values were calculated by means of the energies and recoil angles of the stable ^{36}Ar nuclei. This spectrum also shows the threshold energies for other decay modes of ^{24}Mg and demonstrates the dominance of α decay. Another interesting feature, seen in fig. 6, is the upper limit of approximately 50 MeV of the energy loss, which is much larger than expected from the Coulomb-barrier height with $r_0 = 1.5$ fm.

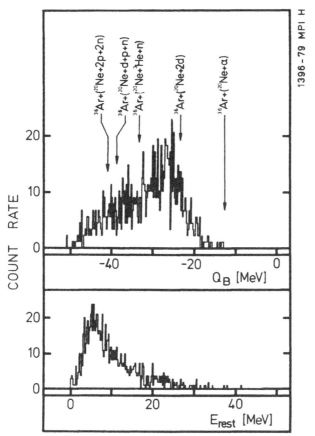

1396-79 MPI H

Figure 6   Energy spectra of the reaction $^{32}S + {}^{28}Si \rightarrow {}^{36}Ar + {}^{20}Ne + \alpha$

The unique identification of the exit channel is in most cases impossible as both fragments observed may be the remnants of the preceding α decay. Because of this ambiguity and because the finite solid angles of the detector induce kinematic correlations into the experimental data the angular and energy distributions of α particles were not transformed into the rest system of the decaying fragment. Instead, the kinematic properties of α particles in the rest system were calculated by means of Monte Carlo calculations and the results were then transformed into the laboratory system. The comparison with the experimental data includes all the restrictions in angular acceptance imposed by the finite dimensions of the detectors.

The Monte-Carlo calculations involve the variation of six parameters which the assumption of a sequential α decay dictates to be Q_B, Θ_{cm}, ϕ_{cm}, E_{rest}, Θ_{rest}, and ϕ_{rest}. The first three quantities characterize the primary collision whereas the last three describe the sub-

sequent α decay. In the Monte-Carlo calculations all quantities were
assumed to be randomly distributed except for Q_B, E_{rest} and θ_{rest}. The
experimental data require an increasing reaction probability with increasing Q_B and decreasing E_{rest}. Furthermore, we observe a strong
out-of-plane anisotropy of the α decay which is described by

$$W(\theta_{rest}) \propto \exp(-A \cos^2 \theta_{rest}).$$

The constant A depends on the properties of the states decaying by α
emission. No attempt was made to substantiate the dependence of the
reaction probability on the various parameters by a more detailed theoretical description as it was felt that the theoretical understanding
of these light-ion reactions is still inadequate.

Figure 7 shows the comparison of the measured results for $\Delta Z = 2$,
$Z_1 - Z_2 = -4$ with the results of two Monte-Carlo calculations. The

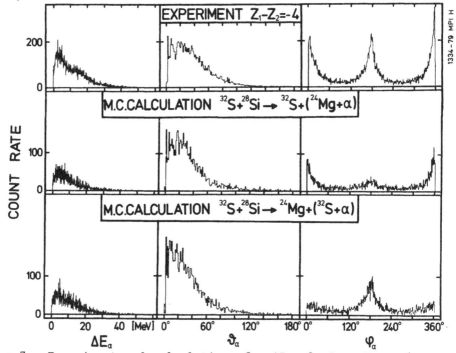

Figure 7   Experiment and calculations for $\Delta Z = 2$, $Z_1 - Z_2 = -4$

Monte-Carlo calculations assume two different fragmentation processes
as indicated, leading to the identical exit channel. It is obvious
from the comparison that both reactions (α decay of ^{28}Si and ^{36}Ar, respectively) contribute to the observed distributions. Furthermore, the
agreement between experiment and calculation indicates that the hypoth-

esis of an intermediate nucleus that exhibits statistical α decay is a valid one. Similar results were also found for the $\Delta Z=2$, $|Z_1-Z_2|=0,8$ exit channels. In order to corroborate these conclusions we performed additional Monte-Carlo calculations by assuming a hot-spot [4] or a random distribution of p_1 and p_2 in the c.m. system. In all cases did we find less agreement with the experimental data than with the assumption of a statistical, sequential decay.

The main conclusions from this work are: (i) Already at low bombarding energies the ^{32}S-induced reactions on light nuclei lead to the production of more than two particles in the exit channel. (ii) The reaction process is best described as a binary reaction followed by the statistical emission of light particles. (iii) The reactions show a selective production of elements with even Z; selective phenomena are also present in the energy transfer of the two-fragment exclusive reactions.

This work was done in collaboration with U. Winkler, R. Novotny and U. Lynen.

1. Bondorf, J.P., Se, J.N., Karvinen, A.O.T., Fai, G., Jacobsson, B.: Phys. Lett. 84B, 162 (1979)
2. Sann, H., Damjantschitsch, H., Hebbard, D., Junge, J., Pelte, D., Povh, B., Schwalm, D., Tran Thoai, D.B.: Nucl. Instr. Meth. 124, 56 (1975)
3. Novotny, R., Hammer, G., Pelte, D., Emling, H., Schwalm, D.: Nucl. Phys. A294, 255 (1978)
4. Ho, H., Albrecht, R., Dünnweber, W., Graw, G., Steadman, S.G., Wurm, J.P., Disdier, D., Rauch, V., Scheibling, F.: Z. Phys. A283, 235 (1977)

EMISSION OF ALPHA PARTICLES IN DEEP INELASTIC
REACTIONS INDUCED BY 148 MeV ^{14}N BEAM

R.K. Bhowmik*, E.C. Pollacco, N.E. Sanderson,
J.B.A. England, D.A. Newton and G.C. Morrison

Department of Physics, University of Birmingham,
Birmingham B15 2TT, England

Correlation measurements for α particles in coincidence with deep
inelastic fragments are presently available on several light heavy ion-
target combinations [1 - 9].   For projectile energies below 8 MeV/A,
these correlations have been interpreted in terms of sequential α
emission from target-like [3,4] or projectile-like [6] fragments.   However, even at these low energies, the angular correlations for α particles show a pronounced enhancement in the forward direction indicating
that the α particles are emitted before the nuclei attain statistical
equilibrium.

As the energy of the incident beam is increased, two effects become apparent.   Firstly, the coincidence cross sections increase
rapidly with increasing energy [7,8].   Large α multiplicities (∿0.5)
are also seen at the higher energies which suggests that α emission
plays an important role in the energy dissipation process for deep
inelastic reactions.   Secondly, the energy spectra of α particles
emitted in the forward direction have a large high energy tail with
velocities approaching the beam velocity.   These two observations
when combined with the strongly forward peaked angular correlations
imply that the α particles are emitted at an early stage of the reaction, with emission times << 10^{-21} sec.

At the University of Birmingham, we have measured correlations of
α particles in coincidence with projectile-like fragments at beam
energies ∿10 MeV/A for different heavy ion-target combinations.   In
the present paper we would like to describe the results obtained with
148 MeV ^{14}N projectiles on ^{12}C, ^{27}Al and ^{58}Ni targets.   The correlation data may be empirically described in a simple factorised form as
a product of the singles α and heavy ion (HI) cross sections.   A reaction model where a prompt α particle is emitted first, prior to the
formation of the deep inelastic fragments, may explain the observed

α-HI correlations.

2. EXPERIMENT

The experiments were performed at A.E.R.E., Harwell using the 148 MeV ^{14}N beam from the Variable Energy Cyclotron. The targets were self-supporting foils of natural carbon, aluminium and isotopically enriched ^{58}Ni of thicknesses 0.6, 3.2 and 2.5 mg/cm^2 respectively.

For the angular correlation measurements the heavy ejectiles (Z = 2 - 8) were detected in a silicon detector telescope consisting of a 30 μm ΔE and a 1.5 mm E counter. The α particles were detected in a second telescope consisting of a 50 μm ΔE and 1.5 mm thick E counter. Two identical α telescopes were used to measure α-HI correlations simultaneously at two angles. The solid angles subtended by the HI and α counters were 2 and 6 msr respectively.

In the HI telescope the ions were detected with a dynamic range of E_{HI} ≈ 7 - 30 MeV for α's and greater than 13, 19, 24, 30, 38 and 45 MeV respectively for Li, Be, B, C, N and O isotopes. The dynamic range of α particles is E_α = 10 - 60 MeV in the α telescopes.

The energy signals from the HI and α telescopes and the coincidence time spectra were recorded event-by-event on magnetic tape using a ND4420 Multiparameter Analyser. The singles events from the HI and α telescopes have also been recorded for normalisation. The coincidence cross sections have been corrected for 'random' events which were typically <10% of real events. To increase the dynamic range of the particles observed, triple counter telescopes were used to detect HI and α particles at selected angles.

For the ^{12}C and ^{58}Ni targets measurements were performed for two HI angles, θ_{HI} = 13°, (grazing angle for ^{58}Ni) and θ_{HI} = 26°. The α telescopes were moved in plane in the angular range -70° < θ_α < 70° where positive θ_α corresponds to HI and α telescopes on opposite sides of the beam. For some θ_α, out-of-plane correlations were measured for E_α > 15 MeV by mounting an α telescope 18° out-of-plane. Only in-plane energy and angular correlations have been measured for the Al target at θ_{HI} = 13°. For all three targets singles cross sections have been extracted in the angular range 6 - 30° for projectile-like fragments and 13° - 70° for α particles.

The amount of carbon and oxygen impurity in the Ni target was estimated to be 8 and 2 μg/cm^2 respectively by observing the elastic

scattering of 25 MeV α particles. For the [27]Al target an upper limit
of 10 μg/cm² of C and 5 μg/cm² of O was deduced. Assuming equal coincidence cross sections from ^{12}C and [16]O targets, light impurity contributions in the α-HI correlations from the [27]Al target have been
estimated to be <8% for light (Z < 6) ejectiles, <10% for Z = 6, <15%
for Z = 7 and <20% for Z = 8. For the [58]Ni targets, the impurity
contributions are estimated to be <10% for Z < 6 at both HI angles.
For Z = 6 ejectiles the impurity contributions are ∿15% and ∿30% at
θ_{HI} = 13° and 26° respectively. No corrections for impurities in
either target have been made.

3. RESULTS

3.1 Angular correlations

The in-plane energy-integrated angular correlations for α particles in coincidence with different ejectiles are shown in Fig.1. For
a comparison between different targets the coincidence cross sections
have been divided by the corresponding singles deep inelastic cross
section (Table 1) to obtain the differential α multiplicities $dM_\alpha/d\Omega_\alpha$
(Z, θ_{HI}). Although the coincidence cross sections at θ_{HI} = 13° and
26° are considerably different, the multiplicities at these two angles
have approximately the same value. For each target shown in Fig.1
the solid curves give the shape of the α singles angular distribution
$d\sigma_\alpha/d\Omega_\alpha$. The dashed curves are drawn through data points to guide
the eye.

For Z < 6 ejectiles the angular correlations are symmetric about
the beam axis with a narrow width ∿40° FWHM. The correlations are
found to be similar in shape to the α singles angular distribution.
The angular correlations for fast α particles $(E_\alpha > 30$ MeV) are
narrower but again have the same shape as the corresponding singles
distribution. It is noted that for Li ejectiles an enhancement of
the coincidence cross section along the direction of the HI telescope
is observed for the [58]Ni target.

For ejectiles heavier than carbon the correlations broaden and
the peaks shift away from the HI telescope. In the case of the [12]C
target the arrows shown in Fig.1 indicate the direction of momentum
transfer to the recoiling nucleus assuming a binary division

$$^{14}N + {^{12}C} \rightarrow HI + Recoil$$

A similar shift in the correlations for heavy ejectiles has also been seen with ^{11}B projectiles [1].

To obtain the integral α-multiplicities the angular correlations have been parametrised in the form

$$\frac{dM}{d\Omega_\alpha}(\theta_\alpha, \phi_\alpha) \sim e^{-\theta_\alpha/\theta_0} \cdot e^{-1/2 \, \phi_\alpha^2/\phi_0^2}$$

where $(\theta_\alpha, \phi_\alpha)$ are the spherical polar co-ordinates of the α counter with respect to the beam axis and $\phi_\alpha = 0$ on the reaction plane. Typical values of the parameters θ_0, ϕ_0 are $\sim 20^\circ$ and 60° respectively [10].

The integral α multiplicities for different ejectiles estimated in this way are given in Table 2. The multiplicities are quite large, in the range 0.4 - 0.8 for ^{27}Al and ^{58}Ni and 0.8 - 1.6 for ^{12}C.

3.2 Energy correlations for ^{58}Ni

In order to obtain detailed information about the reaction mechanism, energy-correlation data with reasonably good statistics were taken for selected angle pairs $(\theta_{HI}, \theta_\alpha)$. Typical two-dimensional energy-correlation spectra for the reaction ^{58}Ni$(^{14}$N, HI$\alpha)$ at $\theta_{HI} = 13^\circ$, $\theta_\alpha = 13^\circ$ are shown in Fig.2. It is found that α particles in coincidence with different ejectiles have quite similar E_α spectra. This can be seen more clearly from Fig.3 where the projected E_α^C spectra in the centre-of-mass frame in coincidence with different ejectiles are' compared with the singles E_α^C spectra at $\theta_\alpha = 26^\circ$ (a) and 51° (b). A similarity between α spectra from α-α and α-HI coincidence data has also been observed.

A comparison of the projected energy spectra for the different ejectiles (E_{HI}^C) with the corresponding singles HI spectra indicate that they are again similar for ejectiles lighter than carbon. Furthermore the HI coincidence spectra do not depend significantly on the angle θ_α.

For C and N ejectiles at $\theta_{HI} = 13^\circ$ the singles spectra show a large bump near the beam energy which is absent from the coincidence energy spectra. We have assumed that the bump arises from quasi-elastic events (QE) that are not associated with coincidence α particles. An estimate of the DI cross section has been made in Table 1 by assuming that the shape of the DI component corresponds to the shape of the projected coincidence HI spectra averaged over all θ_α.

A second feature of the two dimensional spectra is the lack of correlation between HI and α energy.   Comparing the projected HI spectra gated by low and high energy α's it is seen that the centroids of ejectile spectra change by <5 MeV although the average energies of α particles for the two groups differ by ∿25 MeV (as shown, for example in Fig.4).   Similar results are obtained by placing HI energy gates on the projected α particle spectra.

The systematic features of the energy and angular correlations for ejectiles lighter than Z = 6 can be summarized as follows:

(a)   no appreciable correlation between HI energy and α energy is observed.

(b)   the projected energy spectra of α particles are independent of HI type or HI angle and are similar in shape to the singles E_α spectra at the corresponding θ_α.

(c)   the projected energy spectra of different HI's are independent of α angle and are similar to the corresponding singles E_{HI} spectra.

(d)   the α-HI angular correlations are roughly symmetric about the beam axis and have shapes similar to the singles α angular distributions.

3.3   Factorisation of coincidence cross section

The above results suggest that the coincidence cross section may be parametrised in the form

$$\frac{d^4\sigma(Z,E_{HI},E_\alpha,\theta_{HI},\theta_\alpha)}{dE_{HI}\,dE_\alpha\,d\Omega_{HI}\,d\Omega_\alpha} \simeq K \cdot \frac{d^2\sigma(E_{HI},\theta_{HI})}{dE_{HI}\,d\Omega_{HI}} \cdot \frac{d^2\sigma(E_\alpha,\theta_\alpha)}{dE_\alpha\,d\Omega_\alpha}$$

where the R.H.S. is the product of the singles cross sections for HI and α particles at the corresponding angles θ_{HI} and θ_α, Z refers to the charge of ejectile and K is a constant of proportionality.

By integrating over E_{HI} and E_α, the differential α multiplicities are then given by

$$\frac{dM_\alpha}{d\Omega_\alpha} = K \cdot \frac{d\sigma(\theta_\alpha)}{d\Omega_\alpha} \qquad\qquad \ldots\ldots(1)$$

The solid curves in Fig.1 are obtained from the singles α differential cross sections by taking K = 1.0, 0.6 and 0.5 barn^{-1} respectively for the ^{12}C, ^{27}Al and ^{58}Ni target. The agreement between the experimental angular correlations and the predictions of the factorisation procedure (FP) is reasonable for Z < 6. K is found to be $\sim 1/\sigma_R$ where σ_R is the total reaction cross section.

The energy correlations can also be compared point by point for each (E_{HI}, E_α) with the FP calculations. However, in order to improve the statistics of the two dimensional correlations, different kinematic variables Q_3, E_{12} and E_{23} may be calculated and the corresponding one dimensional projections compared. Q_3 is the three body Q value and any correlation along the kinematic line $E_{HI} + E_\alpha = $ constant would show up as an enhancement in the corresponding region of Q_3. The solid curves are the calculated FP correlations normalised to the data, from which contributions from the kinematically forbidden region ($Q_3 >$ 0) have been excluded. E_{12} is the relative energy between the HI and α particles and E_{23} is the relative energy between the α particle and recoiling nucleus assuming only three particles in the final state. The good agreement between the data and the FP calculations confirms the success of factorisation in describing the coincidence cross sections.

For C and N ejectiles at 13° the single HI spectra contain a strong QE component that is absent in the coincidence spectra. For this reason, the coincidence data for these ejectiles could not be reproduced using the singles energy spectra. Reasonable agreement could, however, be obtained by replacing the singles $d^2\sigma_{HI}$ and $d^2\sigma_\alpha$ by the projected coincidence energy spectra (dashed curves in Fig.5).

3.4 Energy correlations for ^{12}C and ^{27}Al

For ^{12}C and ^{27}Al targets, the energy correlation data for ejectiles lighter than C have been reasonably well reproduced by FP. The shapes of the projected energy spectra for different ejectiles approximately correspond to the shapes of the corresponding singles E_{HI} spectra (Fig.6). Similarly, the shapes of the coincident α particle spectra are independent of ejectile type or ejectile angle.

For Z = 6 ejectiles a slight dependence of the projected HI spectra with θ_α is observed. This effect is pronounced for Z = 7 ejectiles where the projected E_{HI} spectra appear to have two components. The angular correlations for α particles in coincidence with the higher energy heavy ejectiles (Z > 6) are broader than the correlations for α's in coincidence with lower energy ejectiles and the peak of the correlation is shifted further away from the heavy ion telescope. Such a shift for higher energy ejectiles has also been seen with ^{11}B projectiles [1].

4. DISCUSSION

The most interesting aspect of the present data is the success of FP in describing the α-HI coincidence cross sections for ejectiles lighter than the projectile. Such a factorisation of coincidence cross section as a product of singles cross sections implies that the probability of HI emission and α emission are approximately independent. This feature of the reaction could be understood if the HI and α particles were formed at different times during the collision process.

The strongly forward peaked angular correlations for α particles indicate that the life-time for α emission is very short and comparable to the formation time of the DI fragment. This suggests that the α particle is emitted in the forward direction at a very early stage by a direct reaction process. The projectile-target combination, after emitting the α particle, undergoes a deep inelastic collision and finally breaks up into two components. Due to the statistical nature of the energy relaxation process, the fragments retain no memory of the momentum transferred to the α particle and the coincidence cross section may be written as the produce of the probabilities for separately detecting the α particle and the ejectile. Since the α multiplicities are close to unity, this would imply that a major fraction of the ejectiles observed in singles are formed by such a three body process. In this way the close agreement between the singles and coincidence ejectile spectra could be explained.

It must be stressed that although the above process implies a time ordering of events, the break up of the dinuclear complex is also a fast process as the energies of the ejectiles are well above the Coulomb barrier. Consequently a weak dependence of the HI energy on the energy carried away by the α particle is expected.

By integrating the α-HI coincidence cross sections over θ_α and θ_{HI}, the total cross section for α particles observed in coincidence with heavy ejectiles (Z = 3 - 8) is estimated to be ∿200 mb. The cross section for α - α coincidences is an order of magnitude larger than the total cross section for the α - HI coincidences, although the energy spectra and angular correlations are again similar in shape to the α singles distributions. Recent measurements on the system $^{12}C + {}^{160}Gd$ at different bombarding energies by the Groningen group [11] appear to indicate that the α-α coincidence cross sections are proportional to the product of the α singles cross section. At the highest bombarding energy studied most of the α particles observed in singles arise from 3α break up of the ^{12}C projectile. In the present experiment, break up of ^{14}N with two or three α particles emitted may explain the large α-α multiplicities observed.

One of the problems in obtaining any systematic information about α-HI correlations is the difficulty in comparing the data from different projectile-target systems. If the factorisation approach provides a qualitative description for a number of systems, a quantitative comparison may become feasible by analysing deviations from such a first order approximation. It would be instructive to test to what extent factorisation is valid for the α-HI correlation data presented by H.Ho and C.Gelbke at the present symposium.

REFERENCES

*paper presented by R.K. Bhowmik.

1. R.K. Bhowmik, E.C. Pollacco, N.E. Sanderson, J.B.A. England, and
    G.C. Morrison, Phys. Lett. 80B, 41 (1978).

2. R.K. Bhowmik, E.C. Pollacco, N.E. Sanderson, J.B.A. England, and
    G.C. Morrison, Phys. Rev. Letts. 43, 619 (1979).

3. J.W. Harris, T.M. Cormier, D.F. Geesaman, L.L. Lee Jr., R.L. McGrath,
    and J.P. Wurm, Phys. Rev. Lett. 38, 1460 (1977).

4. H. Ho, R. Albrecht, W. Dünnweber, G. Graw, S.G. Steadman, J.P. Wurm,
    D. Disdier, V. Rauch, and F. Scheibling, Z. Phys. A.283, 235 (1977).

5. C.K. Gelbke, M. Bini, C. Olmer, D.L. Hendrie, J.L. Laville,
    J. Mahoney, M.C. Mermaz, D.K. Scott, and H.H. Wieman, Phys. Lett.
    71B, 83 (1977).

6. T. Shimoda, M. Ishihara, H. Kamitsubo, T. Motobayashi, and T. Fukoda,
    in Proceeding of the ICPR Symposium, Hakone, Japan, September 1977,
    edited by H. Kamitsubo and M. Ishihara (unpublished).

7. T. Fukuda, S. Tanemo, K. Kondo, M. Tanaka, H. Ogata, I. Miura,
    M. Inoue, T. Yamazaki, T. Yamagata, H. Kamitsubo, T. Shimoda,
    and M. Ishihara, R.C.N.P. Annual Report, 1978, Osaka University,
    Suita, Osaka, Japan.

8. R. Billerey, C. Cerruti, A. Chevarier, N. Chevarier, B. Cheynis,
    and A. Demeyer, (Private communication).

9. H. Ho, this conference.

10. Nuclear Structure Group Annual Progress Report, University of
    Birmingham, Department of Physics, 1979.

11. J. Wilczynski, R. Kamermans, J. van Popta, R.H. Siemssen,
    K. Siwek-Wilczynska and S.Y. van der Werf, to be published, and
    K.V.I. Annual Report, 1978, Groningen, the Netherlands.

Table 1 Singles laboratory cross sections in mb/Sr for
(^{14}N,HI) reactions on ^{12}C, ^{27}Al and ^{58}Ni targets

Target	^{12}C		^{27}Al		^{58}Ni	
ejectile	13°	26°	13°	26°	13°	26°
α	1700	880	1400	730	1500	720
Li	75	37	64	29	59	18
Be	33	15	32	10	37	9
B	140	42	100	25	100	16
C(DI)	500	98	260	50	(190)	36
C(QE)					(176)	
N(DI)	420	42	194	23	(200)	60
N(QE)					(330)	
O	240	20	60	14	17	6

Table 2 Integral α multiplicities for the reaction
(^{14}N,HIα) on ^{12}C, ^{27}Al and ^{58}Ni targets for
different θ_{HI}

Target	^{12}C		^{27}Al	^{58}Ni	
ejectile	13°	26°	13°	13°	26°
α	1.6	1.2	1.1	0.8	0.8
Li	1.5	1.1	0.8	0.7	0.8
Be	1.4	0.7	0.6	0.5	0.5
B	1.3	0.8	0.5	0.4	0.6
C	1.0	1.4	0.4	(0.4)	0.4
N	0.7	1.1	0.2	(0.15)	0.4
O	0.7	1.1	0.3		

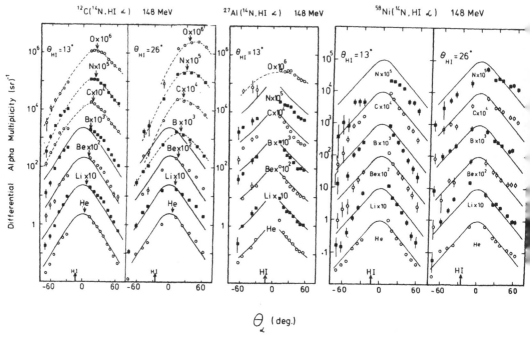

Fig. 1   Differential α multiplicities for ^{12}C, ^{27}Al and ^{58}Ni
targets for the reaction (^{14}N,HIα) at 148 MeV.

^{14}N + ^{58}Ni   148 MeV

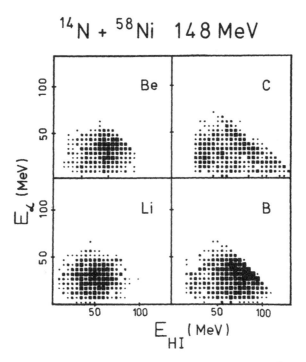

Fig.2 E_{HI} versus E_α energy correlations for Li, Be, B and C ejectiles for the reaction ^{58}Ni(^{14}N,HIα) at θ_{HI} = 13°, θ_α = 13°.

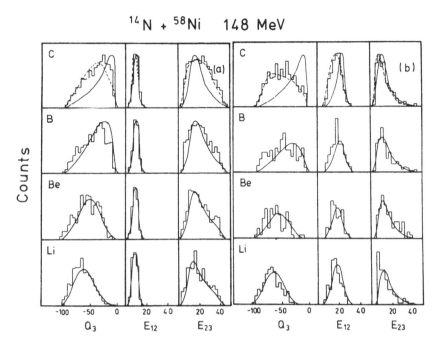

Fig.3 Projected centre-of-mass energy spectra of different HI and
α particles at (a) $\theta_{HI} = 13°$, $\theta_\alpha = 26°$ and (b) $\theta_{HI} = 13°$, $\theta_\alpha = 51°$.
The solid curves are the singles c.m. energy spectra for HI
and α particles at the corresponding angles normalised to
the data.

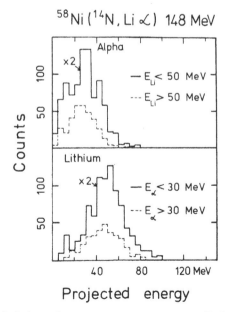

Fig.4 Projected laboratory energy spectra of (a) Li ejectiles gated
by 'low' and 'high' energy α's and (b) α particles gated by
'low' and 'high' energy Li ejectiles at $\theta_{HI} = 13°$, $\theta_\alpha = 13°$.

$^{14}N + ^{58}Ni$ 148 MeV

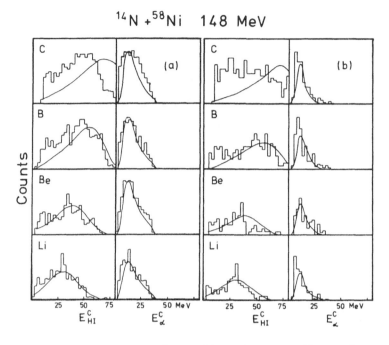

Fig.5   Projected Q_3, E_{12} and E_{23} distributions (in MeV) at (a) $\theta_{HI} = 13^O$, $\theta_\alpha = 26^O$ and (b) $\theta_{HI} = 13^O$, $\theta_\alpha = 51^O$.   The solid and dashed curves are FP predictions normalised to the data.

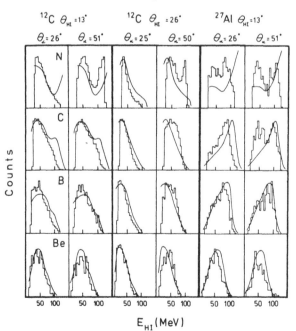

Fig.6   Projected laboratory energy spectra of different ejectiles from ^{12}C and ^{27}Al targets.   The solid curves are the corresponding singles energy spectra normalised to the data.

HEAVY ION REACTIONS AT E/A $\geq$ 10 MEV/NUCLEON

C.K. Gelbke[*]

Heavy Ion Laboratory and Department of Physics
Michigan State University
East Lansing, MI 48824 USA

Abstract

The energy dependence of heavy ion reactions above 10 MeV/nucleon
is discussed. At higher projectile velocities, fragmentation and pre-
equilibrium processes are becoming increasingly more important.
Coincidence experiments performed with ^{16}O ions at 20 MeV/nucleon
indicate that the simple participant spectator model does not yet
apply at this energy even though results from single particle inclusive
measurements show great similarities when compared to data taken at
relativistic energies.

1. INTRODUCTION

Let me start by reminding you of the present status of heavy
ion accelerators either in operation or under construction, see Fig. 1.
Most of our understanding of heavy ion reactions has been obtained
from low energy machines like the UNILAC or the Super HILAC. Some
data are also available from the Berkeley Bevalac at energies above
100 MeV/nucleon (not included in the figure in its present configuration). The energy range between 10 and 200 MeV/nucleon, however,
is practically virgin territory where only a few exploratory experiments have been performed until now. This is the energy range the
new generation of heavy ion accelerators is planned to cover in the
next years to come. In this talk I want to focus on this energy range
and discuss first results that have been obtained until now.

For orientation, Fig. 2 shows a "heavy ion reaction phase diagram"
that was first introduced by Bondorf.[1] The coordinates of this schematic
diagram are the projectile velocity and impact parameter. At low
energies we have the well known processes of grazing collisions (generally associated with inelastic scattering and few nucleon transfer
reactions to low lying states), deeply inelastic and fusion reactions.
For peripheral reactions at high energies one assumes the validity
of the "participant-spectator" picture. An example[2] of such a reaction
is shown in Fig. 3: An incoming ^{14}N projectile of 2.1 GeV/nucleon
interacts with an emulsion nucleus and is disintegrated into several

light fragments that travel with nearly the beam velocity; the target
nucleus does not appear to have suffered any appreciable momentum
or energy transfer - it acts like a spectator. For central collisions
at high energies, the observed emission of a large number of light
particles is taken as the base for the term "total explosion", see
Fig. 4 as an example:[2] An ^{40}Ar projectile of 1.8 GeV/Nucleon collides
with a heavy emulsion nucleus; neither projectile nor target survive
the reaction which results in the emission of a very large number
of light particles over a wide range of angles.

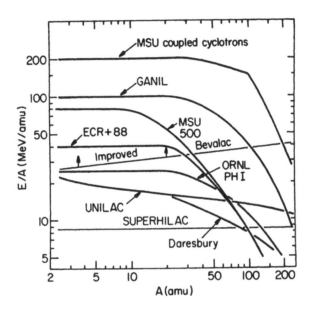

Fig. 1: Overview of heavy ion accelerators in operation or under
construction.

The speculation implied in Fig. 2 is the existence of relatively
well defined boundaries separating the various processes in contrast
to their possible coexistence over large domains of incident energies
or impact parameters. It certainly is one of the most challenging
questions to be answered by machines like MSU Phase II whether such
a picture will find a solid experimental foundation.

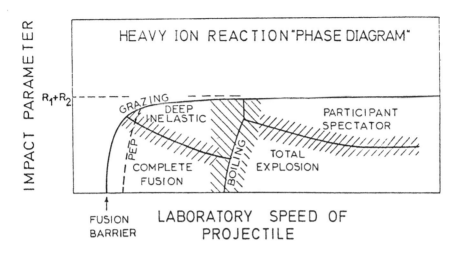

Fig. 2: Schematic "phase diagram" for heavy ion reactions from a few to several hundred MeV per nucleon.[1]

Fig. 3: Photomicrograph[2] of the fragmentation of a 2.1 GeV/nucleon ^{14}N nucleus in nuclear emulsion.

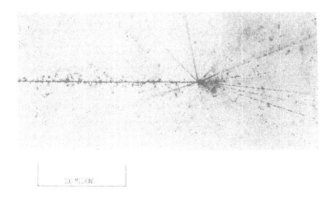

Fig. 4: Central (explosive) collision of an ^{40}Ar projectile of 1.8 GeV/nucleon with a heavy emulsion nucleus.[2]

2. ENERGY SPECTRA OBSERVED IN INCLUSIVE PERIPHERAL REACTIONS

For projectile fragmentation reactions at relativistic energies the energy spectra of the outgoing projectile residues correspond rather closely to Gaussian momentum distributions centered close to the projectile velocity.[3] In the projectile rest frame, the momentum distributions can be parameterized[4] as

$$\frac{d^3\sigma}{dp^3} \sim \exp[-(\vec{p}-\vec{p}_o)^2/2\sigma_f^2] \tag{1}$$

where $\vec{p}_o$ is parallel to the beam momentum $\vec{p}_i$ and, generally, $p_o < \sigma_f < p_i$. An approximately parabolic dependence of the width σ_f on fragment mass is observed (for details about the interpretation see ref. 4)

$$\sigma_f^2 = \sigma_o^2 A_f (A_i - A_f)/(A_i - 1) . \tag{2}$$

Here A_i and A_f are the mass numbers of projectile and projectile fragments, respectively, and σ_o is approximately constant. An example is shown in Fig. 5 for the fragmentation of ^{40}Ar projectiles at 213 MeV/nucleon.[5]

Until now, only very few reactions have been studied over a large dynamic range of energies. One example of such a study is shown in Fig. 6 where the energy spectra, observed close to the grazing angle, for the (^{16}O,^{12}C) reaction on heavy targets are compared for four different projectile energies.[6] Two salient features are immediately apparent (similar observations are made for other exit channels): (i) The energy spectra have maxima corresponding to ejectile velocities rather close to the projectile velocity and (ii) the widths of the energy spectra increase rapidly with energy. If we look at the widths of the energy spectra in a frame that moves with a velocity close to the one of the projectile we obtain the following energy dependence[6,7] (see Fig. 7). Up to beam energies of a few tens of MeV/nucleon the widths of the energy spectra increase quite rapidly and then remain rather constant at energies higher than about 100 MeV/nucleon. This energy dependence is still a subject of speculation and has been suggested as due to the rapid onset of projectile fragmentation processes at energies of a few tens of MeV/nucleon.[6]

Quite recently, there have been several attempts to understand the general shapes of such energy spectra. Within the local-momentum plane-wave approximation[8] narrower widths of the energy spectra are expected for transfer reactions than for projectile fragmentation

reactions, as shown in Fig. 8 by the solid and dashed lines, respectively. The large widths observed for the carbon and nitrogen spectra
at 315 MeV have been explained as due to the dominance of fragmentation
processes. More detailed calculations within the framework of the
DWBA have been performed by the Texas group[9] for the reaction
$^{40}Ca(^{20}Ne,^{16}O)$ at 149 and 262 MeV. An excellent description of the
energy spectra was obtained by assuming the simultaneous occurrence
of transfer and breakup reactions, shown by the dotted and dashed
lines of Fig. 9. In contrast to the results of ref. 8, this approach
yields wider energy distributions for transfer than for breakup reactions.

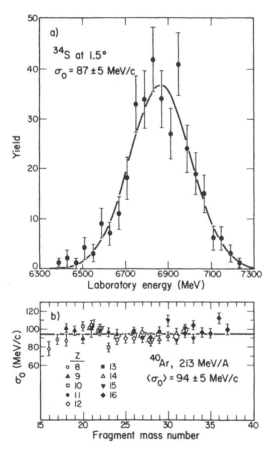

Fig. 5: (a) Measured energy spectrum of ^{34}S at 1.5° from fragmentation
of 213 MeV/nucleon ^{40}Ar on a carbon target. The solid line
corresponds to a fitted Gaussian momentum distribution,
eqs. 1 and 2. (b) Values of σ_0 for the fragments in the
mass range A_f = 16...37. (From ref. 5).

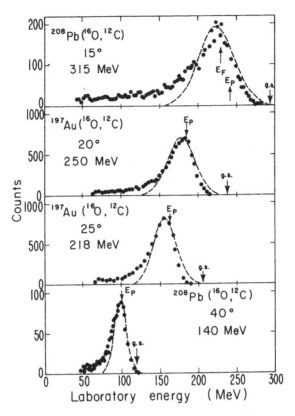

Fig. 6: Energy spectra observed for the (^{16}O,^{12}C) reaction on Pb and Au targets at different energies close to the grazing angle (from ref. 6).

A completely different approach has been used in ref. 10. Using information theoretical concepts, the spectra are described by assuming distributions corresponding to maximal entropy, i.e. minimal informa-tion content, subject to certain constraints. These constraints contain the nontrivial physics that can be obtained from the data. By assuming two constraints, namely on the excitation energy and on the square root of the excitation energy (which may be related to the width of the exciton distribution) and by otherwise assuming a simple Fermi-gas level density for the residual nucleus excellent descriptions of the data are obtained (see, e.g. Fig. 10). This is most remarkable since the analysis assumed the validity of two-body kinematics and no excitation of the ejectile, i.e. no break-up contributions to the cross sections.

From the foregoing discussion it has become apparent that further theoretical and experimental efforts will be necessary if we are to understand the reaction mechanism in more detail.

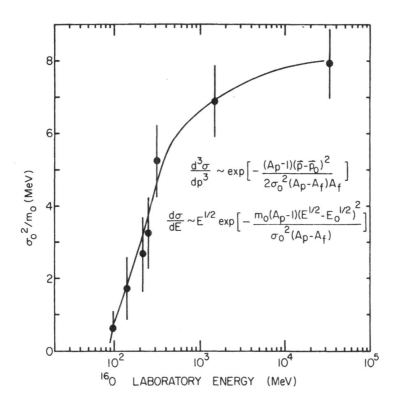

$$\frac{d^3\sigma}{dp^3} \sim \exp\left[-\frac{(A_p-1)(\vec{p}-\vec{p}_0)^2}{2\sigma_0^2(A_p-A_f)A_f}\right]$$

$$\frac{d\sigma}{dE} \sim E^{1/2}\exp\left[-\frac{m_0(A_p-1)(E^{1/2}-E_0^{1/2})^2}{\sigma_0^2(A_p-A_f)}\right]$$

Fig. 7: Energy dependence of the width of the energy spectra of the (^{16}O,^{12}C) reaction. The width has been determined in a rest frame that moves with a velocity close to the projectile velocity (from ref. 6).

3. INCLUSIVE LIGHT PARTICLE SPECTRA

Our present understanding of central heavy ion collisions at relativistic energies is based mainly on the study of light particle emission. Indeed, central collisions are generally associated with high light particle multiplicities that are, quite often, consistent with a complete disintegration of the two interacting nuclei.

Some typical light particle spectra[11] that have been observed for relativistic heavy ion collisions are shown in Fig. 11. The data

can be rather well described by the fireball model which assumes complete momentum transfer and thermal equilibrium in the region of geometrical overlap of the two nuclei and negligible interaction with
the non-overlapping region of the nuclei. Very qualitatively, one
can say that the fireball represents a hot source of nucleons traveling
with a velocity intermediate between projectile and target velocities.
The absolute cross sections are obtained from geometry.

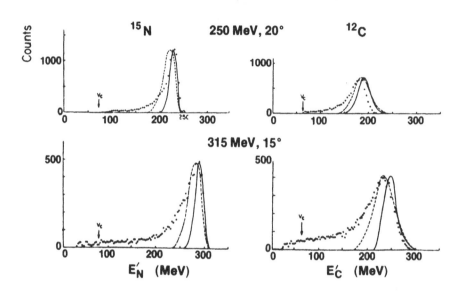

Fig. 8: Comparison of ^{15}N and ^{12}C energy spectra obtained from ^{16}O
         induced reactions on ^{208}Pb and ^{197}Au targets with the pre-
         dictions of the local-momentum plane-wave Born approximation.
         The solid curves are for transfers into the continuum, and
         the dashed ones for projectile fragmentation (from ref. 8).

For comparison, Fig. 12 shows some proton inclusive energy spectra
observed[12] for ^{16}O induced reactions on ^{197}Au at 20 MeV/nucleon.
Surprizingly enough, these data can be very well described by assuming
thermal emission from a hot source (T = 8.1 MeV) moving with half
the beam velocity (see solid lines). Even the fireball model (dashed
lines) is in reasonable agreement with the data. However, also precompound calculations (dotted lines) appear to be quite adequate.
Evidently, these inclusive spectra do not contain sufficient information
to clearly discriminate between the various models—a situation that
has also been encountered at relativistic energies.[13] It should be
clear, however, that the high proton energies observed in this experiment

cannot be explained by emission from the compound nucleus[12] (see also Section 6).

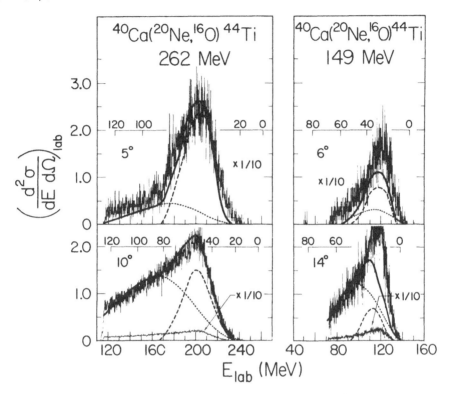

Fig. 9:   Energy spectra observed for the $^{40}Ca(^{20}Ne,^{16}O)$ reaction at 149 and 262 MeV. The dashed and dotted lines represent theoretical cross sections obtained by assuming only break-up and transfer processes, respectively. The solid lines represent the sum of these two contributions (from ref. 9).

4.   ISOTOPE PRODUCTION CROSS SECTIONS

Isotope production cross sections are relatively well understood for deeply inelastic reactions at low energies. Together with the diffusion of nucleons it is generally assumed[14] that the charge asymmetry degree of freedom is rapidly equilibrated and that the excitation energy of the primary reaction products is shared in proportion to their mass (thermal equilibrium). These primary fragments decay by light particle emission to produce the experimentally observed isotope distributions, see e.g. Fig. 13.

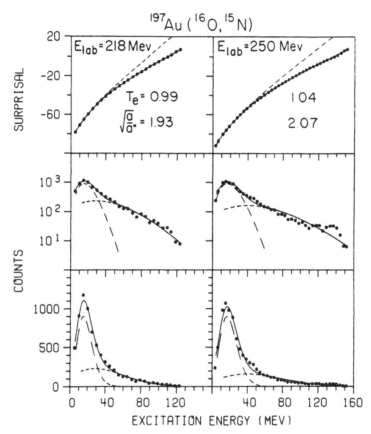

Fig. 10: Surprisal analysis of energy spectra observed for the reactions ^{197}Au(^{16}O,^{15}N) at 218 and 250 MeV (from ref. 10)
The spectra are described by the superposition of a quasi-
elastic and a nearly relaxed component.

At relativistic energies, the abrasion ablation model and cascade
calculations have been quite successful in reproducing isotopic distributions resulting from fragmentation reactions.[5,15,16] In the abrasion
ablation model, the primary mass distributions are calculated from
the geometry of the fireball model. There are, however, large uncertainties about the proton-neutron correlations in the target nucleus
which determine the primary isotopic distributions and about the primary
excitation energies. Quite different assumptions can reproduce the
experimental isotope cross sections which, again, result from the
statistical decay of the primary fragments,[5] see Fig. 14 as an example.
To emphasize the importance of the evaporation stage, Fig. 15 gives

a comparison of two different model calculations which predict very different primary but very similar final fragment distributions.[16]

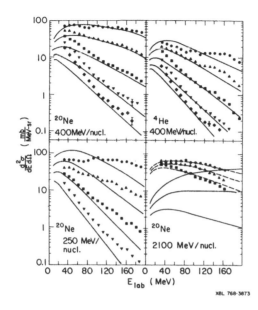

XBL 768-3873

Fig. 11: Measured proton inclusive spectra[11] from a uranium target at 30°, 60°, 90°, 120° and 150° in the laboratory. The solid lines are calculated with the fireball model.[11]

Both deeply inelastic reactions and projectile fragmentation reactions have been used most successfully to produce new neutron rich light isotopes. A recent experiment of ^{48}Ca fragmentation at 212 MeV/nucleon has produced more than a dozen of new nuclei,[17] see Fig. 16. This area of research, the production and, ultimately, the spectroscopy of neutron rich nuclei far from stability has received little attention in recent years--although it might be one of the most promising aspects of future heavy ion research. At present, little is known about the optimum experimental conditions for the production of these "exotic" nuclei and more systematic work will be necessary to study the dependence of isotope production cross sections on projectile-target combination and on beam energy.

First results,[18] comparing element and isotope production cross sections for projectile like fragments are shown in Fig. 17 for ^{16}O induced reactions at 140, 315 and 33600 MeV. Relative element production cross sections change drastically between 140 and 315 MeV and are very similar between 315 and 33600 MeV. This effect has been explained[19] as due to the increasing importance of the geometrical overlap of the colliding nuclei for the determination of the cross sections. However, the isotopic yields are quite different still. The relative abundance of neutron rich isotopes is larger at 315 than at 33600 MeV. It is well possible that the driving force that equili-brates the charge asymmetry degree of freedom at lower energies is still effective at 20 MeV/nucleon and negligible at relativistic energies. Similar conclusions may be drawn by comparing isotope distributions for deeply inelastic[14,20] and projectile fragmentation[5] reactions induced by ^{40}Ar projectiles, see Fig. 18 as an example.

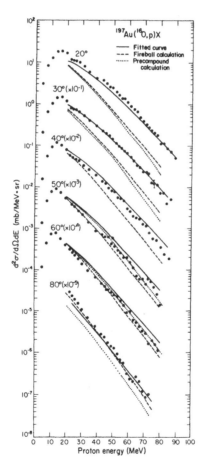

Fig. 12: Cross sections for production of protons in the reaction ^{197}Au(^{16}O,p)X at 315 MeV (from ref. 12).

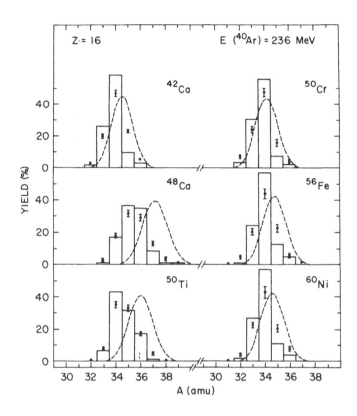

Fig. 13:   Distributions of sulfur isotopes produced in ^{40}Ar induced
           reactions on light targets at 6 MeV/nucleon. Assumed primary
           distributions are shown by dashed lines; calculated final
           distributions are shown by histograms. (From ref. 14).

5.   <u>CORRELATIONS BETWEEN PROJECTILE RESIDUES AND ALPHA PARTICLES</u>

     From the discussion of the previous sections it should be clear
that single particle inclusive experiments are not sufficiently re-
strictive to decide between different models. Therefore, more complex
experiments will be required to make significant progress. I want
to point out, however, that such experimental efforts will only bear
fruit if they are accompanied by more decisive theories that can predict distinctive aspects of many particle exit channels.

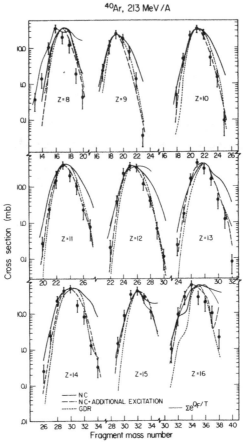

^{40}Ar, 213 MeV/A

Fig. 14:   Experimental isotope distributions for the reaction ^{40}Ar+C
at 213 MeV.   The calculations shown by the thick solid,
dashed and dotted curves are predictions of the abrasion
ablation model using different assumptions on primary excitation energies and proton-neutron correlations.   (From
ref. 5).

Since there are no significant particle correlation data available
at relativistic energies I will restrict myself to a short discussion
of correlations between alpha particles and projectile fragments[21]
observed for ^{16}O induced reactions at 315 MeV.   Fig. 19 shows the
energy spectra observed for ^{12}C + α coincidences for the detection
angle pairs θ_c = 17°, θ_α = 9° and θ_c = 17°, θ_α = 30°.   The kinetic

energy E_{12} associated with the relative motion between the carbon
nucleus and the α particle is very close the α – ^{12}C coulomb barrier
indicating a low excitation energy of the α + ^{12}C system. A double
peaked alpha-particle energy spectrum is clearly observed at θ_α = 9°
and not at θ_α = 30°. These features are qualitatively consistent
with a projectile break-up mechanism. Remarkably different inelastic-
ities are, however, sampled by the two coincidence requirements as
can be seen from the two-body Q-value spectra, Q_3. Whereas rather
high excitation energies of the target residue prevail for θ_α = 9°,
a strong quasi-elastic breakup peak is observed for θ_α = 30°. This
peak corresponds to a break-up process that involves essentially no
target excitation, i.e. a pure projectile fragmentation reaction.

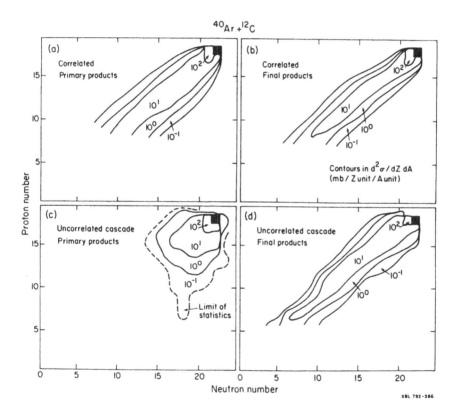

Fig. 15: Comparison of primary and final fragment distributions for
^{40}Ar + ^{12}C at 213 MeV predicted by abrasion-ablation and
cascade calculations.[16]

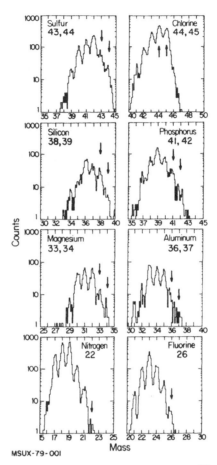

MSUX-79-001

Fig. 16: Isotopes produced by fragmentation of ^{48}Ca at 212 MeV/nucleon. The following new isotopes have been observed for the first time: ^{22}N, ^{26}F, ^{33,34}Mg, ^{36,37}Al, ^{38,39}Si, ^{41,42}P, ^{43,44}S, and ^{44,45}Ar.

Some examples of in-plane angular correlations for α-particles detected in coincidence with carbon and nitrogen nuclei are shown in Fig. 20. For coincident particles detected on opposite sides of the beam axis θ_α is defined to be negative. The angular correlations are shown for different regions of the three-body reaction Q-value Q_3 (see caption of Fig. 20). Only for the most quasi-elastic events of the ^{12}C + α channel (group I, see figs. 20a,d,e) two maxima are observed in the angular correlations which are expected from the kinematics of quasi-free break-up of an excited projectile. For more negative Q-values (groups II and III, see Figs. 20b,c) the angular correlations observed in the ^{12}C - α channel are very similar to those

observed in channels that cannot be trivially attributed to a projectile break-up mechanism. It should also be noted that, at 315 MeV,
the ^{13}C -α and ^{12}C -α cross sections are very comparable and, at 140
MeV, the ^{12}C -α cross sections are smaller than the ^{13,14}C -α cross
sections indicating that pure projectile break-up is not the dominant
reaction mechanism. These observations indicate that interactions
with the target nucleus are very important and that the reaction is
more complicated than a pure break-up reaction. The development of
a proper theory is a prerequisite for a better understanding of such
correlations.

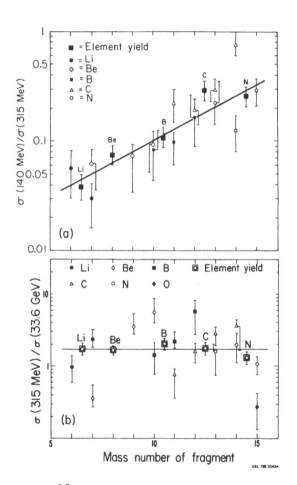

Fig. 17: Comparison[18] of element and isotope distributions of projectile residues for ^{16}O induced reactions at 140, 315 and
33600 MeV beam energies.

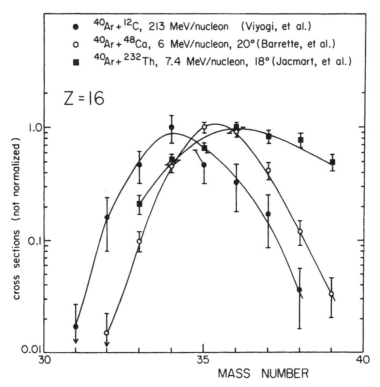

Fig. 18: Comparison of relative abundance of sulphur isotopes for ^{40}Ar induced deeply inelastic and fragmentation reactions.

6. CORRELATIONS WITH FISSION FRAGMENTS

Coincidence experiments like the one discussed in Sect. 5 are kinematically very selective; however, they are still kinematically incomplete for large negative values of Q_3. We have performed a different kind of coincidence experiment that is less selective kinematically but, at the same time, allows some general conclusions about the role of the target nucleus. Light particles $(p,...,0)$ are detected at forward angles in coincidence with two fragments resulting from the sequential fission of the target residue. By measuring the angle θ_{AB} between the two outgoing fission fragments one mainly determines the parallel component of the momentum transfered to the target residue prior to fission, $p_R^{||}$. (For a more detailed discussion of the analysis, see Ref. 22).

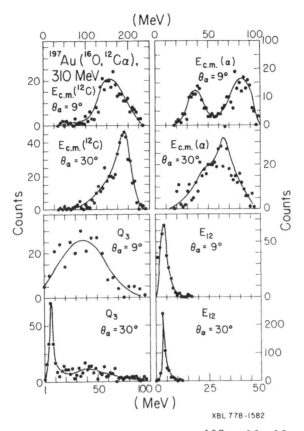

Fig. 19: Energy spectra for the reaction $^{197}Au(^{16}O,^{12}C\alpha)$ at 310 MeV. The energy E_{12} denotes the kinetic energy of relative motion between the ^{12}C nucleus and the alpha particle; Q_3 is the three-body reaction Q-value, $Q_3 = E_{12_C} + E_\alpha + E_{Au} - E_{beam}$.

By setting gates on the various outgoing particles and energies, we have determined the relation between the mean parallel momentum transfered to the target nucleus, $< p_R^{||} >$, and the mean parallel momentum $< p_3^{||} >$ of the coincident projectile residue. The results are summarized in Fig. 21. For orientation, the pure two-body transfer limit, $p_{tr}^{||} = p_1 - p_3^{||}$, is shown by the solid line (p_1 is the momentum of the projectile). The general trend of the data cannot be described by two body kinematics. Instead, a substantial amount of momentum, the "missing momentum", $p_m^{||}$, is carried away by light particles that are not observed in this experiment. A roughly linear dependence of the missing momentum (or the recoil momentum, $p_R^{||}$) on the momentum of the projectile residue is observed; only about two thirds of the momentum lost by the projectile are transferred to the target residue.

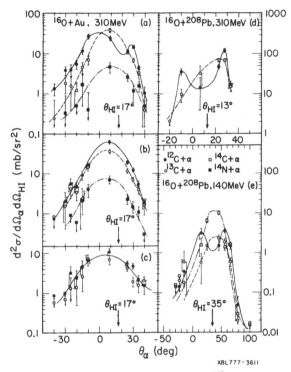

Fig. 20: In-plane angular correlations for ^{16}O induced reactions on ^{197}Au at 310 MeV (parts a-c) and on ^{208}Pb at 310 MeV (part d) and 140 MeV (part e). Three different regions of Q_3-values are displayed. Group I (parts a, d, e): $Q_3(C-\alpha) \geq -20$ MeV, $Q_3(N-\alpha) \geq -30$ MeV. Group II (part b): -60 MeV $\leq Q_3(C-\alpha) < -20$ MeV, -80 MeV $\leq Q_3(N-\alpha) < -30$ MeV. Group III (part c): -100 MeV $\leq Q_3(C-\alpha) < -60$ MeV.

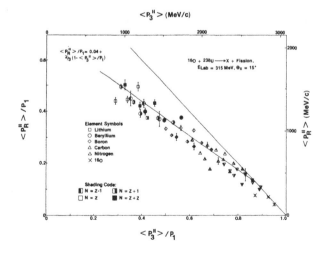

Fig. 21: Dependence of the average parallel recoil momentum of the target residue, $< p_R^{\parallel} >$, on the average parallel momentum, $<p_3^{\parallel}>$, of the projectile residue (from ref. 22).

The large amount of momentum that is transferred to the target residue, however, clearly excludes projectile fragmentation as the dominant reaction mechanism, since, for projectile fragmentation, a negligible amount of momentum should be transferred to the target nucleus, comparable in magnitude to that transferred by inelastically scattered ^{16}O ions. This experiment, however, cannot exclude sequential decay of the projectile residue which, of course, is always allowed kinematically. The fission yields as a function of the folding angle $\theta_{AB} = \theta_A + \theta_B$ are shown in Fig. 22. The momentum scale was obtained by assuming symmetric fission of ^{254}Fm. The inclusive distibutions shows two groups. The high momentum transfer group has a maximum at $\theta_{AB} \simeq 148°$ corresponding to a momentum transfer to the fissioning nucleus of about 92% of the beam momentum. We associate this group with central collisions. The low momentum transfer group of the inclusive θ_{AB} distributions has a maximum at about $173°$ which is close to the maximum of the distribution observed in coincidence with projectile residues (Li,..., O) at $15°$. This group is associated with peripheral collisions. To be definite, we denote reactions with $\theta_{AB} > 160°$ as peripheral and reactions with $\theta_{AB} < 160°$ as central.

There is a sizable probability for the emission of light particles both in central and in peripheral collisions. The arrows in Fig. 22 indicate the position at which the recoil momentum is equal to the difference between the mean momentum carried by the coincident light particle and the beam momentum. These differences agree very well with the maxima observed in the corresponding folding angle distributions in the region of central collisions. This feature of the data suggests that the emission of only one energetic light particle into the forward direction is a most probable mechanism. The motion of the center of mass of any remaining light particles that are emitted during the reaction must carry a very small momentum. On the other hand, the picture of isotropic light particle emission from a source of hot nucleons moving with half the beam velocity implies a sizeable momentum carried by this source if it consists of more than just one or two nucleons. Such a mechanism is unlikely to be dominant, at least for central collisions. We cannot, however, exclude precompound emission or the possibility of the existence of a "hot spot"[23] that remains attached to the target nucleus while it decays either towards equilibrium of the entire nucleus or by the emission of energetic light particles.

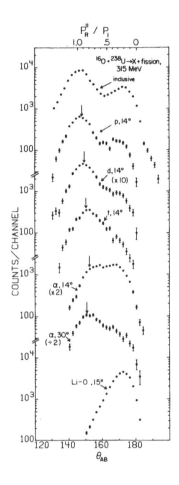

Fig. 22: Folding angle distributions θ_{AB} for coincident fission fragments observed in coincidence with light particles. The
arrows in the figure mark the values of θ_{AB} that are expected
if the mean recoil momentum of the fissioning nucleus is
equal to the difference between the beam momentum and the
average momentum carried by the coincident light particle.

Fig. 23 shows the light particle energy spectra observed at $14°$
in coincidence with central and peripheral collisions. Although there
are differences in the low energy regions it is most remarkable that
very similar slopes are observed in the high energy region of the
spectra. Within the present statistics these slopes show little variation for p, d, t, and α-particles or for central and peripheral collisions. Recently, it has been suggested[1] that the "prompt emission"
of energetic light particles should become an important aspect of

heavy ion reactions at energies of a few tens of MeV per nucleon.
These "promptly emitted particles" should be produced at a very early
stage of the reaction which could yield energy spectra that are rather
independent of the final fate of the colliding nuclei.

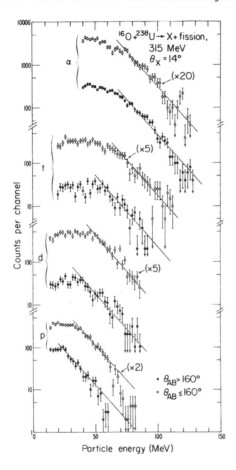

Fig. 23: Energy spectra of light particles detected at 14° in coin-
          cidence with peripheral ($\theta_{AB} > 160°$) and central ($\theta_{AB} < 160°$)
          collisions. The solid lines correspond to the spectral
          shape $\exp(-E/T)$, with $T=13$ MeV.

It is quite clear that we are still far away from understanding
heavy ion reactions in the energy range of a few tens of MeV/nucleon.
Right now, we can, at best, see only the tip of the iceberg. It
should have been clear that I could only cover a few aspects of the
vast area of physics that will be explored by the new generation of
heavy ion accelerators now under construction.

ACKNOWLEDGEMENT

Last, not least, I would like to acknowledge the help and discussions of my collaborators to whom I owe an enormous debt: T.C. Awes, K. van Bibber, B.B. Back, M. Bini, H. Breuer, M. Buenerd, H.J. Crawford, P. Doll, P. Dyer, D.E. Greiner, H.H. Heckman, D.L. Hendrie, J.L. Laville, P.J. Lindstrom, J. Mahoney, G. Mantzouranis, C. McParland, A. Menchaca-Rocha, M.C. Mermaz, A.C. Mignerey, C. Olmer, D.K. Scott, T.J.M. Symons, V.E. Viola, Jr., Y.P. Viyogi, G.D. Westfall, H. Wieman and K.L. Wolf. I also would like to thank the authors of refs. 8, 9, 10, 16 for the permission to use their work prior to publication.

This material is based upon work supported by the National Science Foundation under Grant No. Phy-7822696.

*Alfred P. Sloan Fellow

REFERENCES

1. J.P. Bondorf, Proceedings of the "Workshop of High Resolution Heavy Ion Physics at 20-100 MeV/A", Saclay, May 31-June 2, 1978, p. 37.
2. H.H. Heckman, et al., Phys. Rev. C17 (1978) p. 1651 and p. 1735.
3. D.E. Greiner, et al., Phys. Rev. Lett. 35 (1975) 152.
4. A.S. Goldhaber, Phys. Lett 53B (1974) 306.
5. Y.P. Viyogi, et al., Phys. Rev. Lett. 42 (1979) 33.
6. D.K. Scott, et al., Lawrence Berkeley Laboratory Report LBL-7729 (1978), unpublished.
7. K. van Bibber, et al., Phys. Rev. Lett. 43 (1979) 837.
8. K.W. McVoy and M.C. Nemes, preprint 1979.
9. T. Udagawa, et al., preprint 1979.
10. Y. Alhassid, et al., preprint 1979.
11. J. Gosset, et al., Phys. Rev. C16 (1977) 629.
12. T.J.M. Symons, et al., to be published.
13. See, e.g., Proceedings of the "4th High Energy Heavy Ion Summer Study" held at Berkeley, July 24-28, 1978, Lawrence Berkeley Laboratory Report LBL-7766.
14. J. Barrette, et al., Nucl. Phys. A299 (1978) 147.
15. L.F. Oliveira, et al., Phys. Rev. C19 (1979) 826.
16. D.J. Morrissey, et al., Phys. Rev. C18 (1978) 1267, Phys. Rev. Lett. 43 (1979) 1139
17. G.D. Westfall, et al., to be published.
18. C.K. Gelbke, et al., Phys. Lett. 65B (1976) 227.
19. J. Hüfner, et al., Phys. Lett. 73B (1978) 289.
20. J.C. Jacmart, et al., Nucl. Phys. A242 (1975) 175.
21. C.K. Gelbke, et al., Phys. Lett. 71B (1977) 83, M. Bini, et al., to be published.
22. P. Dyer, et al., Phys. Rev. Lett. 42 (1979) 560; T.C. Awes, et al., Phys. Lett. (to be published).
23. H. Ho, et al., Z. Physik A283 (1977) 235.

Reaction Fragments Resulting from Collisions of ^{20}Ne on ^{197}Au at Energies between 7.5 and 20 MeV/N

H. Homeyer, H.G. Bohlen, Ch. Egelhaaf, H. Fuchs, A. Gamp*, H. Kluge
Hahn-Meitner-Institut für Kernforschung
Bereich Kern- und Strahlenphysik, Berlin, Germany
*Fachbereich Physik der Freien Universität Berlin

Introduction

Very different reaction mechanisms can lead to the production of projectile-like fragments and/or light particles in heavy-ion reactions.
Their relative contributions should depend on bombarding energy, target to projectile mass ratio A_T/A_P, and the structure of the projectile and the target. Without trying to exhaust the list of possibilities, we like to summarize the reaction types which have been observed experimentally and which may be responsible for a considerable contribution to the cross section. This is mainly done to clarify the
terminology for the future discussions in this paper.

1. Quasi-elastic and deeply inelastic collisions

These collisions are connected with the exchange of nucleons and mutual excitation of both partners in the binary process, followed by de-
excitation via γ- or light-particle emission depending on excitation
energy. It has been shown (1) that at low bombarding energies this
process prevails. The total kinetic-energy loss is converted into internal excitation energy of the target- and projectile-like fragment
according to their mass ratio. At higher bombarding energies this process may still be present. The evaporation from excited fragments may
contribute significantly to the yield of fast light ions, in particular in systems with not too different target and projectile mass,
where the projectile-like fragment receives sufficient excitation
energy, even for the quasi-elastic (low total kinetic energy loss)
forward-peaked part of the cross section. The vector addition of
source and evaporation velocity gives large energies for light particles in the forward direction.

2. Massive-transfer or incomplete-fusion reactions

The light particles emerging from massive transfer reactions are very
strongly correlated to the incident energy with respect to their energy and their cross section. Their strength can at least qualitatively

be understood in terms of a generalized concept of critical angular momentum (2).

3. Projectile break-up reactions

Break-up reactions are very closely related to the internal structure of the projectile, namely to the existence of a spatially separated nucleon or cluster in the projectile nucleus, or to say it with a German word, the existence of a "Sollbruchstelle". The word Sollbruchstelle comes from mechanical engineering and literally means a built-in break point where a mechanical system is supposed to break when too much force or tension is applied, in order to avoid catastrophic damage.

Break-up reactions have so far been studied extensively with deuterons and ^{6}Li. Referring to the ^{6}Li data, it has been found in α-d coincidence measurements that at low bombarding energies the main part of the cross section is due to a "sequential break-up" process (3); i.e. the excitation of ^{6}Li into particle-unbound states and subsequent decay into α+d. In our terminology we do not consider this as a break-up process because it does not reflect the cluster structure of the projectile. We would rather put this type of reaction into category 1. of the above list of reaction types. The observation that the main decay channel is the α+d channel has in principle nothing to do with a cluster structure of the projectile in its ground state, but just reflects the fact that the energetically most favoured channel is chosen.

At high bombarding energies (60-150 MeV) the immediate break-up of the ^{6}Li projectile into its ground state cluster constituents plays a dominant role (4, 5). In these experiments strong evidence has been found that a large portion of the break-up fragments experience final-state interaction with the target nucleus.

4. Projectile fragmentation

In fragmentation reactions - i.e. catastrophic damages - the internal structure of the projectile nucleus plays a minor role. We will observe fragmentation when the internal binding energies or binding forces of the projectile are comparable or small with respect to the force applied to or the energy transferred into the projectile during the collision (6). Two alternatives are possible:

i. The target nucleus behaves like a knife and cuts away parts of the projectile nucleus (abrasion-ablation).

ii. The projectile gets very highly excited and then decays in a statistical way. This process may be interpreted by means of a temperature.

Gelbke, Scott and collaborators (7) found a similarity in the element
yields of projectile-like fragments when they bombarded various target nuclei with ^{16}O of 20 MeV/N and of 2 GeV/N. They proposed that
the high-energy limit might at least partially be reached at 20 MeV/N.

In order to get more information on the evolution of the reaction
mechanism with energy we studied the ^{20}Ne + ^{197}Au system at bombarding energies of 7.5, 11, 14.5, and 20 MeV/N. Light-particle and projectile-like fragment emission was investigated. Nearly all data are
inclusive measurements. So far only at 20 MeV/N some coincidence
events between light ions and projectile-like fragments have been
measured.

Experiments

The experiments were performed at the Hahn-Meitner-Institut Berlin
using the 150, 220, 290, and 400 MeV ^{20}Ne beam of the Van-de-Graaff-
isochronous cyclotron accelerator combination VICKSI. Light and heavy
reaction fragments were detected in silicon detector telescopes. For
the projectile-like fragments mass resolution up to mass 20 was a-
chieved. Light particles were stopped in a detector stack of a total of 5 mm. The α-particles with energies up to 120 MeV could be
stopped, protons up to 30 MeV.

a) Spectra of α-particles

In fig. 1 α-energy spectra are shown for three different energies,
220, 290,and 400 MeV at various angles. Two components can be distinguished: one at 20 MeV, the Coulomb barrier between α-particles and a
gold-like partner, the other slightly below the energy corresponding
to the beam velocity. This component is rapidly decreasing at larger
angles. At 150 MeV (not shown) we can barely identify this second component and if it is there, it has a very low cross section.

The mean energy of the fast component in the α-spectra has a linear
dependence on bombarding energy (fig. 2a). The cross section rises
strongly with incident energies between 11 and 14.5 MeV/N and only
slightly between 14.5 and 20 MeV/N (fig. 2b). The two components in
the α-spectra resulting from heavy ion bombardment have first been
observed by Britt and Quinton (8). They proposed that either direct

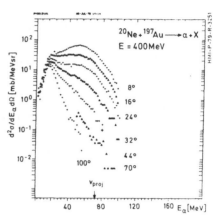

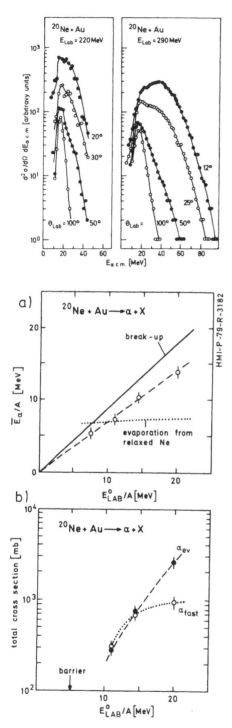

Fig. 1 (top): Center of mass α-energy spectra for three bombarding energies and different angles. At all bombarding energies two components in the spectra can be observed. The low energy component around 20 MeV corresponds to the Coulomb barrier of an α-particle and an Au-like fragment. The high energy part centered below the energy corresponding to beam velocity is rapidly decreasing with detection angle.

Fig. 2 (left): The most probable energy of the fast α-component (2a) and total cross section of the fast and slow components (2b) as a function of bombarding energy. The energy of the fast component depends linearly on incident ^{20}Ne energy. Its cross section rises rapidly with the incident energy.

break-up reactions or projectile excitation followed by decay may be responsible for the high yield of fast α-particles. From the α-spectra alone no further conclusions can be drawn at the present state of the analysis. On the other hand the mechanism proposed by Britt and Quinton should manifest itself also in the heavy-ion spectra.

b) Heavy-ion spectra and cross sections

To illustrate the behaviour at low bombarding energies a ΔE-E experimental scatter plot for θ_{lab} = 50° is shown in fig. 3. Hardly any cross section for $Z > Z_{proj}$ is observed. This is true for all angles The cross section drops rapidly with increasing charge removed from the target. The energies are all well above the Coulomb barrier of the final partner indicated by the solid line. The energy spectra - fig. 4 shows as an example the nitrogen and oxygen isotopes - have the well-known nearly Gaussian shape. The most probable kinetic energy depends on the charge removed from the projectile but not on the

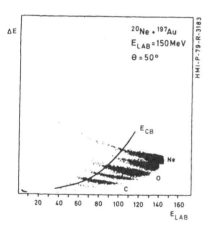

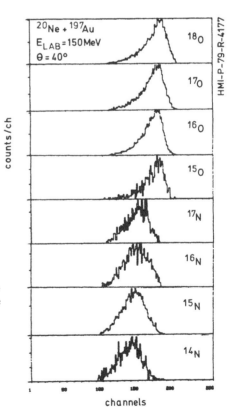

Fig. 3: ΔE-E scatter plot for E_{lab} = 150 MeV.
The cross section for projectile-like fragments decreases rapidly with the number of charge removed from the target. The energies of all ejectiles are well above the Coulomb barrier of the fragments and an Au-like residue.

Fig. 4: Energy spectra of nitrogen and oxygen isotopes at 150 MeV bombarding energy. The spectra of all isotopes have approximately the same width. The scale is normalized to the maxima of the spectra.

mass of the observed element. This is a well-known fact often observed in various systems at low bombarding energies. It can be understood in terms of particle exchange in partially equilibrated systems. The optimum total kinetic-energy loss is essentially determined by Coulomb trajectory matching (9).

The cross section for lower-Z heavy ions increases strongly with bombarding energy. This can be seen in fig. 5 where the relative intensities for the elements with $Z < Z_{proj}$ normalized to the fluorine cross section for the different incident energies are plotted. The plot also roughly indicates the absolute cross section ratios, since the fluorine cross section stays constant within 25 % for all bombarding energies. The strong rise of the cross section for the lighter elements with bombarding energy might be connected with the observation of the increasing yield of the fast α-particles. At 400 MeV the total cross section for fast α-particles amounts to 970 ± 150 mb. If we assume that all α-particles originate from the projectile and add up the heavy fragment cross sections that are connected with one or more α-particles removed from the projectile, weighted with the so defined multiplicity, we get 1050 ± 150 mb at 400 MeV, which is comparable to the measured α-yield.

The energy spectra of projectile-like fragments gradually develop to the behaviour we find at 400 MeV which is different from what we found at 150 MeV. At the highest energy - fig. 6 shows a ΔE-E scatter-plot at $\Theta_{lab} = 22^{\circ}$ - we observe two components, one at high energy with hardly any events for $Z > Z_{proj}$ and a completely damped component centered around or below the Coulomb barrier of the final partners with a very broad Z-distribution. To give an example for the overall cross section dependence on angle and energy a Wilczynski plot for an isotope in the middle of the spectrum, ^{13}C, is shown in fig. 7. The cross section $d\sigma/d\Theta_{CM}$ is plotted in linear steps indicated by the size of the squares. Apart from a stronly forward-peaked component at an energy rougly 40-50 MeV below the one corresponding to the beam velocity we recognize ine-

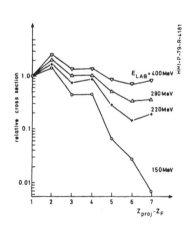

Fig. 5: Z distribution for final products with $Z < Z_{projectile}$ normalized to $Z_{proj} - Z_{fragm.} = 1$. The cross section for low Z fragments rises strongly with bombarding energy.

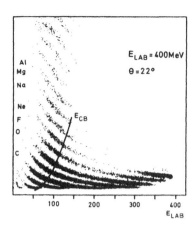

Fig. 6: Δ E-E scatter plot at 400 MeV. Apart from the quasi-elastic the fully damped component shows up near E_B , the Coulomb barrier between the projectile-like and target-like partner in a binary process.

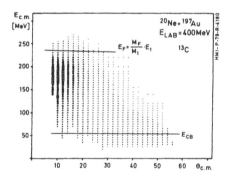

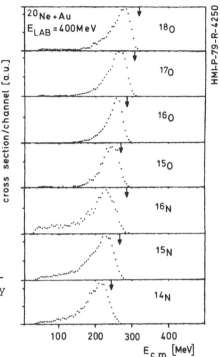

Fig. 8: Angle integrated center-of-mass energy spectra of N and O isotopes. The scale is normalized to the peak in the spectra. The peak energy shifts with mass number. The arrows indicate the energy corresponding to beam velocity.

Fig. 7: Wilczynski plot of ^{13}C at 400 MeV. The sizes of the squares are proportional to the cross section. Most of the cross section is concentrated in the very forward-peaked high energy component of about 50 MeV below the energy corresponding to beam velocity.

elastic contributions extending to the full relaxation (Coulomb barrier) of the system for events at backward angles.

The energy spectra of some isotopes show features different from those at low bombarding energies. Fig. 8 displays the total angle integrated center-of-mass energy spectra of the oxygen and nitrogen isotopes. We observe that

1) most of the cross section is concentrated in the high energy part of the spectrum,

2) the ^{16}O spectrum has the narrowest width,

3) the peak energy shifts with the atomic number and the mass number,

4) their peaks are below the energy corresponding to beam velocity depending on charge and neutron number of the isotope,

5) the shape of the spectra is different for all isotopes. Generally, a stronger shoulder towards lower energies is observed with increasing neutron number.

The accumulated information from the inclusive data unfortunately does not yield direct experimental evidence for the reaction mechanisms involved. On the other hand theoretical model predictions to compare with are scarce. McVoy and Nemes analysed the ^{16}O on ^{208}Pb data at 10-20 MeV/N (10) in the framework of direct transfer- and fragmentation reactions using the local momentum PWBA. From the comparison of calculated and experimental spectral widths they suggest a change in reaction mechanism from transfer to fragmentation for ^{15}N at 18 MeV/N. The ^{12}C data are best fit by fragmentation throughout the energy range. Our data allow some further speculations:

1. The yields of the fast α-particles and projectile-like fragments with more than one charge unit removed from the projectile increase with bombarding energy. At 400 MeV the total inclusive cross sections become comparable. This may imply that the α-particles arise to a substantial degree from the projectile either by excitation of the projectile or a projectile-like fragment above the threshold for α-emission or by direct break-up or by fragmentation processes.

   The effect that at high bombarding energies the most probable energy in the spectra for projectile-like fragments scale with their mass number, especially in the strongly forward peaked component, favours the assumption that reaction mechanisms of a direct type contribute substantially.

2. From the inspection of systematics in the isotopic yields we claimed (11) that break-up or fragmentation processes cannot fully explain the observed distribution. It is not unreasonable to assume that, apart from structure factors, the fragmentation Q-value Q_F, i.e. the separation energies of clusters or nucleons from the projectile, strongly influence the cross section for the respective heavy fragments (12, 13). This implies that for example the ^{16}O and ^{12}C channels (Q_F = 4.7 and 11.9 MeV) are strongly favoured with respect to ^{17}O and ^{13}C (Q_F = 21.1 and 27.4 MeV). In the experiment we find that in both cases the isotopes with N-Z=0

and N-Z=1 are produced with comparable strengths ($\sigma_{16_O}/\sigma_{17_O} = 2.5$, $\sigma_{12_C}/\sigma_{13_C} = 1.3$). This fact and the high yield of ^{15}N and ^{11}B led to the conclusion that final state interaction (mainly proton stripping and neutron pick-up) of the energetically favoured fragments produced in a fast process might be responsible for the relatively high yield of isotopes with N-Z=1.

c) Coincidence data

Direct experimental evidence in favour or against the above mentioned speculations can be drawn from coincidence experiments between light particles and projectile-like fragments. In principle complete angular correlation and multiplicity measurements are necessary to finally obtain quantitative results. However, simple considerations show that qualitatively the essential features can be extracted from a rather small set of coincidence spectra:

The above mentioned reaction mechanisms imply some general predictions for the qualitative angular correlations and the combination of the coincident fragments to be observed.

It is obvious that for coincident events of light and projectile-like fragments in the final channel three different reaction types yield roughly similar experimental observations that can be disentangled by very careful angular and energy correlation measurements only:

a) fragmentation reactions

b) projectile excitation and subsequent statistical decay

c) quasi-elastic transfer reactions with excitation of the projectile-like reaction product above its threshold for light particle emission.

In all cases the light particles are emitted into a narrow cone around the direction of the heavy partner. This focussing arises either from the primary reaction mechanism or just from the transformation of the frame of references since the particles are emitted more or less isotropically from a fast moving source (process b and c).

In particular the combination of coincident fragments are more or less predictable in all three cases. Process a) precisely defines the combination of coincident ejectiles, i.e. $^{19}F+p$, $^{18}F+d$ or $^{18}F+n+p$, $^{17}O+^3He$, $^{16}O+\alpha$, etc. This strict correlation is weakened if the heavy fragments are produced in particle-unstable states and in the processes b and c. Inspecting the particle thresholds for the possible primary reaction products we find that for most of the isotopes involved the α-thresholds are much lower than the thresholds for the emission of hydrogen isotopes. This amounts to the fact that, though we may expect α-coincidences with all isotopes observed in the inclusive data, coincident events with protons, deuterons, and tritons should be essentially restricted to isotopes of F, N, and B.

If the reaction mechanism proceeds via a fast fragmentation or break-up mechanism followed by final-state interaction of the primary products with the target nucleus angular correlations and combinations of coincident ejectiles, different from those mentioned above, are expected*. In the extreme case that <u>all</u> primary products experience final-state interaction, i.e. that we don't observe primary fragments at all, the angular correlations should be essentially a product of the angular distribution of the single events. The combinations of coincident ejectiles, i.e. the distributions of different elements and isotopes for light and heavy partners, should roughly scale with the distributions of the inclusive data. In the less extreme case, where the original fragments only partially suffer final-state interaction the assumed time order of events predicts that within a narrow cone around the heavy partner, we should expect coincidences reflecting the primary process, whereas for larger relative angles between the light-ion and heavy-ion detector the above-mentioned strong correlations in the combinations of coincident ejectiles should be washed out by final-state interaction.

In the experiment we consequently placed the light-particle detectors at very forward angles symmetrically to the beam axis, $\pm 8^{\circ}$, and the detector for projectile-like fragments at $+16.5^{\circ}$ which is approximately at the grazing angle at 400 MeV incident energy. The combination $+8^{\circ}$, $+16.5^{\circ}$ (both detectors on the same side with respect to the beam axis) we define as the "near side", the combination -8°, $+16.5^{\circ}$ as "far side".

---

* By final-state interaction we mean in this context in principle any interaction of the primary fragments with the target nucleus, ranging from quasi-elastic transfer through deep inelastic and fusion reactions in their relative strengths depending on the impact parameter and energy of the fragment at the point where it is created.

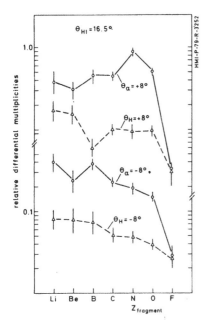

Fig. 9: Relative differential multiplicities for projectile-like fragments in coincidence with α-particles and hydrogen isotopes for both detector positions.

Fig. 9 shows the relative differential multiplicities for projectile-like elements coincident with α-particles and hydrogen isotopes for the far- and near- side detector positions. The relative differential multiplicity is defined as $C \times N_{CO}/N_{FS}$ where C is an arbitrary overall constant, N_{CO} is the number of coincident events and N_{FS} is the number of single events for projectile-like fragments. On the far side the Z-distributions for projectile-like fragments coincident with α-particles or with hydrogen isotopes look very similar. Moreover, we find that the strengths of the different isotopes of the projectile-like fragments (not shown) in both cases correspond within the rather large statistical errors to their relative intensities we observe in the inclusive spectra. According to the arguments given above this means that the primarily produced fragments have changed their identity due to interaction with the target nucleus. The increase of relative multiplicities for elements further away from the projectile can be understood in terms of the increased number of open channels as well in the primary as in the secondary process.

Does the primary process show up in the "near side" coincidences? Firstly, we observe hardly any fluorine isotopes coincident with either α-particles or hydrogen isotopes. Secondly, no enhancement of coincident events of hydrogen isotopes with the isotopes of nitrogen and boron is found. This excludes projectile fragmentation and high projectile or projectile-like fragment excitation followed by light particle emission as the dominant processes.

The relatively high yield of coincident events of α-particles with isotopes of oxygen, nitrogen and carbon on the "near side" compared to the "far side" rather implies that ^{20}Ne break-up reactions into ^{16}O+α and possibly ^{12}C+2α are the essential contributions to the primary process.

Comparison with other experiments

Though the preceding explanation for the coincident spectra and, consequently, parts of the inclusive cross sections, imply the assumption of rather specific reaction mechanisms, especially the time ordering of events several recently performed experiments support the basic ideas. A strong contribution of the quasifree break-up reaction has been observed in coincidence measurements of $^{16}O + ^{208}Pb$ at 20 MeV/N (14) and in $^{32}S + ^{197}Au$ at 11.6 (15). Bhowmik and collaborators (16) who bombarded ^{12}C, ^{27}Al, and ^{58}Ni with ^{14}N at 10 MeV/A and measured correlations between α-particles and projectile-like fragments were able to describe the energy- and angular correlations for elements below boron empirically in a simple factorised form as a product of singles-α and heavy-ion cross sections. As mentioned in the introduction, the target-like residue yields from ^{6}Li induced reactions (4, 5) could be explained by the assumption that the ^{6}Li dissociation fragments, α and d, interact independently with the target nucleus.

Conclusions

The inclusive data have shown that the yield of fast light particles increasing with higher bombarding energies is connected with increasing yields of elements with charge lower than the projectile. This may indicate that most of the light particles arise from the projectile. The coincidence data at 20 MeV/N show that projectile fragmentation or high excitation of the projectile or projectile-like fragments cannot explain the coincident events at the angles measured so far. It seems more likely that the projectile nucleus dissociates into two fragments mainly determined by its internal structure in a fast process giving rise to subsequent interaction of the dissociation products with the target nucleus. More detailed angular correlation measurements are, of course, necessary to prove this conclusion.

References

(1) D. Hilscher, J.R. Birkelund, A.D. Hoover, W.U. Schröder, W.W. Wilcke, J.R. Huizenga, A.C. Mignerey, K.L. Wolf, H.F. Breuer, V.E. Viola, Phys. Rev. C, Vol. 20, No. 2 (1979), 576

(2) K. Siwek-Wilczynska, E.H. du Marchie van Voorthuysen, J. van Popta, R.H. Siemssen, and J. Wilczynski, Phys. Rev. Lett. 42, 1599 (1979)

(3) D. Scholz, H. Gemmeke, L. Lassen, K. Bethge, Nucl. Phys. A288, 351 (1977)

(4) C.M. Castaneda, H.A. Smith, T.E. Ward, T.R. Nees, Phys. Rev. C, 16, 1437 (1977)

(5) B. Neumann, J. Buschmann, H. Klewe-Nebenius, H. Rebel, H.J. Gils, Nucl. Phys. A329, 259 (1979)

(6) A.S. Goldhaber, Phys. Lett. 50B, 211 (1974)

(7) C.K. Gelbke, C. Olmer, M. Buenerd, D.L. Hendrie, J. Mahoney, M.C. Mermaz, D.K. Scott, Phys. Rep. 42, 311 (1978)

(8) H.C. Britt, A.R. Quinton, Phys. Rev. 124, 877 (1961)

(9) J.P. Bondorf, F. Dickmann, D.H.E. Groß, P.J. Siemens, Journal de Physique, C6, 145 (1971)

(10) K.M. McVoy, M.C. Nemes, Proceedings of the Symp. on Heavy Ion Physics from 10 to 200 MeV/A, Brookhaven (1979), to be published

(11) H. Homeyer, C. Egelhaaf, H. Fuchs, A. Gamp, H.G. Bohlen, H. Kluge, Proceedings of the Symposium on Heavy Ion Physics from 10 to 200 MeV/A, Brookhaven (1979), to be published

(12) V.K. Lukyanow, A.I. Titov, Phys. Lett. 57B, 10 (1975)

(13) G. Baur and D. Trautmann, Phys. Reports, 25C, 293 (1976)

(14) C.K. Gelbke, M. Bini, C. Olmer, D.L. Hendrie, J.L. Laville, J. Mahoney, M.C. Marmaz, D.K. Scott, H.H. Wiemann, Phys. Lett. 71B, 83 (1977)

(15) A. Gamp, J.C. Jacmart, N. Poffe, H. Doubre, J.C. Roynette, Phys. Lett. 74B, 215 (1978)

(16) R.K. Bhowmik, E.C. Pollaco, N.E. Sanderson, J.B.A. England, D.A. Newton, G.C. Morrison, Phys. Rev. Lett. 43, 619 (1979) and contribution to this conference

Local excitation in reactions with α-particles

G. Gaul, R. Glasow, H. Löhner, B. Ludewigt, R. Santo

Institut für Kernphysik der Universität Münster, D-4400 Münster

Measurements of energy spectra and angular distributions of reaction products with $1 \leq A \leq 12$ have been carried out in order to test reaction models which can describe consistently light and heavy fragment emission at projectile energies per nucleon slightly larger than the fermi energy in nuclei. This energy domain of about 25 MeV/nucleon seems to play the role of a transition energy on the way to relativistic energy reactions.[1]

We have performed the experiments at the Jülich cyclotron in continuation of our earlier work with 100 MeV α on Ca-targets[2] and measured fragment energy spectra from a ^{58}Ni target at several angles between 10° and 145° with 100 MeV and 172.5 MeV α-particles. Measurements at 172.5 MeV are still in progress. Sufficient accuracy in the particle separation in a wide energy range was achieved using a 4-detector telescope (Si-surface barrier ΔE- and Ge(Li)-E-detector) for detection of the light ejectiles (p, d, t, ^{3}He, α) and data recording in 4-parameter list-mode on tape. The heavier particles (Li, ..., C) were measured with a 2-detector telescope. The minimum detectable energy due to the energy loss in the ΔE-detector varied between 4 and 30 MeV for protons and carbon respectively.

Energy spectra for different fragment isotopes, averaged in 2 MeV energy steps, are shown in fig. 1. The double differential cross sections plotted versus particle laboratory energy fall off nearly exponentially while the slope becomes steeper with increasing fragment mass. This behaviour is similar to that observed earlier with Ca-tar-

gets. Thus we expect a simple dependence on the fragment mass of the

main contribution to the fragment cross section. More detailed infor-

mation about the emission process is obtained from plots of contour

lines of invariant cross sections $(1/p) \cdot (d^2\sigma/d\Omega/dE)$ in the plane of

longitudinal $(\beta_{||})$ and transverse $(\beta_{\perp})$ particle velocity (fig. 2). Data

points indicating the same magnitude of invariant cross section are

close to parts of circles centered at three different origins on the

$\beta_{||}$-axis. The highest p cross sections are close to circles centred at

$\beta_{||} \approx 0.02$, the compound system velocity. Thus in this part of the fi-

gure we observe nearly isotropic emission from the well known compound

nucleus. The circles are disturbed at larger longitudinal velocities.

Here we observe additional cross section from the projectile fragmen-

tation process at the beam velocity $(\beta_{||} = 0.23)$. The lower cross sec-

tion data however are again close to circles centered at an interme-

diate velocity $(\beta_{||} \approx 0.09)$. The underlying reaction process seems to

be isotropic emission from a system moving faster than the compound

system. The d-data reveal the same velocity of the emitting system.

The highest ^{3}He cross sections are concentrated around the projectile

velocity. Thus we observe projectile fragmentation (observed in the

^{3}He channel in other experiments[3,4])   at forward angles and at lar-

ger transverse velocities again isotropic emission from a system mo-

ving faster than the compound system.

A reasonable description of the fragment energy spectra must include

contributions from the observed three processes. Just for an illustra-

tion of the possible reaction mechanism leading to a system velocity

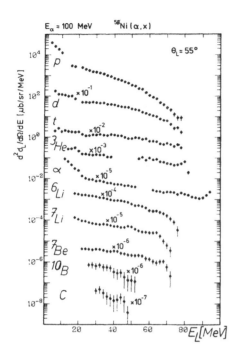

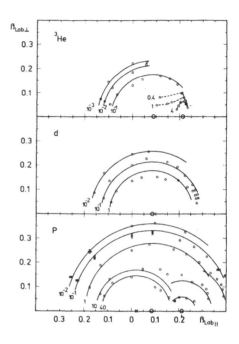

Fig. 1: Fragment energy spectra for various fragment kinds at $\theta_L = 55^O$.

Fig. 2: Contour lines of invariant cross section (in units of $\mu b/(sr \cdot MeV^2/c)$) in the plane of longitudinal and transverse particle velocity for p, d and ^{3}He-ejectiles from 100 MeV α on ^{58}Ni.

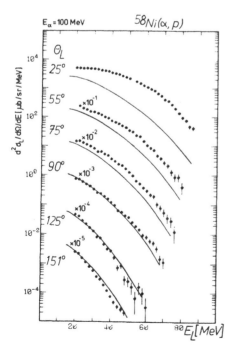

Fig. 3: Experimental proton spectra at various angles compared with local-excitation model calculations, f = 0.1 (equation 2).

intermediate between the projectile and compound system velocity one may look at the time evolution of the nuclear matter distribution in the fluid - dynamics model of the collision process[5] calculated for Ne+U reactions at 250 MeV/nucleons. In the first stage of the reaction at central collisions only a part of the total compound system is strongly deformed and highly excited. In later stages on the reaction time scale the energy spreads over the whole nucleus ending with the decomposition of the target. At larger impact parameters one observes abrasion of nucleons in the projectile fragmentation process. Coming back to bombarding energies of 25 - 43 MeV/nucleon ($0.23 \leq \beta \leq 0.29$) the reaction will end up with compound nucleus fragmentation in central collisions. If however the beam velocity is larger than the speed of sound in nuclear matter ($\beta_s \approx 0.2$[6]) the formation of an intermediate stage - a spatially localised excitation of the target - is expected. Particle emission from states of local equilibrium ("hot spot") has been proposed in connection with a discussion of transport properties in nuclear matter[7] and experimentally investigated in heavy ion reactions[8]. In a first approximation we assume that in the fast step of the interaction only nucleons in the geometrical target-projectile overlap-volume are involved. This rough assumption, from which the system velocity and excitation energy per nucleon is derived, is analogous to the participant spectator model used in relativistic heavy ion collisions.[9] Geometrical considerations have a reasonable basis already at projectile energies discussed in this paper because the α-particle wave-length (λ_α =1.5 fm and 1.1 fm respectively) is smaller than the internucleon distance d $\approx$ 1.8 fm. The overlap region is treated in the statistical model[10] using a Fermi gas level density

$$\rho(U) \sim \frac{1}{U} \exp(2\sqrt{aU}) \tag{1}$$

with excitation energy $U(\varepsilon)$ and kinetic particle energy ε. The level density parameter is $a = N(b)/16$ with $N(b)$ beeing the impact parameter dependend number of nucleons in the overlap region. Using the
classical inverse reaction cross section

$$\sigma_p^* = \pi R^2 (1 - \varepsilon_{coul}/\varepsilon)$$

and integrating over impact parameter b we obtain the double differential proton cross section in the overlap-CM-system

$$d^2\sigma_p / (d\Omega d\varepsilon) = \int 2\pi b db \, f \, N_p(b) \, \sqrt{\varepsilon} \, \sigma_p^* \, \rho(U(b,\varepsilon)). \tag{2}$$

$N_p(b)$ is the number of protons in $N(b)$ and f is a normalisation factor
regarding the emission probability of participant protons. The results
of this straightforward model are compared in fig. 3 with experimental
p spectra. With a normalisation $f = 0.1$ adjusted at $\theta_L = 90^\circ$ we obtain
the correct shape and angle dependence at backward angles. At forward
angles large deviations are observed which are due to projectile fragmentation. This contribution has been included analogous to the Serber-model for α-breakup following Wu et al.[3], who used such a formulation for $(\alpha, ^3He)$ reactions at 140 MeV. The absolute normalisation was
determined by fitting the cross section once at the most forward
angle. Additionally the compound nucleus evaporation spectrum has been
included. The angle integrated spectrum[11] is well approximated by a
simple level density formula (1), which has further been used to describe the double differential cross sections by fitting the cross section height at each angle. The result is a deviation from isotropy at
forward angles by a factor of two. Experimental proton spectra at 100

and 172.5 MeV bombarding energy (absolute cross sections at 172.5 MeV

are preliminary) are compared with calculations containing the three

models in fig. 4a, b. Projectile fragmentation obviously dominates

the shape of the spectra at forward angles. Deviations around 50° at

the higher energy may be related to an overestimated forward peaking in

the breakup cross section, because in the Serber-model the projectile-

target interaction is completely neglected. In spite of certain defi-

ciencies of this model the shape of the breakup spectra is well des-

cribed as has been pointed out in the $(\alpha, {}^3\text{He})$-example[3,12]. Another

problem which needs further investigation arises from the normalisa-

tion f of the local-excitation model. This parameter is lower by a

factor of 2.5 at the higher energy.

For calculations of the complex particle spectra we assume that the

proton energy distributions from our model are a good approximation

for the nucleon energy distribution within the heated compound sub-

system. Therefore we use these proton spectra as input for the coa-

lescence model[9,13]. The basic formula is a power-law of the proton

cross sections for calculations of the complex-particle spectra. For

example an intermediate d-state will be built if the relative momen-

ta of two nucleons are smaller than the coalescence-sphere radius P_o.

This parameter has to be determined by experimental data. Calcula-

tions according to this model are compared to experimental fragment

energy spectra from a ${}^{40}\text{Ca}$ target in fig. 5. Some examples for a

${}^{58}\text{Ni}$-target are given in fig. 6a, b. The general trend of the angle

and energy dependence of the data is well described. The determined

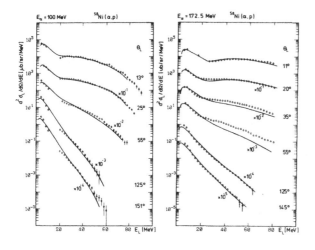

Fig. 4a   Fig. 4b

Fig. 4: Experimental proton spectra compared with calculations containing contributions from compound nucleus evaporation, projectile fragmentation and local excitation
(f = 0.1 (a) and f = 0.04 (b)).

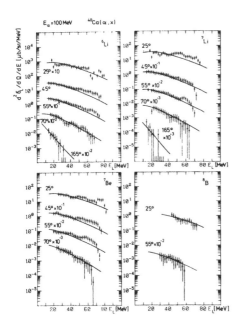

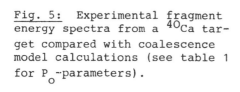

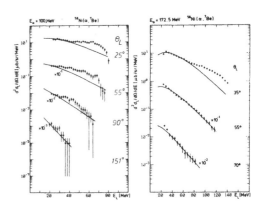

Fig. 6: Experimental ^{7}Be-spectra from a ^{58}Ni target at 100 MeV (a) and 172.5 MeV (b) compared with coalescence model calculations (see table 1 for P_o-parameters)

Fig. 5: Experimental fragment energy spectra from a ^{40}Ca target compared with coalescence model calculations (see table 1 for P_o-parameters).

Table 1: The coalescence-model parameter P_O used for calculating the curves of fig. 5, 6 on the basis of proton-spectra calculations as shown in fig. 4.

P_O [MeV/c]		fragment		
target, energy [MeV]	^{6}Li	^{7}Li	^{7}Be	^{8}B
^{40}Ca, 100	296	297	304	270
^{58}Ni, 100			281	
^{58}Ni, 172.5			440	

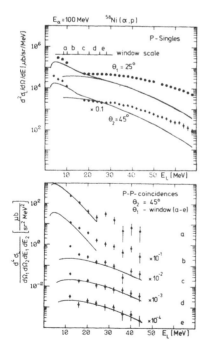

Fig. 7a): Proton spectra from a ^{58}Ni target. The slope of the singles spectra is compared with compound nucleus model plus local-excitation model.

Fig. 7b): The slope of the proton spectra at $\theta_2 = 45^O$ coincident with protons at $\theta_1 = 25^O$

in the energy window a - e is compared with compound nucleus calculations (window a, b) and local-excitation calculations (window c, d, e).

P_o values at E_α=100 MeV (table 1) vary slightly with fragment mass and are of reasonable magnitude around the fermi-momentum.   At   the higher energy we had to use a larger P_o-value indicating a growing importance of complex particle emission at larger excitation energies.

Finally the results of a p-p coincidence experiment shall be mentioned which may help to seperate the different contributions to the nucleon spectra. We measured coincident protons at two   angles separated 20° on the same side of the beam. Fig. 7a shows the singles spectra compared with calculations according to the local-excitation model, once without and once with the compound-nucleus contribution included. In fig. 7b are shown the coincident p-spectra at θ_2 =45° when windows are set in the θ_1-spectrum at the indicated window ener- gy ranges (fig. 7a). With a window set at the compound peak (window a) we observe in the coincident spectrum a more pronounced evaporation peak. If however the trigger energy is increased (window d, e), the compound peak vanishes and a spectrum is observed with a slope signi- ficantly different from the singles spectrum slope. The shape of these coincident spectra is well described by the local-excitation model. Obvioulsy p-coincident spectra are free from direct contributions like projectile fragmentation and display the statistical features of the compound nucleus and the precompound nucleus respectively. Further in- plane and out-of-plane correlation experiments are planned and neces- sary in order to decide unambiguously whether there exists a local thermally equilibrated fast moving source or not.

In conclusion, we have shown that apart from compound-nucleus evapo- ration and projectile fragmentation the shape and angle dependence of the proton spectra measured in α-nucleus reactions at 25 MeV/nucleon

and 43 MeV/nucleon can be explained by a local-excitation model, which

assumes thermal equilibrium in a compound nucleus subsystem.

Complex particle spectra are described in their energy and angle de-

pendence by the coalescence model using reasonable parameters for the

cross section height. First results of a p-p coincident experiment

confirm the p-spectrum shape calculated under the assumption of local

excitation.

References

1) J. Hüfner, C. Sander, G. Wolschin, Phys.Lett. 73B (1978) 289;

2) H. Löhner, B. Ludewigt, D. Frekers, G. Gaul, R. Santo,
       Z. Physik A292 (1979) 35;

3) J.R. Wu, C.C. Chang, H.D. Holmgreen, Phys.Rev.Lett. 40 (1978)
       1013;

4) A. Budzanowski, G. Baur, C. Alderlisten, J. Bojowald, C. Mayer-
       Böricke, W. Oelert, P. Turek, F. Rösel, D. Trautmann,
       Phys.Rev.Lett. 41 (1978) 635;

5) A. Amsden, F.H. Harlow, J.R. Nix, Phys.Rev. C15 (1977) 2059;

6) H.G. Baumgardt, J.U. Schott, Y. Sakamoto, E. Schopper, H. Stöcker,
       J. Hofmann, W. Scheid, W. Greiner, Z.Physik A273 (1975) 359;

7) R. Weiner, W. Weström, Phys. Rev. Lett. 34 (1975) 1523 and
       Nucl. Phys. A286 (1977) 282;

8) H. Ho, R. Albrecht, W. Dünnweber, G. Graw, S.G. Steadman, J.P.
       Wurm, D. Disdier, V. Rauch, F. Scheibling, Z. Physik A283
       (1977) 235;

9) J. Gosset, H.H. Gutbrod, W.G. Meyer, A.M. Poskanzer, A. Sandoval,
       R. Stock, G.D. Westfall, Phys.Rev. C16 (1977) 629;

10) T. Ericson, Adv. in Physics 9 (1960) 425;

11) F. Pühlhofer, Nucl. Phys. A280 (1977) 267;

12) R. Shyam, G. Baur, F. Rösel, D. Trautmann, Phys.Rev. C19 (1979)
       1246;

13) S.T. Butler, C.A. Pearson, Phys.Rev. 129 (1963) 836.

INCOMPLETE FUSION OR MASSIVE TRANSFER ?

J. Wilczyński

Kernfysisch Versneller Instituut, 9747 AA Groningen, The Netherlands

and

Institut of Nuclear Physics, 31-342 Cracow, Poland [*)]

In this report I will discuss some aspects of the mechanism of incomplete fusion reactions. The report is based on recent studies of heavy-ion reactions at the Kernfysisch Versneller Instituut in Groningen.

It was shown in a series of publications [1-5] that a significant part of "fast" α-particles and also protons, deuterons and tritons [4] originate from essentially binary reactions in which a massive fragment of a projectile is captured by a target nucleus. An important information on the energy dependence of the incomplete fusion reactions was obtained by the Groningen group [3]. To explain the main features of the excitation functions the generalized concept of the critical angular momentum was proposed [3]. The sequence of the angular momentum bins dominated by successive incomplete fusion channels, as predicted by the model of ref.3, was confirmed directly via the γ-multiplicity measurements reported recently by the St. Louis - Oak Ridge group [5].

Fig. 1 shows schematically the experimental set-up used by the Groningen group in the first series of experiments in which the ^{12}C + ^{160}Gd reaction was studied at 90, 120, 160, and 200 MeV bombarding energies. Coincidences between charged particles ($2 \leqslant Z \leqslant 8$) and γ-rays were recorded by using a standard ΔE-E solid-state telescope and a Ge(Li) detector. Particle identification of the ejectiles and identification of the target-residue nuclei via characteristic γ-transitions was enough to determine the reaction mode: whether the missing fragment of the projectile was captured by the target nucleus or flew away in a breakup-type process.

The results on the ^{12}C + ^{160}Gd reaction have already been published [3]. Therefore I will recapitulate only the main points and concentrate on the question of the energy dependence of the cross sections, mutual competition between different reaction channels, and some consequences of the model proposed to explain the observed excitation functions.

[*)] Present address.

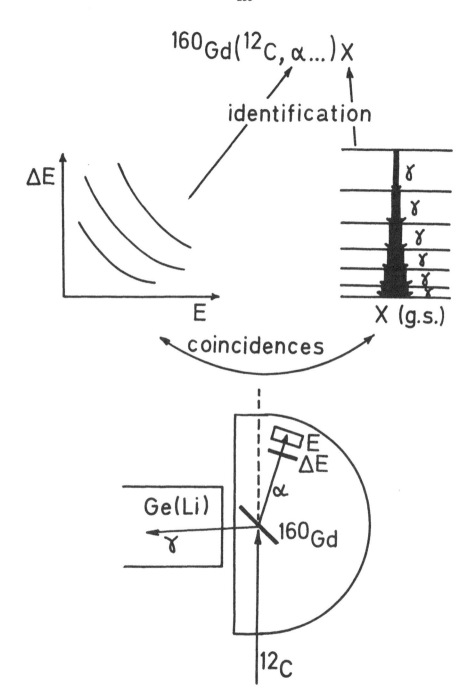

Fig. 1. Scheme of the experimental set-up for studying the incomplete fusion reactions in the ^{12}C + ^{160}Gd system by means of the γ-charged-particle coincidences.

It is seen even from simple inclusive measurements that fast α-particles (i.e. originating from nonequilibrated processes) dominate among projectile-like and lighter ejectiles. Total inclusive cross sections for production of these fast α-particles are of the order of one barn (see fig. 2). Main part of the α-particles is concentrated at forward angles. Their energy spectra at the forward angles extend over a wide range of energies with the centroid close to the beam-velocity energy.

Our study of the incomplete fusion reactions in the $^{12}C + ^{160}Gd$ system was concentrated on the most intensive reaction channels, i.e., those with emission of fast α-particles as ejectiles. In the coincidence α-γ spectra we could identify γ-transitions only in the isotopes of Er and Dy. This means we observed only two reaction modes: ^{160}Gd $(^{12}C,\alpha xn)^{168-x}Er$ and $^{160}Gd(^{12}C,2\alpha xn)^{164-x}Dy$. The first reaction can be interpreted as the capture of a "^8Be" fragment that leads to the formation of the $^{168}Er^*$ compound nucleus and the following emission of x neutrons. Similarly, we interpret the second reaction as the capture of one α-particle.

The energy spectra of the α-particles, gated by selected γ-transitions in the target-residue nuclei, look similarly to those of the singles α-particles: they are peaked approximately at the beam-velocity energy. We conclude therefore that in the incomplete fusion reactions both the captured fragment and the ejectile carry the kinetic energy roughly proportional to the mass of the fragment.

An important information on the incomplete fusion reactions can be deduced from relative intensities of the γ-transitions along the rotational bands in the target-residue nuclei. Specifically, we observed very weak side-feeding to the lowest members of the yrast bands in the isotopes of Er and Dy. This fact leads to the conclusion that low partial waves do not contribute significantly to the cross section for the observed reaction sub-channels. The same feature of the incomplete fusion reactions was observed before by Inamura et al.[1] who suggested, for the first time, that the incomplete fusion reactions are localized in the l-space just above the critical angular momentum for complete fusion reactions.

Total cross sections for two main incomplete fusion channels, $(^{12}C,\alpha)$ and $(^{12}C,2\alpha)$, were obtained by summing up the cross sections for all reaction sub-channels characterized by different numbers of emitted neutrons:

$$\sigma(^{12}C,\alpha) = \sum_x \sigma(^{12}C,\alpha xn), \qquad (1)$$

257

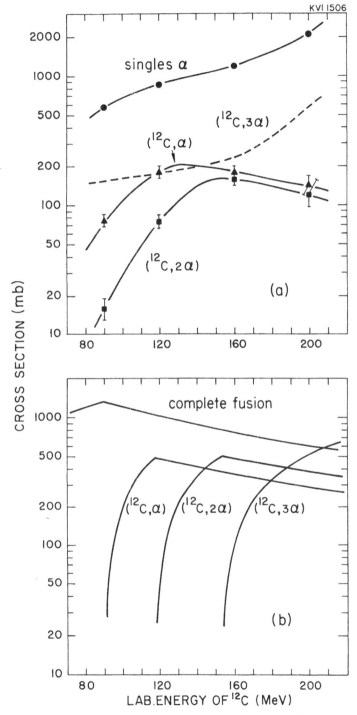

Fig. 2. Energy dependence of the cross sections σ_α(incl.), $\sigma(^{12}C,\alpha)$, $\sigma(^{12}C,2\alpha)$, and $\sigma(^{12}C,3\alpha)$ and comparison with predictions of the model discussed in the text.

$$\sigma(^{12}C,2\alpha) = \sum_x \sigma(^{12}C,2\alpha xn). \tag{2}$$

In addition to the cross sections (1) and (2) we measured the total in-clusive α-production cross section, σ_α(incl.), and the average α-multi-plicity [6]. The results on the average α-multiplicity, combined with the cross sections σ_α(incl.), $\sigma(^{12}C,\alpha)$ and $\sigma(^{12}C,2\alpha)$, allowed us to conclude that only three main processes contribute to the inclusive cross section: the two incomplete fusion channels and the breakup of ^{12}C into three α-particles,

$$\sigma_\alpha(\text{incl.}) \approx \sigma(^{12}C,\alpha) + 2\sigma(^{12}C,2\alpha) + 3\sigma(^{12}C,3\alpha). \tag{3}$$

The energy dependence of all these cross sections is shown in the upper part of fig. 2. The excitation functions for both incomplete fusion channels show rather distinct energy thresholds. The cross sections increase over a limited range of bombarding energies, reach maximum, and then tend to decrease. The energy thresholds are localized signi-ficantly above the entrance channel Coulomb barrier (~ 50 MeV in the lab. system). As it is seen from fig. 2, the ($^{12}C,\alpha$) reaction has the lower energy threshold (at about 80 MeV), while the threshold for the ($^{12}C,2\alpha$) reaction is shifted approximately by 30 MeV towards higher bombarding energies.

These features of the excitation functions can be explained by extending the concept of the angular momentum limitations acting in the entrance channel to a wider class of fusion processes including not only the complete fusion reactions but also the incomplete fusion reac-tions. An idea of such a model is presented in fig. 3. For simplicity considered are only four main reaction modes which are assumed to com-pete mutually in all close contact collisions ($R_{min} \leq C_1+C_2$, where C_1 and C_2 are the half-density radii of the colliding nuclei). Distant collisions ($R_{min} > C_1+C_2$), probably dominated by inelastic scattering, are not considered in this very simple model. We neglect also other incomplete fusion channels, such as ($^{12}C,n$) and ($^{12}C,p$) with emission of fast nucleons.

Basic assumptions of the model are:
(i) Every close-contact collision leads to the reaction which is fa-vored by the strongest "driving force", provided this most favored re-action is not forbidden by the entrance-channel angular momentum limit-ations.
(ii) For l-values above the limiting angular momentum for the most fa-vored reaction the next reaction channel (characterized by the next strongest driving force) dominates.
(iii) The limiting angular momenta for successive incomplete fusion

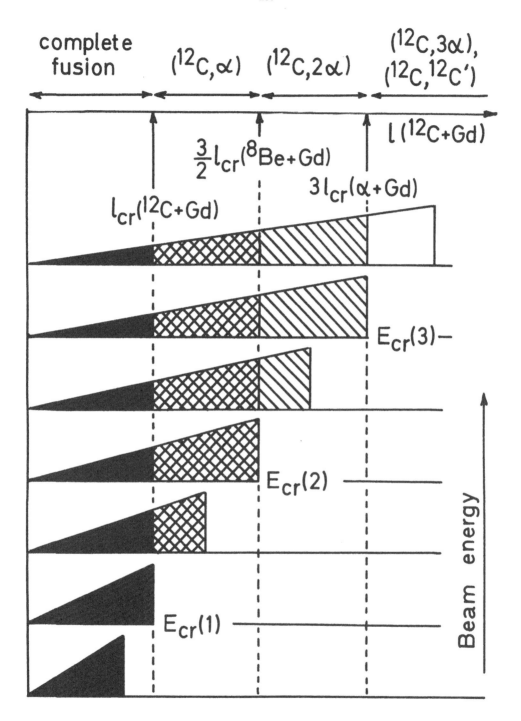

Fig. 3. Figurative explanation of the model discussed in the text.

channels are the critical angular momenta for the target + captured fragment system times the projectile mass to the captured fragment mass ratio. (It is assumed here that the fragments of the projectile share the angular momentum proportionally to their masses.)

With the restraint to the four main reaction modes in the ^{12}C + ^{160}Gd collisions, the sequence of the most favored reactions is as shown in fig. 3: complete fusion, $(^{12}C,\alpha)$, $(^{12}C,2\alpha)$ and $(^{12}C,3\alpha)$. (Later on I will return to this question since the assumption (i) is essential in the proposed model.) The range of the lowest 1-values,

$$0 < 1 < 1_{cr}(^{12}C+Gd) \tag{3}$$

is dominated by the complete fusion reactions. According to the assumptions (i)-(iii), the next 1-bin,

$$1_{cr}(^{12}C+Gd) < 1 < \frac{3}{2} 1_{cr}(^8Be+Gd) \tag{4}$$

is dominated by the $(^{12}C,\alpha)$ reaction. Then, the bin:

$$\frac{3}{2} 1_{cr}(^8Be+Gd) < 1 < 31_{cr}(\alpha+Gd) \tag{5}$$

corresponds to the $(^{12}C,2\alpha)$ reaction. The highest 1-values (accessible only at sufficiently high bombarding energies),

$$1 > 31_{cr}(\alpha+Gd) \tag{6}$$

are open for all "no capture" processes such as deep-inelastic scattering (if ^{12}C remains in the bound states) or the $(^{12}C,3\alpha)$ reaction which is the dominating mode of the breakup of ^{12}C.

As it is seen from fig. 3, the model implies different energy thresholds ($E_{cr}(1)$, $E_{cr}(2)$ and $E_{cr}(3)$ in fig. 3) for different reactions. Such a simple sharp cut-off model gives the energy dependence of the cross sections as shown in the lower part of fig. 2. The calculated cross sections are 2.5 times larger than the measured ones. Certainly, a considerable part of the geometrical cross section within the 1-windows (4)-(6) must be attributed to other reactions, first of all, to such important incomplete fusion channels as $(^{12}C,n)$ and $(^{12}C,p)$ which were beyond the scope of our experimental set-up.

To fit the shapes of the excitation functions exactly one needs to assume a more realistic smooth distribution of the 1-windows (3)-(6). Using the Fermi shape distributions instead of the rectangular windows (3)-(6) we obtained the best fit for the diffuseness parameter $\Delta_1 = 5\hbar$.

The critical angular momenta, $1_{cr}(^{12}C+Gd)$, $1_{cr}(^8Be+Gd)$ and $1_{cr}(\alpha+Gd)$, deduced from the fit agree well with those obtained directly from the complete fusion data and with theoretical predictions. The comparison is shown in table 1. In the last column given are the entrance channel critical angular momenta calculated with the liquid-drop model contact force of ref. 9:

$$(1_{cr} + \tfrac{1}{2})^2 = \frac{\mu(C_1+C_2)^3}{\hbar^2} \left[4\pi\gamma \frac{C_1 C_2}{C_1+C_2} - \frac{Z_1 Z_2 e^2}{(C_1+C_2)^2} \right] . \tag{7}$$

Here $\gamma \approx 0.95$ MeV/fm^2 is the surface tension coefficient, μ is the reduced mass, Z_1 and Z_2 are the atomic numbers, and C_1 and C_2 are the half-density radii. As can be seen from table 1, the estimates of the critical angular momentum based on the liquid-drop model seem to be valid even for as light a projectile as ^{4}He. This surprising fact can be useful for predicting some general trends in the incomplete fusion reactions.

Table 1

Critical angular momenta for ^{12}C+^{160}Gd, ^{8}Be+^{160}Gd, and α +^{160}Gd systems
(in units of $\hbar$)

System	This work	Other data	Eq. (7) [c]
^{12}C + ^{160}Gd	43 ±3	46 ±3 [a]	44.5
^{8}Be + ^{160}Gd	35 ±2	...	35.2
α + ^{160}Gd	21 ±3	25 ±2 [b]	22.4

a) Deduced from the complete-fusion cross section for ^{12}C + ^{158}Gd (Ref. 7).

b) Deduced from the energy dependence of the ^{160}Gd(α,xn) cross section (Ref. 8).

c) Calculated for $\gamma = 0.95$ MeV/fm^2, $C_{1,2} = 1.08$ fm $A_{1,2}^{1/3}$.

As I mentioned at the begining of my report, the model of incomplete fusion reactions based on the generalized concept of critical angular momentum was confirmed by the results of the γ-multiplicity measurements for the $(^{16}O,\alpha)$, $(^{16}O,2\alpha)$ and $(^{16}O,3\alpha)$ reactions in the $^{16}O + ^{154}$Sm system, reported recently by the St. Louis - Oak Ridge group [5]. Thus the existence of the sequence of l-windows for successive incomplete fusion channels has been shown on two independent ways: by the energy dependence of the cross sections [3] and by the γ-multiplicity measurements [5].

Let me now return to the main assumption of the model, i.e., to the assumption (i). We assume that the most favored reaction is characterized by the strongest driving force in the contact configuration. As a measure of the driving force we take the Q-value threshold calculated for a certain relative distance R at which the combined nuclear system is supposed to separate:

$$Q(R) \approx Q_{gg}(r=\infty) + \Delta E_C(R), \tag{8}$$

where Q_{gg} is the ground-state Q-value (for $r=\infty$, as calculated from the ground-state masses), and $\Delta E_C(R) \approx (z_1^i z_2^i - z_1^f z_2^f)/R$ is the change of the Coulomb interaction energy due to the transfer of charge.

Our formulation of the assumption (i), and particularly the choice of the Q-value threshold (8) as the quantity that scales probabilities of different binary reactions seems to be quite natural in the light of the well known dependence of the cross sections on Q_{gg} [10]. It is worth to emphasize that the Q_{gg}-dependence was shown to be valid <u>exclusively</u> for reactions characterized by a large transfer of mass from the projectile to the target, i.e., for ejectiles much lighter than the projectile. But in our terminology this class of binary reactions for which the Q_{gg}-dependence was proven to work is called "incomplete fusion"!

Let us try to continue this line of arguing. Suppose the colliding nuclei had reached the close contact, $R_{min} \leq C_1+C_2$, a kind of a neck between them developed, and at this stage the probabilities for complete fusion and different incomplete fusion processes are proportional to $\exp(Q_{gg}(R)/T)$, as it follows the concept of partial statistical equilibrium [11]. Combining this assumption with the entrance-channel angular momentum limitations (different for different reactions) we can estimate absolute cross sections for all the reactions marked here by the index i:

$$\sigma(i) = \pi \lambda^2 \cdot \sum_{l=0}^{\min[l_{cr}(i), l_{gr}]} (2l+1) \cdot K_l \cdot \exp(Q_{gg}^i(R)/T) . \tag{9}$$

The l-dependent factor K_l is determined by the requirement that all the reaction channels are included, and the total reaction probability is equal to 1:

$$\sum_i K_l \cdot \exp(Q_{gg}^i(R)/T) = 1 . \tag{10}$$

The grazing angular momentum l_{gr} (in eq.(9)) is matched to the close contact distance $R_{min} = C_1+C_2$, and the angular momentum limits $l_{cr}(i)$ are as postulated before (assumption (iii)):

$$l_{cr}(i) = \frac{A(\text{projectile})}{A(\text{captured fragment})} \cdot l_{cr}(\text{captured fragment} + \text{target}). \quad (11)$$

One sees immediately that in case if only few reaction channels are important, and moreover, they differ significantly in the $Q_{gg}(R)$ values, expressions (9)-(11) lead to a simple sequence of the l-windows as discussed before and illustrated in fig. 3.

The Q-value thresholds (8) for reactions in the $^{12}C + ^{160}Gd$ system are given in table 2. According to this list, the complete fusion re-action is strongly favored ($Q_{gg}(R) \approx +37.4$ MeV), and none of other re-action channels can be competitive as long as the critical angular momentum for complete fusion is reached. (The difference in $Q_{gg}(R)$ of 10 MeV makes the reaction probabilities, $\exp(Q_{gg}(R)/T)$, different by a factor of about 150, assuming a typical value of the effective tem-perature $T \approx 2$ MeV.) Above the critical angular momentum for complete fusion there remain only three competitive reaction channels character-ized by sufficiently large $Q_{gg}(R)$ values: $(^{12}C,n)$, $(^{12}C,p)$ and $(^{12}C,\alpha)$, i.e., the incomplete fusion reactions with emission of fast neutrons, protons and α-particles, respectively. First two reactions were not studied in our experiments. As it is seen from the last column in table 2, these two reactions are limited to very narrow l-windows ($\Delta l \approx 0.5\hbar$ and $\Delta l \approx 3\hbar$, respectively). Consequently, the whole range of l-values, $45\hbar < l < 53\hbar$ is open for the $(^{12}C,\alpha)$ reaction, practi-cally without competition of other reactions. (It is worth to note that the hypothetical incomplete fusion reactions with emission of fast neutrons might give very interesting information if their localization in the l-space and the width of the l-window could be determined. The very narrow l-window, as predicted by the model discussed here, would help to distinguish this incomplete fusion reaction from "conventional" preequilibrium emission of fast neutrons.)

Above the limiting angular momentum for the $(^{12}C,\alpha)$ reaction, i.e. for $l > 53\hbar$, only Li, He and lighter fragments can be captured. The list of the $Q_{gg}(R)$ values shows that the $(^{12}C,^{8}Be \rightarrow 2\alpha)$ reaction is the most favored. The probability factor for this reaction is 6 - 20 times larger than for the $(^{12}C,^{6}Li)$, $(^{12}C,^{7}Li)$ and $(^{12}C,^{9}Be)$ reactions which are the only competitors in this range of l-values. Therefore over the whole range, $53\hbar < l < 67\hbar$, the $(^{12}C,2\alpha)$ reaction is predicted to be the strongest reaction channel.

Table 2

Q-value thresholds and limiting angular momenta for different binary reactions for the $^{12}C + ^{160}Gd$ system. The $Q_{gg}(R)$ values are calculated from eq.(8) with R = 12 fm.

ejectile (e)	captured fragment (f)	Q_{gg} (MeV)	$Q_{gg}(R)$ (MeV)	$l_{cr}(e+f)$ ($\hbar$)	$\frac{12}{A_f} \cdot l_{cr}(e+f)$ ($\hbar$)
...	^{12}C	-8.7	+37.4	44.5	44.5
n	^{11}C	-16.7	+29.4	41.3	45.0
p	^{11}B	-16.0	+21.8	43.5	47.5
d	^{10}B	-21.3	+16.5	40.1	48.1
t	^{9}B	-21.6	+16.2	36.6	48.9
^{3}He	^{9}Be	-22.0	+7.8	38.7	51.6
α	^{8}Be	-7.4	+22.4	35.2	52.8
^6Li	^6Li	-19.0	+3.0	29.3	58.5
^7Li	^5Li	-18.0	+4.0	25.2	60.5
$^8Be\ (2\alpha)$	^4He	-6.9	+7.5	22.4	67.3
^9Be	^3He	-12.9	+1.5	17.8	71.2
^{10}B	d	-14.3	-7.2	13.9	83.6
^{11}B	p	-9.2	-2.1	8.0	96.6
^{11}C	n	-13.1	-13.1	8.9	107.0
$^{12}C^* \ (3\alpha)$	...	-7.3	-7.3	...	...

The l_{cr} values are calculated from eq.(7) with $C_{1,2} = 1.08$ fm$\cdot A_{1,2}^{1/3}$, and $\gamma = 0.95$ MeV/fm^2.

I would not like to discuss reactions involving the capture of very light fragments (e.g. d,p,n) in terms of the incomplete fusion mechanism because these reactions may occur in more distant collisions, $R_{min} > C_1 + C_2$, for which the mechanism of direct transfer is more appropriate. With this reservation nearly all binary processes in asymmetric systems can be considered as incomplete fusion reactions. As it is seen from table 2, the sequence of the most probable reactions be-

gins from the capture of the heaviest fragments and ends on the capture
of light fragments. This sequence of reaction channels is correlated
with the sequence of increasing angular momenta which restrict successive reactions. The model implies a certain limitation for the binary
multi-nucleon transfer reactions. Specifically, it follows from eqs.
(7) and (11) that at bombarding energies of about 15 MeV/A (for most
of the colliding systems) none of the projectile fragments (except for
single nucleons or at most ^{3}He) can be captured in peripheral collisions. Consequently, above 15 MeV/A the cross sections for all possible binary multi-nucleon transfer reactions must decrease with increasing bombarding energy, thus making room for inelastic (or deep-inelastic) scattering, projectile breakup and multibody fragmentation processes. This effect is clearly seen in the energy dependence of the
^{160}Gd(^{12}C,3α) cross section (fig. 2). Another argument in support of
this consequence of the present model comes from results of experiments
carried out by Gelbke et al.[12] who observed decided predominance of
the breakup reactions at 20 MeV/A.

Returning now to the question expressed in the title of my report
I would like to emphasize once more the unquestionable fact that the
cross sections of the reactions that we call "incomplete fusion"[1,3,5]
or "massive transfer"[2,4] are correlated with the Q_{gg} values. Within
our present knowledge the only explanation of such correlation is that
these reactions proceed via the stage of a composite system characterized by equilibration of at least those degrees of freedom which are
essential for ensuring statistical probabilities of certain final configurations. With this picture in mind we should consider all the reactions (including the complete fusion) on common grounds. The mechanism of the incomplete fusion reactions can be interpreted then as a
natural extension of the fusion mechanism to the region of high angular
momenta. These are reasons why I prefer the name "incomplete fusion"
rather than "massive transfer".

References

1) T. Inamura, M. Ishihara, T. Fukuda, T. Shimoda and H. Hiruta,
   Phys. Lett. 68B (1977) 51

2) D.R. Zolnowski, H. Yamada, S.E. Cala, A.C. Kahler and T.T. Sugihara,
   Phys. Rev. Lett. 41 (1978) 92

3) K. Siwek-Wilczyńska, E.H. du Marchie van Voorthuysen, J. van Popta,
   R.H. Siemssen and J. Wilczyński, Phys. Rev. Lett. 42 (1979) 1599

4) H. Yamada, D.R. Zolnowski, S.E. Cala, A.C. Kahler, J. Pierce and
   T.T. Sugihara, Phys. Rev. Lett. 43 (1979) 605

5) K.A. Geoffroy, D.G. Sarantites, M.L. Halbert, D.C. Hensley,
   R.A. Dayras and J.H. Barker, Phys. Rev. Lett. (in press);
   M.L. Halbert, reported at this Symposium
6) J. Wilczyński, R. Kamermans, J. van Popta, R.H. Siemssen,
   K. Siwek-Wilczyńska and S.Y. van der Werf, Phys. Lett. (in press)
7) A.M. Zebelman and J.M. Miller, Phys. Rev. Lett. <u>30</u> (1973) 27
8) W.J. Ockels, Ph.D. thesis, University of Groningen, 1978;
   D. Chmielewska, Z. Sujkowski, J.F.W. Jansen, W.J. Ockels and
   M.J.A. de Voigt, to be published
9) J. Wilczyński, Nucl. Phys. <u>A216</u> (1973) 386
10) A.G. Artukh, V.V. Avdeichikov, G.F. Gridnev, V.L. Mikheev,
    V.V. Volkov and J. Wilczyński, Nucl. Phys. <u>A168</u> (1971) 321
11) J.P. Bondorf, F. Dickmann, D.H.E. Gross and P.J. Siemens,
    Journal de Phys. <u>32</u> (1971) C6-145
12) C.K. Gelbke, C. Olmer, M. Buenerd, D.L. Hendrie, J. Mahoney,
    M.C. Mermaz and D.K. Scott, Phys. Rep. <u>42C</u> (1978) 311

ANGULAR MOMENTUM TRANSFER IN INCOMPLETE FUSION REACTIONS[*][x]

K.A. Geoffroy,[1] D.G. Sarantites,[1] M.L. Halbert,[2] D.C. Hensley,[2] R. A. Dayras,[2] and J.H. Barker[3]

Incomplete fusion[4,5] is a peripheral process in which part of the projectile fuses with the target while the rest proceeds forward with little disturbance. Indirect evidence for its occurrence has been inferred from excitation functions[4] and from sidefeeding patterns.[5,6] We have obtained more direct evidence from γ-ray multiplicity data for specific exit channels in coincidence with charged particles from 10° to 160° from reactions of 153-MeV ^{16}O on ^{154}Sm. For the energetic, very forward-peaked ^{4}He and ^{12}C accompanying capture of "^{4}He", "^{8}Be", or "^{12}C" by the target, the transferred angular momentum increases from 20 to 40 linearly with captured mass. The linear increase supports the picture of projectile fragmentation during incomplete fusion. Assuming that the entrance-channel angular momentum ℓ is divided between the fragments according to their masses, we deduce $\langle \ell \rangle$ = 52, 60, and 74 for capture of ^{12}C, ^{8}Be, and ^{4}He, respectively. The observed multiplicity widths, transformed to ℓ-space, are on the order of 10-20 h FWHM. The observations on $\langle \ell \rangle$ demonstrate the peripheral nature of the process and are quantitatively consistent with a model[4] based on successive critical angular momenta for various degrees of incomplete fusion.

---

[*]Supported in part by the U.S.D.O.E. Division of Basic Energy Sciences.
[1]Washington University, St. Louis, MO 63130.
[2]Oak Ridge National Laboratory, Oak Ridge, TN 37830 (operated by Union Carbide Corp. for U.S. Department of Energy).
[3]St. Louis University, St. Louis, MO 63103.
[4]K. Siwek-Wilczynska et al., Phys. Rev. Letters 42, 1599 (1979).
[5]T. Inamura et al., Phys. Lett. 68B, 51 (1977).
[6]D.R. Zolnowski et al., Phys. Rev. Lett. 41, 92 (1978).

[x]See also Phys. Rev. Lett. 43 (1979/303)

THE CONTRIBUTION OF PERIPHERAL FRAGMENTATION PROCESSES TO CONTINUOUS PARTICLE SPECTRA IN NUCLEUS-NUCLEUS COLLISIONS[+]

G. Baur*, F. Rösel and D. Trautmann
Institut für Theoretische Physik der Universität Basel,
CH-4056 Basel, Switzerland

R. Shyam**
Institut für Kernphysik, Kernforschungsanlage Jülich,
D-5170 Jülich, W.Germany

1. Introduction

A typical spectrum of particles in nuclear reactions shows different
reaction mechanisms : at the high energy end of the spectrum there are
isolated peaks which are due to fast one-step transitions to discrete
states of the residual nucleus. The low energy part of the spectrum
is usually described by more complicated deep-inelastic processes, where
more collisions in the target are necessary to loose energy. For even
lower energies, the spectrum is dominated by the evaporation from the
compound nucleus.

The topic of this talk is another mechanism, the break-up (fragmentation) process. Over the last years this break-up process has been studied rather extensively both theoretically and experimentally. We present here the direct reaction theory of the break-up process. We distinguish two modes, the elastic[1] and inelastic[2-4] break-up, depending on whether the target remains in the ground state or not during
the collision. This formulation of the inclusive break-up, which consists of the elastic and inelastic modes, is physically closely related to the work of Lipperheide and Möhring[5].

Although there is a great activity in this field at the present time,
it should be mentioned that the break-up process has quite a long hi-

---

[+]Invited talk presented by G.Baur at the Symposium on Deep-Inelastic
and Fusion Reactions with Heavy Ions, Berlin,October 23-25, 1979.

*Permanent address: Institut für Kernphysik,Kernforschungsanlage Jülich,
D-5170 Jülich, W.Germany.

**Alexander von Humboldt Fellow.

story in nuclear physics. The break-up of the deuteron in the Coulomb field of the nucleus was first considered theoretically by Oppenheimer[6]. A later and much more accurate investigation was given by Landau and Lifshitz[7]. The early experiments of Helmholtz, McMillan and Sewell[8] in 1947 of 200 MeV deuteron break-up could be well accounted for by the Serber model[9]. Our theoretical framework incorporates the theories of Landau and Lifshitz[7] and Serber[9] as limiting cases.

Nowadays the "abrasion process" also plays a dominant role in high energy heavy ion physics[10]. Deuteron-nucleus collisions in the relativistic region are studied theoretically by Fäldt and Pilkuhn[11] and experimentally by Ashgirey et al.[12].

It is very important to check carefully the reaction mechanism of these continuum spectra. As we shall see in this talk, there may be quite a large fraction of direct processes hidden in these spectra, depending, of course, on the angle and energy of the emitted particles. It is, therefore, important that one can clearly separate the fast one-step processes from the more complicated multistep processes (deep inelastic collisions). We feel that we now understand the direct break-up mechanism well enough to make such a separation possible. In our calculations we prove now directly - as one expects intuitively - the strong localization of the break-up process in the surface region, which is in accordance with the geometrical model of Serber[9]. This suggests a rather simple dependence of the reaction mechanism on the incoming ℓ-value or impact parameter.

After presentation of a simple picture of the break-up process, we give in chapter 2 a review of the theory of the elastic and inelastic break-up modes along with its most important characteristics. In chapter 3 we give a comparison of our theoretical results with experimental data. We study mainly light-ion induced reactions, which serve as a theoretical "playground" for the computationally more involved case of heavy ion reactions. (After all, the α-particle shows all the properties of a heavy ion). As an example for heavy ions, we consider the ^{8}Be-continuum spectra of ^{9}Be induced reactions at subcoulomb energies. In chapter 4 we study the impact parameter dependence for the break-up process, out of which a very simple picture of the gross properties of the break-up process emerges. Our conclusions are given in chapter 5.

2. Theoretical Frame-Work

2.1 Simple picture. Qualitative considerations.

In a simple picture we can already see qualitatively some important features of the process. In fig. 1 particle a (a=b+x) impinges on a target nucleus A with velocity $\vec{v}_a^{\,o}$. In certain cases the constituent b will miss the target nucleus while x interacts (elastically or inelastically) with the target nuclues A. Disregarding binding effects,

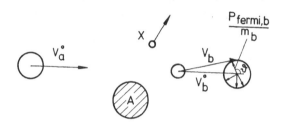

Fig.1 : Simple picture of the break-up process, the spectator model.

particle b will move on essentially undisturbed with the velocity $\vec{v}_b^{\,o} = \vec{v}_a^{\,o}$. This velocity is smeared out by the Fermi motion $\vec{p}_{fermi,b}/m_b$ of particle b inside the projectile a, therefore we expect for the velocity of particle b in the final state $\vec{v}_b = \vec{v}_b^{\,o} + \vec{p}_{fermi,b}/m_b$. The energy of the outgoing particle b will be given by

$$E_b = \frac{1}{2} m_b v_b^2 = \frac{1}{2} m_b \, (\vec{v}_b^{\,o} + \frac{\vec{p}_{fermi,b}}{m_b})^2 \tag{1a}$$

For $v_b^o \gg \dfrac{p_{fermi,b}}{m_b}$ we obtain

$$E_b \cong \frac{m_b}{m_a} E_a + v_b^o \, p_{fermi,b} \, \cos\theta \tag{1b}$$

where θ is the angle between $\vec{v}_b^{\,o}$ and $\vec{p}_{fermi,b}$. We therefore expect a bump in the spectrum at forward angles with a peak energy $E_b^{peak} = m_b/m_a \, E_a$, where E_a is the energy of the projectile. The width of the bump will be given by $v_b^o \, p_{fermi,b} = p_{fermi,b} \sqrt{2m_a E_a}$. It therefore measures directly the Fermi motion of b in the projectile a. It will be seen below how these very simple features emerge from our theoretical approach.

In addition to the "spectator mechanism" considered here, there is also another mechanism which can contribute to the peak at $E_b = \dfrac{m_b}{m_a} E_a$.

The projectile is inelastically excited in the field of the target nucleus to some continuum (resonant) state which decays subsequently into b + x. Such a possibility has recently been advocated mainly by Udagawa, Tamura and coworkers. We feel that this mechanism, although important in special situations, is not the dominant process in the inclusive spectra considered here. Furthermore, it seems difficult to incorporate the inelastic break-up process, which we find to dominate, into this approach. The work of Udagawa et al.[13] is based on a formulation of deuteron break-up by Rybicki and Austern[14]; those authors considered deuteron break-up at E_d = 12 MeV (E/A = 6 MeV), they find complete disagreement with the experimental coincidence data for the (d,pn) reaction.

2.2. Outline of the basic theory.

The theory of the coincidence cross section for the process A(a,bx)A is reviewed in ref. 1. Recently, it has been shown that this theory which is based on the post form of the DWBA can also explain the recently measured ^{58}Ni(α,tp)^{58}Ni coincidence cross section at E_α = 172.5 MeV, details can be found in ref. 15.

Now we want to calculate the inclusive (a,b) cross section, which consists of the "elastic" and "inelastic" modes. The contribution of the elastic mode can simply be obtained by integration of the expression[1] for the A(a,bx)A coincidence cross section over the angle of the unobserved neutron. This can be done analytically by virtue of the orthogonality of the spherical harmonics. The sum over the angular momenta of the neutron then becomes incoherent in the formula for the double differential cross section. For the inclusive type of spectra we also have to consider all kinds of inelastic processes of the type A+a→b+c, where c denotes an open channel of the system B = A+x at the energy in question. In principle it may be possible to calculate all these transitions individually and to sum them up. But this would be very difficult and impracticable if there are many open channels c. There is an approximate procedure which allows us to make use of the unitarity of the S-matrix (for the system B = A+x), and to calculate the inelastic part of the inclusive cross section essentially by means of quantities which were already needed to calculate the elastic break-up. This is described in detail in refs. 2, 3 and 4.

2.3 Transition from bound- to unbound-state stripping.

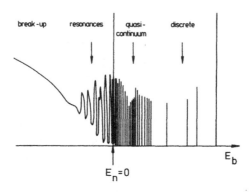

<u>Fig.2</u> : Schematic view of a spectrum of the (a,b) reaction at a given angle. The threshold for emission of a neutron is denoted by an arrow.

In fig. 2 a schematic view of the spectrum of an (a,b) stripping reaction is shown. (For simplicity of presentation we assume that the transferred particle x is a neutron, i.e. we put a=b+n in this section). The discrete well separated states at the high energy end are followed by a region where the level density becomes higher and higher and may not be resolved experimentally any more. This region was called continuum by Cohen et al.[16), yet the levels are still discrete bound states. Above the neutron emission threshold there will be a population of isolated resonances which will go over into an even more structure-less continuum at lower energies E_b.

The continuous transition from single particle bound states to single particle resonances was established in ref.17. There the following relation between the width $\Gamma_{s.p.}$ and the asymptotic normalization $N_{s.p.}$ of a single particle (Gamow) state was given :

$$\Gamma_{s.p.} = \frac{\hbar^2}{m_n q_n} N^2_{s.p.} \tag{2}$$

Let us now deal with the situation where the single particle strength is spread out over very many states. Then we can define an energy averaged double differential cross section for stripping to the bound states[16) ("quasi-continuum") by (see also ref. 17)

$$\frac{d^2\sigma_\ell}{d\Omega_b dE_b} = \frac{1}{2} \frac{m_a m_b}{(2\pi\hbar^2)^2} \frac{q_b}{q_a} D_o^2 \frac{S \cdot N^2_{s.p.}}{D} \sum_m |T^+_{\ell m}|^2 \quad , \tag{3}$$

where we have restricted ourselves to one transferred ℓ-value; $\frac{1}{D}$ is the number of energy levels per energy interval and S denotes the average spectroscopic factor (for states with a given ℓ). The matrix element $T^+_{\ell m}$ is defined by

$$T^+_{\ell m} = \int d^3r \ \chi_b^{(-)}(\vec{r})^* h_\ell Y_{\ell m}(\hat{r}) \cdot \chi_a^{(+)}(\vec{r}) \ . \tag{4}$$

Let us now establish the connection of eq.(3) to the unbound region. Because of the phase space factor, the elastic break-up (see ref.1) tends to zero for $E_n \to 0$, therefore around threshold we will only have to consider the inelastic break-up. It tends to a limit different from zero in the presence of absorption in the neutron channel[2,3,4]. We introduce now the relation between the energy averaged total cross section $<\sigma_\ell>$ and the strength function Γ/D (for the given ℓ) in the threshold region :

$$<\sigma_\ell> \ \stackrel{\sim}{=} \ \sigma_\ell^{reaction} = \frac{2\pi^2}{q_n^2} (2\ell+1) \frac{\Gamma}{D} \tag{5}$$

This allows us to rewrite the inelastic break-up cross section[2-4] in the following way :

$$\frac{d^2\sigma_\ell}{d\Omega_b dE_b} = \frac{1}{2} \frac{m_a m_b}{(2\pi\hbar^2)^2} \frac{q_b}{q_a} D_o^2 \frac{q_n m_n \Gamma}{D\hbar^2} \sum_m |T^+_{\ell m}|^2 \tag{6}$$

With the help of eq. (2) we immediately establish the continuous transition. Hereby we have introduced a natural definition[18] of a spectroscopic factor for resonant states : $\Gamma = S \cdot \Gamma_{s.p.}$. It is gratifying to see how these apparently unrelated formulations of stripping to bound and unbound states have indeed a common origin. This is expected, because usually nothing dramatic happens in the experimental spectra around neutron threshold.

2.4 Plane wave limit, connection to the momentum distribution of particle b inside projectile a.

It is interesting, although dangerous for quantitative calculations[19], to insert plane waves in the break-up matrix element for the particles a and b. With such an approximation we can express the T-matrix (we specialize to the $(\alpha, ^3He)$ reaction) in terms of the off-shell n-A T-matrix and the momentum distribution ϕ_n of a neutron in the α-particle (see e.g. ref. 20).

$$T = - \frac{\hbar^2}{2m_n} (2\pi)^{3/2} t(\vec{q}_\alpha - \vec{q}_3, \vec{q}_n) \phi_n (\frac{3}{4} \vec{q}_\alpha - \vec{q}_3) \tag{7}$$

This formula shows the qualitative features given in section 2.1 : to the extent that we can neglect the energy variation of the off-shell matrix element t and of the phase space factor, the cross section is determined entirely by ϕ_n (Fermi motion). We expect a peak in the spectrum at $\vec{q}_3 = \frac{3}{4} \vec{q}_\alpha$, i.e. $\vec{v}_3 = \vec{v}_\alpha$; and the width of this peak is given by the momentum distribution of the neutron in the α-particle. This is in accord with the prediction of the Serber model[9].

3. Comparison of the Theory with Experiment

3.1 (d,p) and (α,^{3}He) continuum spectra

Quite recently, proton spectra in E_d = 25.5 MeV deuteron induced reactions have been measured for many target nuclei and analyzed with the present theory[2]. In these proton spectra prominent bumps at around half of the incident deuteron energy were found, especially at forward angles. In the bump region the agreement is quite remarkable. At lower proton energies E_p, the experimental points are higher because of additional contributions from preequilibrium and evaporation processes. In refs. 2 and 15 experimental and theoretical angle integrated break-up cross sections are compared for ^{27}Al, ^{62}Ni, ^{119}Sn and ^{181}Ta. The break-up of 80 MeV deuterons incident on a wide range of target nuclei is studied experimentally and theoretically in ref. 21. Again a pronounced peak in the proton spectra in forward direction around $E_p \cong 40$MeV is found, which is well explained by our theory.

The (α,^{3}He) break-up process was discovered in Jülich and Maryland and is described in refs. 3 and 22.

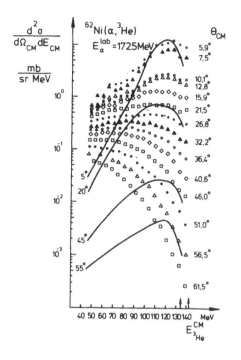

Fig.3 : Double differential cross section for the ^{62}Ni$(\alpha,{}^{3}$He$)$ reaction at E_{α} = 172.5 MeV. Full lines indicate theoretical calculations.

In fig.3, taken from ref. 3, the experimental results for the experimental double differential cross section for the ^{62}Ni$(\alpha,{}^{3}$He$)$ reaction are compared with our calculations. In the energy region of the broad peak our calculations are in reasonable agreement with experiment. The strong decrease in the angular distribution over three orders of magnitude is well reproduced. In the deep inelastic region $(E_{3He}<90MeV)$ our calculations underestimate the $(\alpha,{}^{3}$He$)$ cross section by at least one order of magnitude. This is expected since the calculations did not include multistep processes, which become more important for lower ^{3}He energies. For these processes, a multistep direct analysis[23] ,using statistical concepts should be tried; a complementary and very interesting attempt to explain these spectra was put forward by the Münster group[24] , who use the fireball concept of partial thermal equilibrium together with the coalescence model to describe the emission of complex particles. Because only ^{3}He is observed, various partial waves of the neutron add up incoherently. In fig.4 (ref.3) the contributions of different ℓ_{n}-values to the break-up peak at θ_{3He} = 5° are shown separately. It is seen that higher ℓ_{n}-values dominate the peak with de-

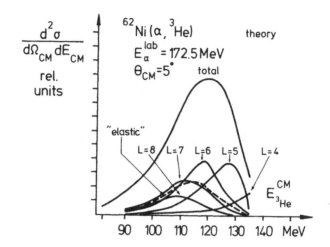

Fig.4 : Calculated contributions of various neutron partial waves to the total cross section for the $(\alpha,^3He)$ break-up at $E_\alpha=172.5$ MeV and $\theta_{^3He}= 5^\circ$. The elastic break-up contribution is indicated by the dashed line.

creasing ^{3}He energy. Thus the reaction selectively excites favoured ℓ_n-values. This makes the $(\alpha,^3He)$ reaction a useful tool for extending studies of the single neutron strength distribution in nuclei made so far by means of the (d,p) reaction[20] towards higher ℓ_n-values. We note also from fig. 4 that the elastic break-up accounts also for about 25% of the total inclusive cross section, a prediction which can be considered to be verified by now[15].

3.2 The break-up of a heavy ion : the $(^9Be,^8Be)$ reaction.

Our formalism can also be applied to heavy ion break-up reactions. The theory of $(^9Be,^8Be)$ continuum spectra is dealt with in ref. 4; it is very important to consider finite range and recoil effects, as can be seen in fig. 4, taken from ref. 4. Further details can be found there.

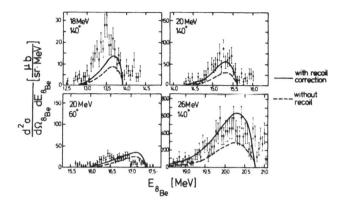

Fig.5 : Comparison of the theoretical inclusive spectra with experiment for the reaction ^{197}Au(^{9}Be,^{8}Be). The theoretical curves are the sum of the elastic and inelastic break-up modes. The importance of the recoil effect, which is taken into account in the Buttle and Goldfarb[25] approximation, is shown.

4. The Dependence of the Break-Up Cross Section on the Impact Parameter and Gross Properties

It is instructive to represent the total cross section $\sigma_{total}^{break-up}$ as a sum over the partial waves ℓ_a of the incident projectile with wave number q_a :

$$\sigma_{total}^{break-up} = \frac{\pi}{q_a^2} \sum_{\ell_a} (2\ell_a+1) \, T_{\ell_a}^{break-up} \quad . \tag{8}$$

We have defined a probability for break-up $T_{\ell_a}^{break-up}$ in analogy to the usual transmission coefficient T_{ℓ_a} which determinates the total reaction cross section

$$\sigma^{reaction} = \frac{\pi}{q_a^2} \sum_{\ell_a} (2\ell_a+1) \, T_{\ell_a} \quad . \tag{9}$$

With an optical model potential we may e.g. calculate the transmission coefficients. With the usual smooth (or sharp) cut-off models for T_{ℓ_a} we can see that $\sigma^{reaction}$ is proportional to R^2, i.e. to $A^{2/3}$, where A is the mass number of the target nucleus. Direct reactions, which are centered around $\ell_a \simeq \ell_{grazing}$, are expected to be proportional to R, i.e. to $A^{1/3}$. For the transmission coefficient T_{ℓ_a} we obtain a smooth cut-off behaviour, whereas $T_{\ell_a}^{break-up}$ shows a distinct peak around a grazing angular momentum[2,26]. This is a characteristic feature of

a peripheral process. In this region the break-up process is the dominant reaction channel for grazing angular momenta.

In our calculation for the α-particle energy the shape of the break-up
probability curve remains unchanged, only its magnitude increases,
with some kind of "saturation"[26]. This shows the geometrical nature
of the break-up process. Thus it is established that the break-up is
an important absorption effect in the surface region for α-induced reactions (at least for α-particle energies much larger than the break-
up threshold). Therefore, all kinds of theories which disregard the
possibility of the break-up of the α-particle must lead to quite appreciable deviations of the theoretical transmission coefficients as
compared to the experimental ones. Since the grazing angular momenta
determine the angular distributions, the possibility of break-up has
to be included in all optical model theories for composite particles.
This holds especially for α-nucleus optical potentials calculated in
ref.[27] where absorption is assumed to be only due to the excitation
of target states treating the α-particles as an elementary particle.

Finally, we try to extract some "gross properties" of the break-up reaction. As suggested by our numerical results, a simple parametrization of the break-up process seems possible which shows a remarkable
similarity to the Serber picture. We introduce the parametrization

$$T_\ell^{b-up}(a,b) = \beta(E_a,E_{bind})e^{-\frac{(\ell-\ell_o)^2}{(\Delta\ell)^2}} = \beta(E_a,E_{bind})e^{-\frac{(b-b_o)^2}{(\Delta R)^2}} \qquad (10)$$

With $b = \ell/q_a$, $b_o = \ell_o/q_a$ and $\Delta R = \Delta\ell/q_a \cdot E_a$ and E_{bind} correspond to the
incident and binding energies of the projectile, respectively. The
break-up probability has a peak at a partial wave ℓ_o with a width $\Delta\ell$.
The impact parameters b_o and ΔR are almost independent of the incident
energy. The factor β describes the strength of the break-up process,
which is expected to show a saturation for sufficiently high incident
energies and vanish for small incident energies. We relate b_o to the
size of the target and projectile by $b_o = r_o(A^{1/3}+a^{1/3})$. The numbers
are remarkably independent of A and E_a (ref.26). This supports the
Serber picture. With the simple parametrized form of $T_\ell^{b-up}(a,b)$ we
can directly calculate the total (a,b) break-up probability

$$\sigma_{total}^{b-up}(a,b) = 2\pi\beta\int_0^\infty db\, b\, e^{-\frac{(b-b_o)^2}{(\Delta R)^2}} \simeq 2\pi^{3/2}\beta\, b_o\, \Delta R \qquad (11)$$

This formula corresponds directly to the Serber formula[9]. It also shows the factorization property found in heavy ion fragmentation reactions[10]. The total cross section factorizes into a part ΔR, which depends only on the projectile and fragment, and a target factor $b_0 = r_0(A^{1/3}+a^{1/3})$ with $r_0 \simeq 1.2$ fm, which is directly related to the size of the colliding systems.

5. Conclusion

We have seen that break-up (fragmentation, abrasion) is an important reaction mechanism for continuous spectra. It is most important for peripheral collisions. In those cases, where the cross section for break-up is a substantial part of the total reaction cross section its influence on other channels (e.g. the elastic one) cannot be neglected. Furthermore, the break-up may act as a doorway mechanism for more complex nuclear cascades. The inelastic break-up mode is responsible for a substantial part of the experimental break-up yield. The stripping reaction to unbound states is also an interesting tool to study hitherto unexplored high lying single particle properties, much in the same way as inelastic scattering probes the giant resonance and pick-up reactions the deep-lying hole structure of nuclei.

References

1) For a review see G.Baur and D.Trautmann, Phys.Rep.25C (76) 293 and further references contained therein.

2) J.Pampus,J.Bisplinghoff,J.Ernst.T.Mayer-Kuckuk,J.Rama Rao, G.Baur, F.Rösel and D.Trautmann, Nucl.Phys.A311 (78) 141.
G.Baur,F.Rösel,D.Trautmann,J.Pampus,J.Bisplinghoff,J.Ernst,T.Mayer-Kuckuk,J.Rama Rao, A.Budzanowski,R.Shyam, Proc.Workshop on Reaction Models for Continuous Spectra of Light Particles,Bad Honnef,Nov.15-17, 1978, p.27.

3) A.Budzanowski,G.Baur,C.Alderliesten,J.Bojowald,C.Mayer-Boericke, W.Oelert,P.Turek,F.Rösel and D.Trautmann, Phys.Rev.Lett.41 (78)635.

4) G.Baur,M.Pauli,F.Rösel and D.Trautmann, Nucl.Phys.A315 (79) 241.

5) R.Lipperheide and K.Möhring, Nucl.Phys.A211 (73) 125.

6) J.R.Oppenheimer, Phys.Rev.47 (35) 845.

7) L.Landau and E.Lifshitz, JETP 18 (48) (English translation in: Collected Papers of L.Landau).

8) A.C.Helmholtz,E.M.McMillan and D.C.Sewell, Phys.Rev.72 (47) 1003.

9) R.Serber, Phys.Rev.72 (47) 1008.

10) H.H.Heckmann,D.E.Greiner,P.J.Lindström and F.S.Bieser,Phys.Rev.Lett. 28 (72) 926.

11) G.Fäldt and H.Pilkuhn,Ann.Phys.58 (70) 454.

12) L.S.Ashgirey,M.A.Ignatenko,V.V.Ivanov,A.S.Kusnetsov,M.G.Mescherya-kov,S.V.Razin,G.D.Stoletov,I.K.Vzorov and V.N.Zhmyrov, Nucl.Phys. A305 (78) 404.

13) T.Udagawa, Tamura,T.Shimoda,H.Fröhlich,M.Ishihara and K.Nagatami, Phys.Rev.C, to be published.

14) F.Rybicki and N.Austern, Phys.Rev.C6 (72) 1525.

15) A.Budzanowski,G.Baur,R.Shyam,J.Bojowald,W.Oelert,G.Riepe,M.Rogge, P.Turek,F.Rösel and D.Trautmann, Z.Phys., in press.

16) K.C.Chan,B.L.Cohen,L.Shabason,J.R.Alzona and T.Congeno,Phys.Rev. C12 (75) 1844.

17) G.Baur and D.Trautmann, Z.Phys.267 (74) 103.

18) J.P.Schiffer, Nucl.Phys.46 (63) 246.

19) R.Shyam,G.Baur,F.Rösel and D.Trautmann, Phys.Rev.C19 (79) 370.

20) G.Baur, Z.Phys.A277 (76) 147.

21) U.Bechstedt,H.Machner,G.Baur,R.Shyam,C.Alderliesten,A.Djaloeis, P.Jahn,C.Mayer-Boericke,F.Rösel and D.Trautmann, to be published.

22) J.R.Wu,C.C.Chang and H.D.Holmgren, Phys.Rev.Lett.40 (78) 1013.

23) H.Feshbach in Proc.of the Intern.Conf.on Nucl.Reaction Mechanism, Varenna,Italy,13-17 June 1977, Cooperativa Libraria Universitaria Editrice Democratica, Milano, Italy 1977, p.1.

24) A.Frekers, G.Gaul, H.Zöhne,B.Ludewigt and R.Santo, Proc.Workshop on Reaction Models for Continuous Spectra of Light Particles, Bad Honnef, Nov.15-17,1978, p.119.

25) B.J.Buttle and L.J.B.Goldfarb, Nucl.Phys.A176 (71) 299.

26) G.Baur,R.Shyam,F.Rösel and D.Trautmann, to be published.

27) N.Vinh Mau.Proc.Workshop on Microscopic Optical Particles, Hamburg 1978, ed.H.V.von Geramb, Springer Verlag, Heidelberg, 1978.

FUSION REACTIONS: SUCCESSES AND LIMITATIONS
OF A ONE-DIMENSIONAL DESCRIPTION

R. Bass

Institut für Kernphysik,
University of Frankfurt am Main, Germany.

The basic ingredients of various "one-dimensional" classical fusion
models are reviewed. Effects of the conservative potential and of energy dissipation on the fusion cross section are considered, and a method
for the deduction of an empirical nucleus-nucleus potential from experimental fusion cross sections is discussed. Finally attention is drawn
to limitations of the simple one-dimensional models which become important at high energies and for heavy systems.

1. Introduction: "one-dimensional" fusion models

The term "one-dimensional" will be used here for any model which considers the relative vector between two interacting nuclei as the only significant degree of freedom, and neglects more complicated effects such
as shape changes or mass transfer. Relative angular momentum will be
included, however, and in this sense the models in question are actually three-dimensional. Clearly such a simple picture can only apply to
the fast, initial stages of a nucleus-nucleus collision. Its success depends therefore, in large measure, on the fact that in many cases of
practical interest a "point of no return" is passed on the ingoing part
of the trajectory, where the eventual transition into the fusion channel is decided.

In addition, I shall use a strictly classical description. Thus the
sharp cut-off approximation in angular momentum space is implied, and
the energy region near or below the fusion barrier is excluded.

Models with these characteristics have been applied with considerable
success during the last six years. They may be grouped into two basic
categories, namely

I) "friction models", where explicit assumptions are made on distance-
dependent friction forces[1,2], and

II) "critical distance models", where complete damping of the relative
motion at a well-defined critical distance is implied[3-6].

In the former models trajectories are calculated by numerical integration of the Schrödinger equation, and all trajectories which do not re-
emerge are presumed to fuse. In contrast, the latter models postulate

that the critical distance must be reached as a necessary condition
for fusion. By virtue of this somewhat schematic criterion they yield
analytical formulae for the fusion cross section (see eqs.(1),(2)),
which simplify matters considerably.

Additional criteria may be used in the critical distance models, in
order to decide whether a system at the critical distance will evolve
towards fusion or towards fission. The consequences of such criteria
are illustrated in Fig. 1 where I have plotted the product $E\sigma_{fu}$ versus
energy E. I believe that this type of plot has several advantages compared to the popular way of plotting σ_{fu} versus 1/E: the ordinate is
directly related to the limiting angular momentum (prop. L^2) and a mo-
notonically increasing function of E, while the abscissa is more easily interpretable and emphasizes the physically interesting high-energy
region rather than the region around the barrier. As in the other type
of plot, the slope of the data is determined by a characteristic distance (see eq. (3)).

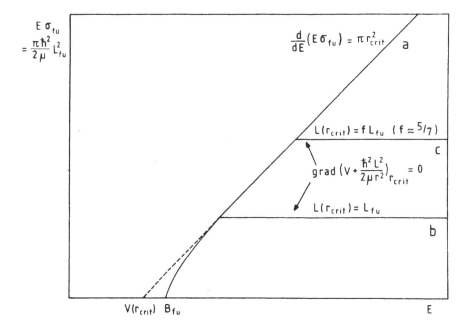

Fig. 1 Schematic energy dependence of the limiting angular momentum
for fusion(L_{fu}) in different versions of the critical distance
model.

As shown in Fig. 1, three different versions of the critical distance model may be distinguished. The relevant assumptions are:

a) that fusion occurs for all trajectories which reach r_{crit} [5,6];

b) that fusion only occurs if the effective potential, as calculated with the <u>asymptotic</u> angular momentum, has a "pocket" at $r \geq r_{crit}$ [3];

c) that the relative angular momentum is reduced by "tangential friction" at $r = r_{crit}$, and that fusion occurs whenever the effective potential, as calculated with the <u>reduced</u> angular momentum, has a "pocket" at $r \geq r_{crit}$ [4].

At this point I should like to emphasize that all models need dissipation of energy (and angular momentum) in order to produce fusion, and hence there is no such thing as a "friction-free model". The physical differences between different models arise, in fact, from the different assumptions which are made either explicitly or implicitly on friction, and not from the use of different conservative potentials (although different authors may have different favourite potentials).

Coming back now to Fig. 1 one can state that the cases a), b) and c) all imply complete radial damping, but different degrees of angular momentum loss at the critical distance. With increasing loss of angular momentum the centrifugal forces are reduced; hence, the number of partial waves which experience an attractive interaction at $r = r_{crit}$ is increased, and so is the saturation limit of the fusion cross section. The latter is most severe in case b), which disregards angular momentum transfer, and absent in case a), which may be interpreted as corresponding to a complete loss of angular momentum at $r = r_{crit}$. The intermediate case c) seems most plausible on theoretical grounds and also agrees best with more elaborate calculations (see Figs. 6,7). However, as discussed in section 4, experimental tests meet with difficulties in defining an appropriate experimental fusion cross section in the presence of strong dissipative effects.

2. The nucleus-nucleus potential

It is now well established that the fusion cross sections for relatively light systems and low energies are mainly determined by the conservative nucleus-nucleus potential, and rather insensitive to the quantitative details of dissipative processes. Physical reasons for this behaviour will be discussed in somewhat more detail in section 3. A number of potentials have been suggested in the literature which are capable of reproducing the overall systematics of experimental fusion cross sections in this domain of "potential-fusion". A well-known example is the so-called "proximity-potential" proposed by the Berkeley group[7].

Rather than reviewing the various theoretical and semi-theoretical potentials I should like to discuss now the problem how to deduce an empirical nucleus-nucleus potential from experimental fusion cross sections. Such a potential can serve several purposes, namely: a) to test theoretical potentials, b) to predict fusion cross sections for various practical applications, c) to provide a reliable basis for the discussion of anomalies and nuclear structure effects in fusion reactions.

The following discussion is based on the critical distance model, i.e. disregards dissipation at $r > r_{crit}$. This may be justified partly by considerations given in section 3, and partly by the success of the method in correlating a large amount of experimental and theoretical information. Within our model the fusion cross section at energy E is given by

$$\pi r_{fu}^2 \left\{ E - V(r_{fu}) \right\} = E \sigma_{fu} \tag{1}$$

$$r^2 \left\{ E - V(r) \right\} = \text{Min! for } r = r_{fu} \leq r_{crit} \tag{2}$$

where $V(r)$ is the sum of the Coulomb and nuclear potentials, and r_{fu} is an energy-dependent "fusion distance" defined by (2). The latter quantity marks the position of an effective barrier which must be passed by the system in order to reach the critical distance (see Fig. 2).

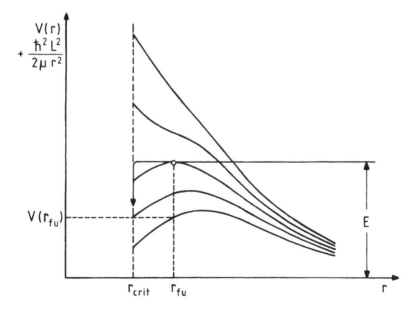

Fig. 2 Effective two-body potentials relevant to the fusion problem

As a consequence of (2) the cross section for a given energy is insensitive to r_{fu}, $V(r_{fu})$ separately, but is mainly determined by the functional dependence $V(r)$. Conversely this means that experimental fusion data can be used to deduce the function $V(r)$, although individual values r_{fu}, $V(r_{fu})$ for a given energy may not be defined accurately.

By differentiating (1) with respect to energy and inserting (2) the following equations can be derived, which relate corresponding pairs r_{fu}, $V(r_{fu})$ to measured fusion excitation functions[8)]

$$\pi r^2_{fu} = \frac{d}{dE}(E\sigma_{fu}) \qquad (3)$$

$$V(r_{fu}) = E - (E\sigma_{fu})\Big/\frac{d}{dE}(E\sigma_{fu}) \qquad (4)$$

The analysis is carried out most conveniently with a simple graphical technique which is illustrated in Fig. 3. As there has been some confusion in the literature concerning the accuracy of the method I should like to give some quantitative details. The experimental errors in σ_{fu} are shown as a hatched band in Fig. 3 and give rise to the following uncertainties in r for given values of V, V_N (see Fig. 3) or in V, V_N for a given value of r:

$$\left(\frac{\Delta r}{r}\right)_V = \frac{1}{2}\frac{\Delta\sigma}{\sigma} \qquad (5)$$

$$\left(\frac{\Delta r}{r}\right)_{V_N} = \frac{1}{2}\left(\frac{dV_N}{dr}\right)^{-1}\left(\frac{dV}{dr}\right)\frac{\Delta\sigma}{\sigma} = \frac{1}{2}\left\{-\frac{V_C}{r}\left(\frac{dV_N}{dr}\right)^{-1} + 1\right\}\frac{\Delta\sigma}{\sigma} \qquad (6)$$

$$\left(\frac{\Delta V}{V}\right)_r = \frac{1}{2}\frac{r}{V}\frac{dV}{dr}\frac{\Delta\sigma}{\sigma} \qquad (7)$$

$$\left(\frac{\Delta V_N}{V_N}\right)_r = \frac{1}{2}\frac{r}{V_N}\frac{dV}{dr}\frac{\Delta\sigma}{\sigma} = \frac{1}{2}\left\{-\frac{V_C}{V_N} + \frac{r}{V_N}\frac{dV_N}{dr}\right\}\frac{\Delta\sigma}{\sigma} \qquad (8)$$

Here V_C, V_N denote the Coulomb and nuclear part of the potential, respectively. In order to make the analysis meaningful, the experimental data must cover a sufficiently large energy range and should have errors of not more than about ± 10 %. Typical errors in r (for fixed V_N) are then within ± 0.2 fm, while V_N (for fixed r) may be uncertain by ± 10 % to ± 25 %. These estimates do not include, of course, errors which may be inherent in the model and arise, for example, from the neglect of friction at $r > r_{crit}$.

A global analysis of experimental fusion data has been performed, based on a representation of the nuclear part of the potential in accordance with the "proximity theorem"[4,7,8)]

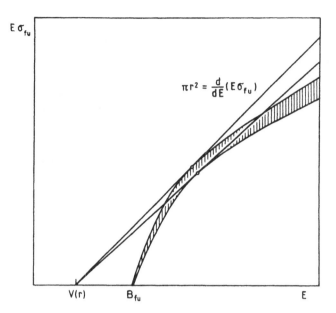

Fig. 3 Graphical method for deriving the function V(r) from experimental fusion excitation functions.

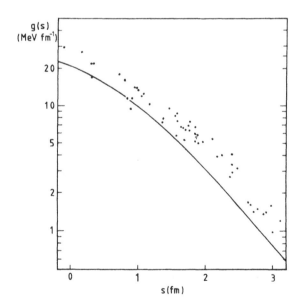

Fig. 4 "Experimental" points representing the function g(s) as defined
by (5). The proximity potential of Blocki et al.[7) is shown for
comparison.

$$-V_N(s) = 4\pi\gamma \frac{R_1 R_2}{R_1+R_2} f(s) = \frac{R_1 R_2}{R_1+R_2} g(s) \tag{9}$$

where γ is the specific surface energy, s measures the separation between the half-density surfaces of the interacting nuclei, and $f(s),g(s)$
are universal functions. The half-density radii were calculated with a
formula given by Blocki et al.[7] and the Coulomb potential was approximated by that of two point charges. Corresponding pairs s,g(s) deduced
in this manner for a large number of systems are shown in Fig. 4 together with the theoretical function g(s) as obtained by the Berkeley group[7].
The empirical points are seen to reproduce very closely the trend of
the theoretical potential but are, on the average, about 50 % higher
(or, alternatively, shifted by about 0.3 fm towards larger values of s).
The empirical potential is well reproduced by the following function

$$g(s) = \left\{ A \exp(\frac{s}{d_1}) + B \exp(\frac{s}{d_2}) \right\}^{-1} \tag{10}$$

with $A = 0.033$ MeV^{-1}fm, $B = 0.007$ MeV^{-1}fm, $d_1 = 3.5$ fm, $d_2 = 0.65$ fm.
It should be noted that these parameter values differ slightly from those published previously[8] due to the use of a different formula for the
half-density radii and the inclusion of more recent experimental data.
Nevertheless the potential V(r) for any given system is practically unaffected by these changes.

The potential given by (9) and (10) summarizes our present knowledge of
fusion cross sections and should therefore provide a reliable basis for
predictions and comparisons with theoretical calculations. It can, however, not be expected to reproduce accurately individual fluctuations
from system to system, which may be due to shell effects in nuclear radii or other reasons of nuclear structure. Practical experience has
shown that in most cases complete agreement between experimental and predicted fusion cross sections or fusion barriers can be obtained, if the
potential radii (R_1+R_2) are adjusted by less than $\pm$ 0.3 fm.

In closing this discussion of the conservative potential I should like
to mention that the empirical potential presented here not only agrees
remarkably well with theoretical expectations and reproduces intermediate energy fusion cross sections. In addition, it has been used successfully to fit sub-barrier fusion and elastic scattering data, i.e. processes which are sensitive mainly to the tail region at large distances.

3. Dissipation of energy and angular momentum (friction).

In discussing the influence of dissipation on fusion cross sections it is important to realize that only the limiting trajectory ($L = L_{fu}$) is of interest, which is usually a grazing trajectory in the cases considered here. Thus it is irrelevant whether the assumption of negligible dissipation at $r > r_{crit}$ breaks down for more central collisions ($L < L_{fu}$). In the following I consider separately the effects of tangential friction and of radial friction on the limiting trajectory (see Fig. 5).

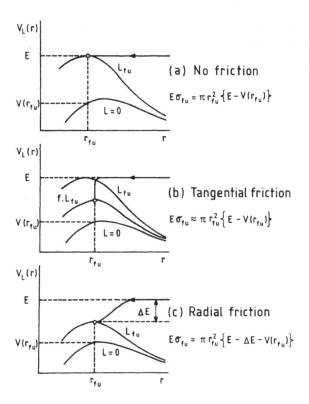

Fig. 5 Influence of tangential and radial friction on the limiting trajectory, and resulting expressions for the fusion cross section.

If the tangential and radial friction coefficients are comparable - as, for example, in the "proximity friction model"[9,10] - one expects the limiting trajectory to be practically unaffected by radial friction for $r > r_{crit}$. The energy loss is then due to tangential friction and localized in a narrow radial region close to the stationary point of the trajectory (case (b) in Fig. 5). In this case the fusion cross section is given approximately by the same expression as in the absence of friction. However, the effective barrier position r_{fu} now refers to a reduced angular momentum ($f \cdot L_{fu}$) and may be significantly shifted compared to the

case without friction, assuming the same potential (case (a) in Fig. 5).
Nevertheless, the fusion cross section is only slightly increased as a
consequence of the previously discussed minimum condition (compare (1),
(2)).

The effect of radial friction is to reduce the incident energy without
loss of angular momentum. This implies either close collisions ($r \approx r_{crit}$)
or a much stronger radial than tangential friction force (as postulated,
for example, in the model of Gross and Kalinowski[1]). In the latter case
the energy loss is not well localized and will extend well beyond the
stationary point of the limiting trajectory (case (c) in Fig. 5). In contrast to tangential friction, radial friction always reduces the fusion
cross section for a given conservative potential. However, there is an
ambiguity in the sense that the effect of radial friction can be com-
pensated by the use of a more attractive potential.

The consistent picture emerging from numerous analyses of experimental
data based on the critical distance and proximity models may be taken
as a strong indication that fusion cross sections are not sensitive to
dissipative effects, as long as the limiting trajectory remains outside the half-density distance. This condition is satisfied for comparatively light systems and moderate energies. For heavier systems and
higher energies the evidence is less clear, although the results seem
to support the concept of a critical distance. Examples of higher energy data are discussed in the following section.

4. Limitations of the one-dimensional models

In going to heavier systems or higher energies, where the position of
the effective barrier approaches the half-density distance (r_{crit}), one
encounters the following complications:

a) dissipative effects - and our incomplete knowledge of these effects -
become more important;
b) additional degrees of freedom, like precompound emission, mass and
charge transfer, neck formation and other types of deformation will increasingly influence the evolution of the system on the limiting trajectory;
c) properties of the compound nucleus, like the position of the y-rast
line or the existence of a fission barrier, may affect the fusion cross
section;
d) the analysis of the experimental cross sections in terms of complete
fusion and surface reactions may be ambiguous due to a strong overlap
of the final mass and energy distributions, as expected for the differ-

ent mechanisms.

It should be realized that the points listed above are not completely unrelated. Moreover, they are probably not just technical points, which can be settled simply by the use of more elaborate models or refined experimental techniques. A more plausible view seems to be, that each of these points has to do in one way or another with the beginning disappearance of "complete fusion" as a distinct physical process. The reason is that compound nuclei, if formed under such circumstances (i.e. with large mass, charge and (or) excitation energy), are expected to have lifetimes of the same order as typical collision times.

It appears, therefore, that in most cases where fusion is clearly definable, the one-dimensional description is appropriate. The problems with the simple models at higher energies or for heavier systems are symptomatic of the more fundamental problem of finding unambiguous and experimentally recognizable criteria for "fusion". Unfortunately, the experimental evidence concerning this interesting regime is rather limited so far. In the following I should like to discuss briefly a few examples of recently published results.

Figure 6 refers to the system ^{12}C + ^{14}N. The high energy data have been obtained at Oak Ridge and are the result of a careful analysis involving kinematic considerations and evaporation calculations[11]. It should be realized, however, that at these energies the distinction between compound nucleus residues and scattering products is not straightforward. The theoretical curves are derived from trajectory calculations by Birkelund et al.[2] and from the critical distance model including angular momentum transfer[4,8]. At the highest energies the experimental cross sections are seen to be significantly larger than predicted by both calculations. At low energies, on the other hand, the calculations agree well with each other and with the experimental data[12].

Figure 7 shows results for the system ^{16}O + ^{27}Al in the same representation. In this case the experimental data[13-15] are in reasonable accord with the model calculations over the complete energy range covered by the experiments. Additional measurements in the high energy region would be clearly desirable, however.

Finally Fig. 8 illustrates the situation for heavier systems, and is representative for a number of studies where medium or heavy target nuclei were bombarded with ^{40}Ar projectiles. In these cases one usually observes a strong yield of reaction products corresponding to an approximately symmetric division of the compound system. The interpretation of these symmetric yields as arising either from compound nucleus fission

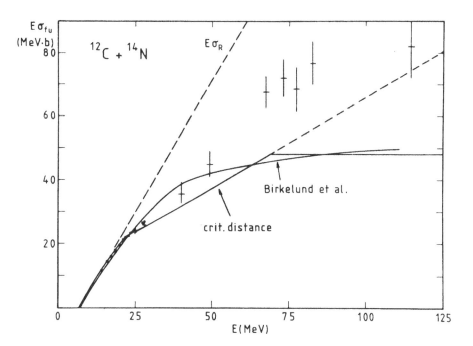

Fig. 6 Comparison of fusion data for the system $^{12}C + ^{14}N$ with model predictions.

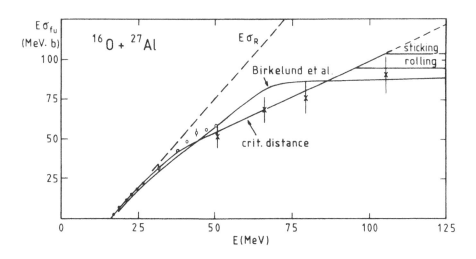

Fig. 7 Comparison of fusion data for the system $^{16}O + ^{27}Al$ with model predictions.

or from a surface diffusion mechanism is an open problem, which is discussed in other contributions to this conference.

The experimental results shown in Fig. 8 refer to the system ^{40}Ar + ^{109}Ag
and were obtained by Britt et al. at Berkeley[16]. In addition to the
cross section for evaporation residue formation (σ_{er}), the authors at-
tribute a part of the symmetric yield to fusion ("fusion-fission", σ_{ff}),
whereas the rest is interpreted as "quasi-fission" (σ_{qf}). The fusion
cross section deduced in this manner agrees with critical distance predictions (curves a,b) at the lower energies, but remains lower than the
predicted saturation limits at the highest energy. A different conclusion would be reached, however, if part or all of the quasi-fission component was included in the "fusion cross section".

It seems clear from these few examples that more experimental as well as
theoretical work is required in order to establish useful criteria of
"fusion" at high energies and for heavy systems. The next step should
then be the refinement of the existing models in accordance with such
criteria.

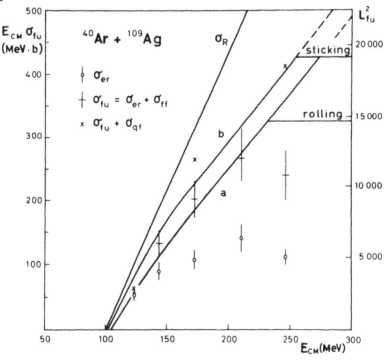

Fig. 8 Comparison of fusion data for the system ^{40}Ar + ^{109}Ag with model
predictions. The curves marked a and b are calculated in the
critical distance model with potentials from refs. 4 and 8, respectively. For further explanation see text.

References

1) D.H.E. Gross and H. Kalinowski, Phys. Lett. 48 B (1974) 302
   D.H.E. Gross, H. Kalinowski and J.N. De, in Classical and Quantum
   Mechanical Aspects of Heavy Ion Collisions, ed. by H.L. Harney et al.,
   Lecture Notes in Physics Vol. 33, p. 194, Springer-Verlag Berlin-
   Heidelberg-New York 1975
2) J.R. Birkelund et al., Phys. Rev. Lett. 40 (1978) 1123
   J.R. Birkelund et al., University of Rochester Report UR-NSRL-193
   (1979), to be published in Physics Reports
3) J. Wilczynski, Nucl. Phys. A216 (1973) 386
4) R. Bass, Phys. Lett. 47 B (1973) 139
   R. Bass Nucl. Phys. A231 (1974) 45
5) J. Galin et al., Phys. Rev. C9 (1974) 1018
6) D. Glas and U. Mosel, Phys. Rev. C10 (1974) 2620
   D. Glas and U. Mosel, Nucl. Phys. A237 (1975) 429
7) J. Blocki et al., Ann. Phys. (N.Y.) 105 (1977) 427
8) R. Bass, Phys. Rev. Lett. 39 (1977) 265
9) R. Bass, Proc. Europ. Conf. on Nucl. Phys. with Heavy Ions, Caen
   1976, ed. by J. Fernandez et al., Communications p. 147
10) J. Randrup, Ann. Phys. (N.Y.) 112 (1978) 356
11) R.G. Stokstad et al., Phys. Rev. Lett. 36 (1976) 1529
    R.G. Stokstad et al., Phys. Lett. 70 B (1977) 289
    J. Gomez del Campo et al., Phys. Rev. C19 (1979) 2170
12) M. Conjeaud et al., Nucl. Phys. A309 (1978) 515
13) B.B. Back et al., Nucl. Phys. A285 (1977) 317
14) Y. Eisen et al., Nucl. Phys. A291 (1977) 459
15) R.L. Kozub et al., Phys. Rev. C11 (1975) 1497
16) H. C. Britt et al., Phys. Rev. C13 (1976) 483

HEAVY-ION FUSION: A CLASSICAL TRAJECTORY MODEL

J. R. BIRKELUND, L. E. TUBBS AND J. R. HUIZENGA
Departments of Chemistry and Physics
and
Nuclear Structure Research Laboratory
University of Rochester
Rochester, New York 14627 U.S.A.

and

J. N. DE AND D. SPERBER
Department of Physics
Rensselaer Polytechnic Institute
Troy, New York 12180 U.S.A.

I.   INTRODUCTION

The fusion data of a wide variety of heavy nuclei at incident energies a few MeV/nucleon above the Coulomb barrier can be reproduced quite well by models based on the classical motion of the nuclei[1] in a potential field, including the nuclear proximity potential[2], and dissipative forces based on the one body friction of Randrup[3]. The present model, based on systematic nuclear properties, is only expected to reproduce the general trends of fusion excitation functions, and may lack precise agreement with data in individual cases. This paper discusses the extent to which such classical models lead to fusion excitation functions in agreement with currently available data[3], and the predictions of such models at higher energies where little data currently exists. In addition, some discussion will be given of the limitations expected to be found in the model, as the projectile energies are raised into the region of 10-20 MeV/u.

Early analyses of fusion excitation functions[4] were based on 'friction free' models which assumed negligible friction for nuclei on trajectories up to the fusion barrier, and complete fusion of any trajectories which crossed the barrier. The simplest models of this type assume the barrier to be fixed independent of the orbital angular momentum, while more sophisticated models of the 'friction free' type calculate the barrier positions according to an assumed form for the nuclear potential[5]. As more data became available, the shape of the fusion excitation functions led to the proposal of the critical distance models[6,7], which require the nuclei to reach a critical separation before fusion can occur. However, the

J. R. BIRKELUND, ET AL.

known importance of damped reaction mechanisms[8] for non-fusing trajectories suggests that friction effects on the fusing trajectories need to be considered in models of heavy nucleus fusion. Hence, information about the conservative and dissipative forces may be obtained from fusion reactions as well as deep-inelastic collisions.

II. THE MODEL

The equations of motion for the heavy nuclei are solved numerically[1]. Explicit account is taken of the transfer of angular momentum from orbital to intrinsic spin. Several possible choices for the conservative potentials have also been tested. The degrees of freedom used in the model are shown in Fig. 1, and include the radial separation of the nuclear mass center r, the angular orientation of the radius vector θ, and the angular orientation of the target and projectile nuclei θ_T and θ_P. All the nuclear radius parameters of the model have been taken from the liquid drop model systematics of Myers[4].

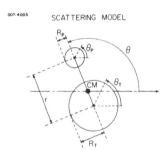

007-4003   SCATTERING MODEL

FOUR DEGREES OF FREEDOM $(r, \theta, \theta_P$ AND $\theta_T)$

FIGURE 1

III. THE COULOMB POTENTIAL

Several possible choices of Coulomb potential have been examined. These include the point charge approximation, a less repulsive potential suggested by Bondorf et al.[10], based on the Coulomb self energies of the nuclei, and the potential derived from the interaction of a point charge with a uniform spherical charge distribution. The relevant expression for the point charge approximation is

(1) $$V_C(r) = 1.438 \ Z_T Z_P / r \text{ MeV}.$$

For the Bondorf potential, at radii less than the sum of the nuclear charge radii, $R_{CT} + R_{CP}$, the expression in Eq. (1) is replaced by

HEAVY-ION FUSION: A CLASSICAL....

(2)
$$V_C(r) = V_0 - kr^n$$

where

(3)
$$V_0 = 0.6 \left[\frac{(Z_T + Z_P)^2}{(R_{CT}^3 + R_{CP}^3)^{1/3}} - \frac{Z_T^2}{R_{CT}} - \frac{Z_P^2}{R_{CP}} \right] (1.438) \text{ MeV.}$$

The parameters n and k are obtained by smoothly matching expressions 1 and 2 at
the matching radius $R_C = R_{CT} + R_{CP}$. The nuclear charge radii are taken from the
systematics of Myers[10]. The point charge plus uniform charge distribution leads
to the following equation for $r < R_C$

(4)
$$V_C(r) = 1.438 \; (Z_T Z_P / 2R_C)(3 - r^2/R_C^2) \quad .$$

In this case $R_C = r_0(A_T^{1/3} + A_P^{1/3})$, where we have taken $r_0 = 1.3$ fm.

The various choices of Coulomb potential are shown in Figs. 2 and 3 for the
^{62}Ni + ^{35}Cl and ^{116}Sn + ^{35}Cl systems. Also shown in the figures are Coulomb
potentials calculated for a Fermi charge density distribution. The point charge
approximation is the most repulsive of the potentials and is an overestimate of
the Coulomb potential at small separations. The Bondorf potential is correct in
the limit of target and projectile merged into a single nucleus, and the functional
form at larger radii gives a smaller potential than the calculation based on the
Fermi charge distributions, where no account is taken of enhancement of the nuclear
density in the overlap region. The point plus uniform charge distribution is
much softer than the other three potentials, with the parameters chosen for Eq. (4).

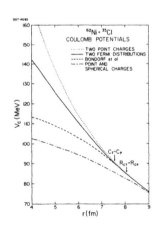

FIGURE 2

J. R. BIRKELUND, ET AL.

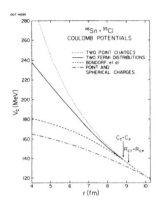

FIGURE 3

IV. THE NUCLEAR POTENTIAL

Most calculations have been performed with the nuclear proximity potential of Blocki[2] et al. and a modification suggested by Randrup[11]. The potential is given by

(5)
$$V_N(\zeta) = 4\pi \ \gamma \ \bar{R} \ b \ \phi(\zeta)$$

where

$$\bar{R} = \frac{C_T C_P}{C_T + C_P} \quad .$$

The surface energy $\gamma = 0.9517 \ [1 - 1.7826\{(N-Z)/A\}^2]$, where Z, N and A refer to the combined system. The value of $\zeta = r - C_T - C_P$, and is the surface separation of the target and projectile. The universal proximity function $\phi(\zeta)$ has been tabulated by Blocki et al., from a Thomas-Fermi model calculation. The surface diffuseness parameter b was taken to be 1 fm.

The modified version of the potential removes the hard core which appears in the potential at small values of ζ because the standard proximity potential allows the nuclear density to rise above the bulk density in the overlap region. This potential is equivalent to ensuring that the nuclear density never rises above its central value, and hence is an approximate procedure accounting for deformations occurring during the interaction.

The modified proximity potential is given by Eq. (5) for $\zeta > 0$, and for $\zeta < 0$ is given by

HEAVY-ION FUSION: A CLASSICAL....

(6) $$V_N(\zeta) = 4\pi \gamma \bar{R} b [\phi(\zeta = 0) + \zeta] .$$

The choices of nuclear potential are shown in Fig. 4 for the systems $^{27}Al + ^{16}O$ and $^{165}Ho + ^{56}Fe$. Also shown in Fig. 4 is a potential recently suggested by Krappe et al.[12], which also leads to excitation functions in agreement with the data when the hard core is removed.

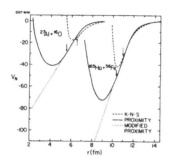

FIGURE 4

    The total effective potentials, including nuclear, Coulomb and centrifugal terms, are shown in Figs. 5 and 6 for the $^{65}Ni + ^{35}Cl$ and $^{165}Ho + ^{56}Fe$ systems. These figures show that the various choices of potentials vary the strengths of the repulsive core in the effective potentials, but leave the barriers essentially unchanged. This characteristic will be seen in the calculated excitation functions, as an insensitivity to the changes in potential for low projectile energies.

V.   DISSIPATIVE FORCES

    The friction form factors $f_r(r)$ and $f_\theta(r)$ have been calculated from the nuclear one body friction model of Randrup[3]. The form factors are given by

(7a) $$f_r(r) = 4\pi n_0 \bar{R} b \psi(\zeta)$$

(7b) $$f_\theta(r) = 2\pi n_0 \bar{R} b \psi(\zeta)$$

where $f_r(r)$ and $f_\theta(r)$ are the radial and tangential friction form factors, and where n_0 is the nucleon bulk flux within the nucleus, taken to be 0.264×10^{-22} MeV sec fm^{-4}. The proximity flux function $\psi(\zeta)$ was taken from the

J. R. BIRKELUND, ET AL.

tabulated values of Randrup[3].

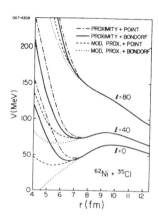

FIGURE 5

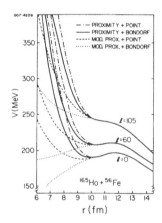

FIGURE 6

VI.   THE MODEL DEFINITION OF FUSION

Within the context of the model described in this paper, fusion is defined to occur whenever the target and projectile are trapped within the inter-nuclear potential. This may occur either by reflection from inside the potential barrier in the exit channel, or in the case of the softest potential choices, by motion of the trajectory inside the arbitrarily chosen separation of 0.2 fm. This definition of fusion may not correspond in all cases to more generally used

HEAVY-ION FUSION: A CLASSICAL....

definitions, which will be discussed below.

VII. COMPARISON WITH DATA

Calculated fusion excitation functions are compared with data for several
systems in Figs. 7, 8, 9, 10 and 11. Four possible combinations of Coulomb and
nuclear potentials have been tested in the calculations. The combination of the
standard proximity potential with the point charge Coulomb potential (model SP) is
denoted by short-long dashed lines. The model SP is the most repulsive of all the
potentials used at small radial separations. Other combinations in order of
increasing softness, are the standard proximity potential, with the Bondorf Coulomb
potential (model SB), shown as dots; the modified proximity potential with the
point charge Coulomb potential (model MP), shown as dash-two dots; and the modified
proximity potential with the Bondorf Coulomb potential (model MB), shown as dashes.

For the comparatively light system ^{27}Al + ^{16}O $^{12-16}$ shown in Fig. 7, the four
choices of potentials lead to the same excitation function, which is shown in the
figure by a solid line. The excitation function is plotted as a function of $E_{c.m.}^{-1}$,
and shows the general characteristics of all the excitation functions calculated
with the model. Near the threshold the cross section increases linearly as $\frac{1}{E}$
decreases, reaches a maximum, and decreases in the higher energy region. Within
the errors, and the variation between the different experimental measurements,
there is good overall agreement between the measurements and the calculated excitation functions.

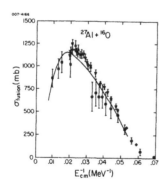

FIGURE 7

As the mass of the system is increased, the trajectories calculated with the

J. R. BIRKELUND, ET AL.

different choices of potential lead to different excitation functions in the high energy region. This is seen in Fig. 8 for the system ^{62}Ni + ^{35}Cl [17,18] where the excitation functions for models SP and MB are shown. This figure shows an additional characteristic of the calculations, which is the equality of the excitation functions from the various models in the low energy region below the cross section peak. Such behavior occurs because the potential modifications tested do not alter the barriers or potentials at small overlap of the projectile and target. Thus, at low energies, where penetration of the nuclei is small, all models lead to the same excitation function. The sensitivity to the potential arises only in the high energy region. However, it is not possible to isolate the effects of Coulomb and nuclear potentials in an unambiguous fashion, unless some assumption is made about the form of one of the potentials. The excitation functions are not very sensitive to the friction strength, provided that the friction is always adequate to reduce the relative motion to the rolling condition, and provided that rolling friction is not introduced into the equations of motion. If a rolling friction term is included, the projectile and target will 'stick' and the cross section will be further enhanced in the high energy region for asymmetric systems.

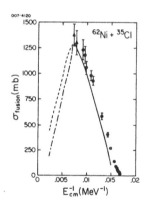

FIGURE 8

In Fig. 9, the ^{62}Ni + ^{35}Cl system is again shown, with excitation functions calculated using nuclear radii increased by 0.16 fm. This radius increase is within the accuracy of the Myers' systematics, and leads to an improvement in the agreement between calculations and the data. This figure indicates the sensitivity of the calculations to nuclear radius changes at low energies and illustrates the

HEAVY-ION FUSION: A CLASSICAL....

improvement in the fit that can be obtained if one allows for parameter changes for individual reactions.

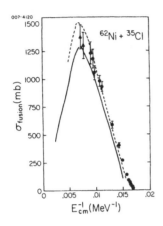

FIGURE 9

The data and calculations for the system ^{116}Sn + ^{35}Cl [18,19] are shown in Fig. 10. The sensitivity of the calculated excitation functions to the potentials is even greater for the heavier ^{116}Sn + ^{35}Cl system than for the calculations of the ^{62}Ni + ^{35}Cl system of Fig. 8. In Fig. 10, three model calculations are shown, but as yet no data exists in the region of greatest sensitivity to the potentials.

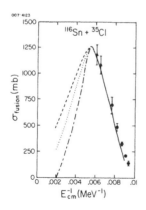

FIGURE 10

J. R. BIRKELUND, ET AL.

In Fig. 11 the data and calculations are shown for ^{165}Ho + ^{56}Fe [20]. In this case, the excitation functions for all four models are well separated. This condition is only true for the very heaviest systems. The single data point suggests that the most appropriate potential combination lies between models MB and MP, with the standard proximity potential being too repulsive when combined with either of the Coulomb potentials.

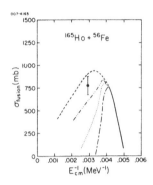

FIGURE 11

Little data currently exists in the energy region where the excitation functions are most sensitive to the potentials. However, current data do indicate that some account must be taken of the angular momentum transfer, if agreement is to be obtained between data and calculation. This is illustrated in Fig. 12 where the 'friction free' calculation is compared to the data for the ^{62}Ni + ^{35}Cl system. This type of calculation uses the barrier positions from the potential combination SP, but simply assumes that the system fuses on a given trajectory, if a pocket exists in the effective potential, and if the energy is sufficiently high to take the system over the barrier. By comparison with the data and Fig. 8 it is seen that the simple friction free model underestimates the data and is below the model calculations in the region near the cross section peak. The explanation for the improved agreement between the model calculation including friction and the data, lies in the calculation of the angular momentum transfer during the reaction. For trajectories which show no 'pocket' in the effective potential at the asymptotic value of the orbital angular momentum, a pocket may be produced at some point on the trajectory by transfer of orbital angular momentum to intrinsic spin. Thus, it is possible for the model calculation to produce fusion from a higher number of

HEAVY-ION FUSION: A CLASSICAL....

trajectories than 'friction free' models which take no account of angular momentum transfer. It is possible that nuclear potentials deeper than the proximity poten- tial would produce a sufficient number of pockets in the effective potential for a friction free model to reproduce the data. However, inclusion of friction in such models would then cause the calculation to overestimate the fusion cross sections.

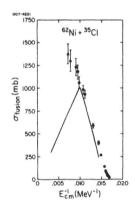

FIGURE 12

VIII. ANGULAR MOMENTUM LIMITS TO FUSION

The angular momentum limitation on fusion is included in the model described here, by the inclusion of the centrifugal potential. This is an entrance channel limitation which is not calculated in the same manner as the liquid drop angular momentum stability limits of Cohen, Plasil and Swiatecki[21]. That these limits are not identical can be seen from Table 1, where some measured angular momenta from fusion cross sections in agreement with the model, are compared with the liquid drop stability limits. It can be seen from Table 1 that measured values of ℓ_f consider- ably exceed the liquid drop angular momentum limits for some systems. These observa- tions need not indicate that the liquid drop limits are too low, but rather that interaction times on trapped trajectories are so long that the considerable relaxa- tion of the mass asymmetry degree of freedom makes very difficult an experimental distinction of fusion at angular momenta below the liquid drop limits. The model regards trapped trajectories as leading to fusion, but consideration of the evolu- tion of the trapped system may indicate that although the interaction time is long, complete equilibration is not achieved before fission of the system. In this

J. R. BIRKELUND, ET AL.

TABLE 1   Comparison of experimentally measured values of ℓ_f with the liquid-drop
model limiting values of angular momenta ℓ_{LDM}

Reaction	E_{cm} (MeV)	σ_{fus} (mb)	ℓ_f	ℓ_{LDM}
^{116}Sn + ^{35}Cl	130.1	695	89	78
^{109}Ag + ^{40}Ar	246.5	975	109	83
^{121}Sb + ^{40}Ar	222.5	1130	107	82
^{165}Ho + ^{40}Ar	241.5	1450	126	79
^{238}U + ^{40}Ar	256.8	1030	117	40
^{165}Ho + ^{56}Fe	344.0	763	129	66
^{56}Fe + ^{136}Xe	225.0	1118	125	80

respect for heavy systems at higher energies this model will lead to higher fusion
cross sections than predicted by the liquid drop model. Considerable interaction
times are also predicted by the model for a few impact parameters above the maximum
value of angular momentum for which trapping of the system occurs. This is illustrated in Fig. 13 where the interaction times for untrapped trajectories are plotted
for several projectile energies for the systems ^{27}Al + ^{12}C and ^{165}Ho + ^{56}Fe. As can

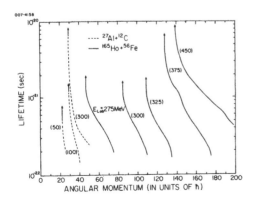

FIGURE 13

HEAVY-ION FUSION: A CLASSICAL....

be seen from this figure, at the higher energies, interaction times are achieved
which are considerably in excess of 10^{-21} sec.   However, the model retains a sharp
distinction between these relatively long lived trajectories and trapped trajectories.

IX.   SYSTEMATICS OF FUSION CROSS SECTION MAXIMA

There has been some interest in the systematics of fusion cross sections for
target-projectile systems involving 1p-shell and 2s - 1d-shell nuclei[22]. Such
systems show in some cases oscillatory structure in the excitation functions[23], and
variations in the maximum fusion cross sections from one system to another. In
Fig. 14 are compared the data and the results of the calculation for fusion cross
section maxima for some of the light systems. As can be seen from Fig. 14, except
for the system $^{10}B + ^{16}O$, the fusion cross section maxima are reasonably well reproduced in view of the experimental uncertainties. However, it should be noted
that in some cases there is disagreement between calculated and measured energies of
the peak in the excitation functions. This general agreement in cross section
maxima suggests that the fusion cross section maxima depend principally on the macroscopic structure of the nuclei and not in any systematic way on the microscopic details of the nuclear structure. Further, as can be seen from Fig. 9 small changes in
nuclear radii, within the accuracy of the radius systematics, will make significant
differences in the calculated maximum fusion cross section.

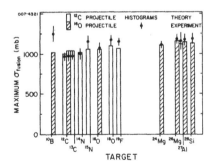

FIGURE 14

In addition, measurements of fusion cross sections for light systems are
difficult because of the confusion which can arise between evaporation residues and
the products of few nucleon transfer reactions. This can be seen from Fig. 7 where
the data for the system $^{27}Al + ^{16}O$ is shown. The different data symbols indicate
measurements made by different groups or techniques. The variation in the measured

J. R. BIRKELUND, ET AL.

cross sections illustrate the experimental difficulties in this mass range.

X.   LIMITATIONS OF THE MODEL

The model as described in this paper, with parameters taken from liquid drop model systematics, is in good agreement with most available fusion data, for a wide mass range of target and projectile. However, in assessing the validity of the model calculations in energy regions for which no data are available, it is neces- sary to consider likely limitations of the model. These limitations fall into three categories, including effects unaccounted for by the model, such as deformations which are expected to be important in the reaction; conceptual problems in defining complete fusion, which are expected to become more important at higher energies; and the related category of experimental problems in the measurement of complete fusion.

Deformations of the target and projectile and neck formation are approximately accounted for in the model by the variations in the nuclear potential which have been tested. However, the relationship between the potentials and the dynamically pro- duced deformations is not clear in the model and no account is taken of the effect of deformation on the friction. This is a general limitation of the model which will become increasingly important as the projectile energy increases and the inter- action time and nuclear interpenetration increase on the fusion trajectories. In addition, the one-body friction is expected to become less dominant as the tempera- ture of the system rises, thus reducing the nucleon mean free path in the nucleus. A further limitation of the model arises from the neglect of the effects of the mass transfer, other than the one-body friction itself. The deformation and mass trans- fer can be accounted for in a more sophisticated model, which also makes use of the one-body friction[11,24]. However, as yet only preliminary calculations of the fusion cross section have been made with the more sophisticated model.

Problems concerned with the definition of an experimentally measurable con- cept of fusion, may also limit the applicability of the model. A commonly used conceptual definition of fusion requires that the fused system contain essentailly all of the nucleons of the target and projectile, have deformations inside the saddle point for fission on the deformation potential surface, and be in statistical equilibrium in all its degrees of freedom. However, for heavy ion reactions, especially at high energies it may not be possible to experimentally verify that all these conditions have been met. Further, since a composite system will be formed in a highly excited state, it is necessary to decide which de-excitation mechanisms are allowable within the context of fusion. It seems reasonable to allow

HEAVY-ION FUSION: A CLASSICAL....

the inclusion of de-excitation through statistical evaporation of nucleons or alpha
particles, and the subsequent emission of γ-rays. There is evidence that heavy-ion
reactions, including reactions leading to fusion-like products[25], may emit pre-
equilibrium light particles[26,27,28]. Such particles may arise from hot spots in the
nuclei formed during the interaction[29], or from jets of particles which are expected
to arise because of the one-body nature of the interaction[30,31]. In addition, such
fast light particles may arise from the breakup of the projectile. The presence of
pre-equilibrium particles in the reaction should not be taken to preclude the possibility of fusion. If the pre-equilibrium particles arise on the trajectory after the
system is trapped, or if they arise from processes which are fundamentally related
to the energy loss mechanisms, then any resulting composite nucleus may be regarded
as fused. Conceptually, if the fast particles arise from the breakup of the projectile early in the interaction before trapping, with subsequent capture of the
remaining projectile fragment, then such events should be excluded from the fusion
cross section. However, it may be experimentally difficult to distinguish
such reactions from those in which the pre-equilibrium particles are emitted at a
later stage of the reaction. The model described in this paper does not account for
mass change on fusion trajectories in which pre-equilibrium particles are emitted
before trapping. Further, if the pre-equilibrium particles come from a hot spot,
this may indicate that the long mean free path assumption of the one-body friction
is not valid.

   Composite systems which become trapped will undergo relaxation of the mass
asymmetry degree of freedom, and will then lead either to evaporation residue-like
fragments, or fission-like fragments from heavy systems at sufficiently high excitation energies. However, even assuming that such fragments can be experimentally distinguished from damped and transfer reaction products, the observation of evaporation residue or fission-like fragments does not invariably indicate that fusion
occurred. Incomplete momentum transfer from projectile to target has been observed
for heavy systems in cases where fission-like products are observed[32]. In lighter
systems such incomplete momentum transfer processes have been called incomplete
fusion[33] or massive transfer[34]. If the incomplete momentum transfer results from
the emission of pre-equilibrium α-particles or nucleons of the type described
above, then such processes may be part of the fusion cross section. However, the
loss of nucleons or more massive fragments during the reaction before trapping may
be related to angular momentum instabilities[33] rather than energy loss mechanisms,
and such processes should not be counted as part of the fusion cross section, even
though they may result in fission-like or evaporation residue-like products.

   There may be no experimental method of determining whether a heavy system

J. R. BIRKELUND, ET AL.

which produces fission-like products ever reached a configuration inside the fission
saddle point during the reaction. The relaxation of the mass asymmetry on a trapped
trajectory is probably sufficient to produce fission-like fragments which are difficult to distinguish from those resulting from systems which move inside the fission
saddle point. In addition, the relaxation of the slowly equilibrating mass asymmetry
degree of freedom may be the best observable indicaiton that some of the degrees of
freedom of the system have approached statistical equilibrium before fission. There
are some data which indicate an increase in the width of the mass distribution[35] of
fission-like fragments, whenever the fusion cross section contains angular momenta
above the liquid drop stability limits. Such variations in width may possibly reflect the interaction time and path of the trapped system on the deformation potential surface, but are unlikely to show clearly whether the system passed behind the
fission saddle point. Thus, except for the limitations on fast-particle emission
discussed above, the definition of trapped systems as fusion, may be the only viable
one for heavy systems and high bombarding energies.

XI.   CONCLUSION

The calculation of heavy ion fusion excitation functions with a model based on
the proximity potential and one-body friction reproduces most available data over a
wide mass range. The calculations suggest that transfer of angular momentum from
orbital motion to intrinsic spin is necessary to produce the observed fusion cross
sections. In addition, the model predicts a peaking of the fusion excitation functions with a decrease in the cross section as the projectile energy increases. The
fusion excitation functions are shown to be most sensitive to the potentials and
friction at higher energies, although no completely unambiguous separation of
potential and friction effects can be made. The predictions of the model at higher
energies should be treated with some reservation, however, since the one-body
friction is likely to be less applicable at higher energies, and the deformations
occurring in the reaction are unaccounted for by the model. In addition, care
should be taken in comparing the model with data at the higher energies, since the
measurement of fusion at high energies presents considerable problems.

This research was supported by the U.S. Department of Energy.

HEAVY-ION FUSION: A CLASSICAL....

REFERENCES

1. J.R. Birkelund, J.R. Huizenga, J.N. De, and D. Sperber, Phys. Rev. Lett. 40, (1978) 1123; J.R. Birkelund, L.E. Tubbs, J.R. Huizenga, J.N. De and D. Sperber, Physics Reports, 56, (1979) 107.
2. J. Blocki, J. Randrup, W.J. Swiatecki, and C.F. Tsang, Ann. Phys. (N.Y.) 105, (1977) 427.
3. J. Randrup, Ann. Phys. (N.Y.) 112, (1978) 356.
4. H.H. Gutbrod, W.G. Winn, and M. Blann, Nucl. Phys. A213, (1973) 267.
5. R. Bass, Nucl. Phys. A231, (1974) 45.
6. J. Galin, D. Guerreau, M. Lefort, and X. Tarrago, Phys. Rev. C9, (1976) 1018.
7. D. Glas and U. Mosel, Nucl. Phys. A237, (1975) 429; A264, (1976) 268.
8. W.U. Schröder and J.R. Huizenga, Ann. Rev. Nucl. Sci. 27, (1977) 465.
9. W.D. Myers, Nucl. Phys. A204, (1973) 465.
10. W.D. Myers, Phys. Lett. 30B, (1964) 451.
11. J. Randrup, Nucl. Phys. A307, (1978) 319.
12. Y. Eisen, I. Tserruya, Y. Eyal, Z. Fraenkel, and M. Hillman, Nucl. Phys. A291, (1977) 459.
13. B. Back, R.R. Betts, C. Gaarde, J.S. Larsen, E. Michelsen, and Tai Kuang-Hsi, Nucl. Phys. A285, (1977) 317.
14. J. Dauk, K.P. Lieb, and A.M. Kleinfeld, Nucl. Phys. A241, (1975) 170.
15. R.L. Kozub, N.H. Lu, J.M. Miller, D. Logan, T.W. Debiak, and L. Kowalski, Phys. Rev. C11, (1975) 1497.
16. R. Rascher, W.F.J. Muller, and K.P. Lieb, Preprint (1979).
17. W. Scobel, H.H. Gutbrod, M. Blann, and A. Mignerey, Phys. Rev. C14, (1976) 1808.
18. B. Sikora, W. Scobel, M. Beckerman, M. Blann, and L. Tubbs, Bull, Am. Phys. Soc. 22, (1977) 1019.
19. P. David, J. Bisplinghof, M. Blann, T. Mayer-Kukuk, and A. Mignerey, Nucl. Phys. A287, (1977) 179.
20. A.D. Hoover, L.E. Tubbs, J.R. Birkelund, W.W. Wilcke, W.U. Schröder, D. Hilscher, and J.R. Huizenga, (to be published).
21. S. Cohen, F. Plasil, and W.J. Swiatecki, Ann. Phys. (N.Y.) 82, (1974) 557.
22. J.D. Schiffer, Proceedings of International Conference on Nuclear Structure, Tokyo, Japan (1977) p. 13 (T. Marumori, Editor); D.G. Kovar, D.F. Geesaman T.H. Braid, Y. Eisen, W. Henning, T.R. Ophel, M. Paul, K.E. Rehm, S.J. Sanders, P. Sperr, J.P. Schiffer, S.L. Tabor, S. Vigdor, B. Zeidman, and F.W. Prosser, Jr., Phys. Rev. C20, (1979) 1305.
23. P. Sperr, T.H. Braid, Y. Eisen, D.G. Kovar, F.W. Prosser, J.P. Schiffer, S.L. Tabor, and S.E. Vigdor, Phys. Rev. Lett. 37, (1976) 321.
24. J. Randrup, Nucl. Phys. A327, (1979) 490.
25. L. Westerberg, D.G. Sarantites, D.C. Hensley, R.A. Dayras, M.L. Halbert, and J.H. Barker, Phys. Rev. C18, (1978) 796.
26. T. Nomura, H. Utsunomiya, T. Motobayashi, T. Inamura, and M. Yanokura, Phys. Rev. Lett. 40, (1978) 694.
27. H. Ho, R. Albrecht, W. Dünnweber, G. Graw, S.G. Steadman, J.P. Wurm, D. Disdier, V. Rauch, and F. Scheibling, Z. Phys. A283, (1977) 235.
28. J.M. Miller, G.L. Catchen, D. Logan, M. Rajagopalan, J.M. Alexander, M. Kaplan, and M.S. Zisman, Phys. Rev. Lett. 40, (1978) 100.
29. P.A. Gottschalk and M. Weström, Phys. Rev. Lett. 39, (1977) 1250.
30. J.P. Bondorf, J.N. De, A.O.T. Karvan, G. Fai, and B. Jakobsson, Phys. Lett. 84B, (1979) 162.
31. D.H.E. Gross, and J. Wilczyński, Phys. Lett. 67B, (1977) 1.
32. P. Dyer, T.C. Awes, C.K. Gelbke, B.B. Back, A. Mignerey, K.L. Wolf, H. Breuer, V.E. Viola, and W.G. Myer, Phys. Rev. Lett. 42, (1979) 560.
33. K. Siwek-Wilczyńska, E.H. DuMarchie Van Voorthuyzen, J. Van Popta, R.H. Siemssen, and J. Wilczyński, Phys. Rev. Lett. 42, (1979) 1599.

J. R. BIRKELUND, <u>ET AL</u>.

34.   D.R. Zolnowski, H. Yamada, S.E. Cala, A.C. Kahler, and T.T. Sugihara, Phys. Rev. Lett. <u>41</u>, (1978) 92.
35.   C. Lebrun, F. Hanappe, J.F. Lecolley, F. Lefebvres, C. Ngô, J. Péter, and B. Tamain, Nucl. Phys. <u>A321</u>, (1979) 207; B. Heusch, C. Volant, H. Freiesleben, R.P. Chestnut, K.D. Hildenbrand, F. Pühlhofer, and W.F.W. Schneider, Z. Phys. <u>A288</u>, (1978) 391; M. Lefort, Proceedings of this symposium.

REVIEW OF MODELS FOR FUSION

H. J. Krappe

Hahn-Meitner-Institut für Kernforschung Berlin
D 1000 Berlin 39, Glienicker Straße 100

1. Introduction

It is the intention of these notes to compare the physical assumptions on which the various models for heavy-ion fusion are based and to discuss the limits of their applicability rather than to present an outline of these theories. A more detailed description of some fusion models is given in other contributions to this conference.

The starting point for the description of heavy-ion fusion is the formula

$$\sigma_{fus} = \pi \lambda^2 \sum_{l=0}^{\infty} (2l + 1) T_l \qquad (1)$$

borrowed from the theory of compound-nucleus formation in nucleon-nucleus scattering. The simplest form for the l-dependence of the transmission coefficients $T_l(E)$ follows from the picture that all classical trajectories which lead to contact of the two sharp nuclear surfaces result in fusion. The remaining trajectories shall not contribute at all. This leads to the concept of a critical angular momentum l_{cr} such that

$$T_l = \begin{cases} 1 & l \le l_{cr}(E) \\ 0 & l > l_{cr}(E) \end{cases} \qquad (2)$$

and the fusion cross section has the classical form

$$\sigma_{fus} = \pi R^2 \left(1 - \frac{V(R)}{E}\right), \qquad (3)$$

where R is the sum of the nuclear radii and V(R) the interaction potential at the centers-of-mass distance R. From the form of (3) follows the practice of plotting σ_{fus} versus E^{-1} as shown schematically in fig. 1. The intercepts of the straight line representing σ_{fus} with the coordinate axes immediately yield R and V(R). From (2) follows

the break-up of the cross section into contributions from individual
angular-momentum bins as drawn schematically in fig. 2.

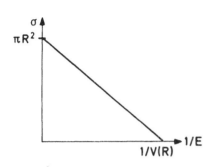

Fig. 1   Schematic drawing of the
          fusion excitation func-
          tion according to (1)
          against the inverse ener-
          gy.

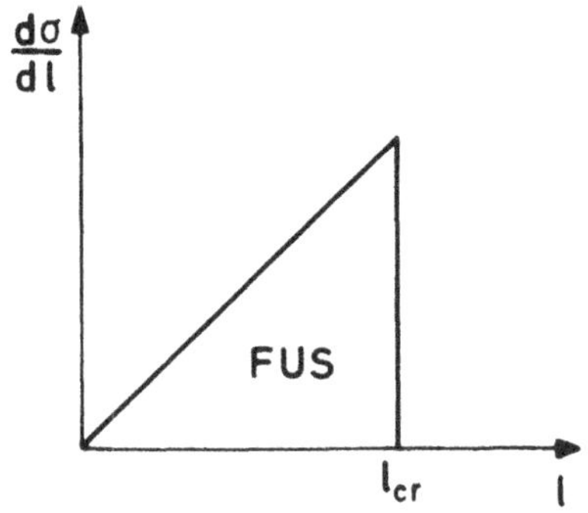

Fig. 2   Contribution of different
          partial waves to the fusion
          cross section according to
          (2) at a given energy.

In the following we present criticism and some refinements of these
simple pictures.

2.   Barrier Penetration

In the first refinement to be discussed, fusion is considered as a barrier-penetration process of two mass points without internal degrees of
freedom moving in a static, local potential. An ingoing-wave boundary
condition [1] is required inside the potential barrier. It expresses
the assumption that all the flux that passes the barrier is somehow
trapped inside and leads to fusion. Clearly such an assumption is plausible only if there is a sufficiently deep pocket in the effective potential for all partial waves which contribute to fusion at the energy
considered. Since for large values of Z_1Z_2 or large angular momenta the
pocket disappears, the barrier-penetration model is restricted to lighter systems and to energies not too high above the barrier.

Often the potential barrier is approximated by an inverted parabola, which leads to the Hill-Wheeler formula for the transmission coefficients [2]

$$T_1 = [1 + \exp(2\pi \; \frac{V_B(1) - E}{\hbar \, \omega_B(1)})]^{-1}$$ (4)

with

$$V_B(1) = \max_r V(1,r)$$ (5)

and

$$\omega_B^2(1) = -\frac{1}{\mu} \; (\frac{\partial^2 V(1,r)}{\partial r^2}) \; r = r_{max},$$ (6)

where

$$V(1,r) = V_{nucl}(r) + V_{Coul}(r) + \frac{1(1+1)\hbar^2}{2 \, \mu \, r^2}$$ (7)

is the effective potential for the 1th partial wave. The parabolic approximation limits the applicability of (4) to energies which are not more than a few MeV below the s-wave barrier. In fig. 3 fusion data for the system $^{16}O + ^{27}Al$ from various measurements [3 - 7] are compared with the barrier-penetration model. In this calculation the generalized liquid-drop potential of ref. [8] has been used. However other potentials satisfying conditions (15) - (18) would lead to fits of similar quality. The crosses in fig. 3 are the calculated points. They do not exactly lie on a straight line. However within the experimental uncertainties and for energies somewhat above the barrier a straight line is a sufficiently accurate representation of the calculation.

The barrier is most sensitively probed by the fusion reaction at energies around and below the barrier, where the straight-line prediction of the classical model (3) does not hold. The plot shown in fig. 4 is therefore more appropriate for a comparison of data with model predictions. All curves in fig. 4 were calculated with the generalized liquid-drop potential [8] assuming spherical nuclear shapes and electric point charges for the calculation of the interaction potential. The two lightest systems are described in a very satisfactory way by the barrier penetration model. However for heavier systems the model prediction is seen to be systematically below the experimental data, in particular for sub-

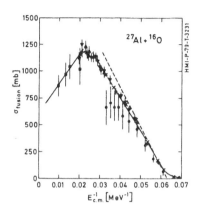

Fig. 3　Fusion excitation function for $^{16}O + ^{27}Al$. Dots represent data from ref. [3], squares from ref. [4], diamonds from ref. [5] and triangles from refs. [6,7]. The full line at lower energies follows from inserting (4) into (1) and using the generalized liquid-drop potential [8]. The full line at high energies is taken from Mosel's work [9]. The dashed line is the prediction of (3) using the s-wave barrier height and position of the generalized liquid-drop potential.

barrier energies. The model fails not only to reproduce the barrier height, but also the slope of the excitation functions. One may therefore conclude that in these cases a quantitative understanding of fusion requires the inclusion of other degrees of freedom, in particular the formation of a neck. In table 1 relevant parameters, on which the onset of neckformation may depend, are collected for the five systems of fig. 4. Since there is some uncertainty in the relative normalisation of different measurements of the fusion cross section (as seen in fig. 3) and since excitation functions at and below the barrier are only available for a few heavier systems no systematic investigation of these effects has been made.

It has been noticed [10] that agreement between the data for heavier systems and the barrier-penetration model can be achieved by substituting the transmission coefficients $T_1(E)$ in (1) by energy-averaged values. Averaging widths of a few MeV are required. The averaging is meant to simulate the effects of deformation degrees of freedom. The fits are however not easily converted into more specific statements about changes of the nuclear shape during fusion.

Resonance structures observed in fusion cross sections of a few light systems involving α-nuclei [16 - 19] are of course beyond the scope of the simple barrier-penetration model.

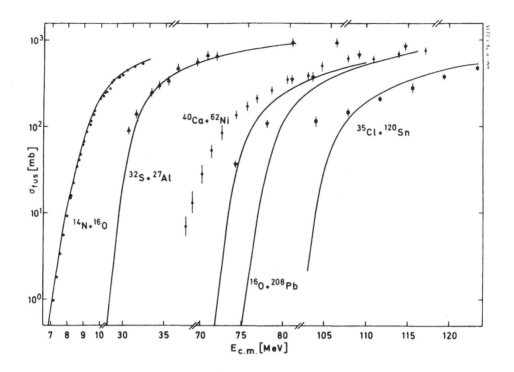

Fig. 4   Fusion excitation functions calculated with the barrier-penetration model. The generalized liquid-drop potential has been
used. The dots are data for ^{14}N + ^{16}O from ref. [11], upward
pointing triangles for ^{32}S + ^{27}Al from ref. [12], dots with
error bars for ^{40}Ca + ^{62}Ni from ref. [13], downward pointing
triangles for ^{16}O + ^{208}Pb from ref. [14], and squares for
^{35}Cl + ^{120}Sn from ref. [15].

For energies well below the barrier the Hill-Wheeler approximation for
the transmission coefficients (4) is insufficient. Instead the Schrödinger equation has to be integrated with an ingoing-wave boundary condition. Therefore the knowledge of the effective, local, heavy-ion potential is required not only around the barrier, but also further inside.
As will be discussed later, there is considerable uncertainty about the
nuclear potential at close distances. Therefore we are severly limited
in our ability to extrapolate measured fusion rates well below the barrier [11, 20, 21] into the still lower energy range of astrophysical interest below 1 MeV center-of-mass energy [22, 23]. The situation is of
course even worse when the excitation function is influenced by intermediate resonances [19]. The measurement of reaction rates far below

System	$Z_1 Z_2$	d [fm]
$^{14}N + ^{16}O$	56	2.31
$^{27}Al + ^{32}S$	208	1.76
$^{40}Ca + ^{62}Ni$	560	1.49
$^{16}O + ^{208}Pb$	656	1.27
$^{35}Cl + ^{120}Sn$	850	1.14

Table 1 $Z_1 Z_2$ values and distances d between equivalent sharp surfaces at the position of the s-wave barrier for the systems shown in fig. 4.

the barrier is on the other hand a unique tool to determine the potential on the "back" side of the interaction barrier.

Compared to the classical model (2), (3) two refinements are introduced by the barrier-penetration model. First, the sharp cutoff by a critical angular momentum in fig.2 is replaced by a smooth transition region with a width of a few l-values. Second, the inward shift of the barrier position with increasing l and the l-dependence of the curvature of the barrier top are properly incorporated. In fig. 5 the l-dependence of these two quantities is shown for a typical light system and a heavier one. It is seen that for light systems the changes are substantial, but that they can be neglected in heavy systems.

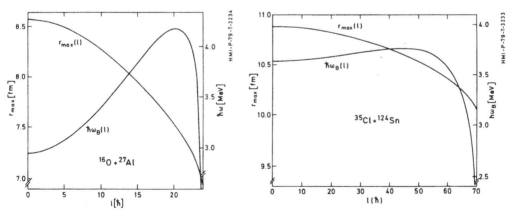

Fig. 5   The barrier position and curvature for $^{16}O + ^{27}Al$ and $^{35}Cl + ^{124}Sn$ as a function of angular momentum. The curvature is given by $\hbar \omega_B$ defined by (6). The nuclear potential is from ref. [8].

Two approximations to the model based on (1) and (4) have been proposed. In Wong's model [24] the l-dependence of the barrier is neglected. If the sum in (1) is approximated by an integral, a closed expression for the fusion cross section in terms of the height, position, and curvature of the s-wave barrier is obtained

$$\sigma_{fus} = \frac{\hbar \omega_B r^2_{max}}{2 E} \ln [1 + \exp(2\pi \frac{E-V_B}{\hbar \omega_B})] \qquad (8)$$

with

$$\omega_B = \omega_B (l=0)$$

$$r_{max} = r_{max} (l=0)$$

$$V_B = V(r_{max}).$$

Fig. 6 shows the prediction of Wong's model in comparison with the calculation based on (1), (4) - (7). As to be expected the calculations agree for energies at and below the barrier, the region dominated by the s-wave. For higher energies (8) approaches asymptotically the classical expression (3). In heavier systems the deviation at high energies is less pronounced.

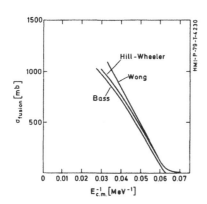

Fig. 6 Fusion excitation function for ^{16}O + ^{27}Al in the Bass and Wong models [24, 25] compared with the full barrier-penetration model (labeled Hill-Wheeler). The potential is from ref. [8].

Bass has pointed out [25] that in the classical model the increase in the fusion cross section with increasing energy $\sigma(E + \Delta E) - \sigma(E)$ is due to the angular momentum bin around $l_{cr}(E)$. This increase should therefore reflect the radius and height of the barrier of the effective potential for $l_{cr}(E)$. Since l_{cr} increases from 0 to some $l_{cr}(E_{max})$ when the energy E varies in the range $V_B (l=0) \leq E \leq E_{max}$ the differential cross section $\partial \sigma / \partial E$ in this energy range should reflect the form of the potential between $r_{max}(1=0)$ and $r_{max}(1_{cr}(E_{max}))$. Fig.6 shows an excitation function obtained by taking the shift of the barrier with increasing angular momentum properly into account, but keeping the sharp-cutoff prescription (2) of the classical model. It is seen that this model deviates from the full barrier-penetration model at all energies because of the sharp-cutoff assumption. Bass proposes [25] to use experimental values for $\partial\sigma/\partial E$ to extract a phenomenological nuclear interaction potential. Apart from the systematic error in the differential quantity $\partial\sigma/\partial E$ introduced by the sharp cutoff, the data do not seem to be accurate enough to extract the derivative from the measured integral quantity σ in order to construct the nuclear potential.

3. Compound-Nucleus and Entrance-Channel Effects

The next step in sophistication of fusion models consists in a more detailed description of the compound nucleus (at high angular momenta) and of the dynamics in the entrance channel. We start with a discussion of limiting effects of compound-nucleus properties on the fusion cross section.

3.1 Limiting Condition for the Compound-Nucleus Formation

Within the classical model an upper limit for the critical angular momentum at a given energy can be obtained [9] by requiring that all the available energy is converted into rotational energy

$$\frac{\hbar^2 \, 1^2_{cr}}{2\,\Theta\,(1_{cr})} = Q + E + E_{def}(1_{cr}), \qquad (9)$$

where $\Theta(1_{cr})$ is the rigid moment of inertia which corresponds to the equilibrium shape of the compound nucleus with angular momentum 1_{cr}, E_{def} is the deformation energy of this shape, and Q is the reaction Q-value. Clearly it is unlikely that all the available energy appears as rotational energy, i.e. that the compound nucleus is formed at its yrast line. However, as pointed out by Mosel [9] (cf. also [68]), (9) can be used together with (1) to obtain an upper limit for the fusion cross section

$$
\sigma_{fus} \leq
\begin{cases}
\pi \dfrac{\theta}{\mu} \left(1 + \dfrac{Q + E_{def}}{E} \right) & E \leq E_{cr} \\[2ex]
\pi \lambda^2 \, (1_{max} + 1)^2 & E > E_{cr},
\end{cases}
$$

where 1_{max} is the angular momentum for which the fission barrier of the compound nucleus vanishes. No higher partial waves can contribute the compound formation. E_{cr} follows from

$$
E_{cr} = \frac{\hbar^2 \, 1^2_{max}}{2 \theta \, (1_{max})} - Q - E_{def} \, (1_{max}).
$$

Nuclear equilibrium shapes and deformation energies have been calculated in the liquid-drop model with [28] and without shell corrections [26, 27, 30] and with the projected Hartree-Fock-Bogoliubov method with a pairing-plus-quadrupole force [29]. Most of these calculations were restricted to spheroidal and ellipsoidal shapes [27] or quadrupoloids, characterized by deformation parameters β, γ [29] or ε, γ, ε_4 [28]. Only in the liquid-drop calculation of ref. [26] more general equilibrium shapes are considered.

One can summarize the results of multi-dimensional potential-energy calculations, which include the mass asymmetry and the neck-size among the deformation parameters [30], in the following way. The compound state is surrounded in this many-dimensional space of shape parameters by a mountain ridge, the lowest pass of which is the fission saddle point. With increasing charge and angular momentum of the system the area inside the mountain ridge shrinks and the area of strong nuclear interaction outside the ridge increases. Heavy-ion collisions leading to such heavier combined systems can therefore result in trajectories which never pass the mountain ridge, but remain in the strong interaction regime (corresponding to a rather large neck cross section) for a fairly long time. There can also be trajectories which correspond to the same reaction time, but which happen to pass the mountain ridge, only to leave it again at a point where the height of the ridge is somewhat lower. There is no qualitative difference between such trajectories and no experimental signature can be found to distinguish between them. The only criterion for a distinction between a deep-inelastic collision and fusion-fission can be on the basis of the reaction time (measured for instance by the relatively slow mode of the mass asymmetry or its spreading width). However, even a sharp distinction between these reactions by means of the reaction time remains necessarily somewhat arbitrary.

3.2 Global Description of Internal Degrees of Freedom in the Fusion Process

The simplest and most schematic description of energy and angular-momentum transfer from the orbital motion into intrinsic degrees of freedom is given in the critical radius model [31 - 34]. The expression (1) for the fusion cross section is substituted by

$$\sigma_{fus} = \pi \lambdabar^2 \sum_{l=0}^{\infty} (2\ l + 1)\ P_l\ T_l,$$ (10)

where P_l is the absorption probability inside the pocket. This expression implies that transmission through the barrier and absorption inside are statistically independent. In the critical-radius model the assumption is made that fusion is described by the dynamics of two point masses in a static potential without coupling to other degrees of freedom until a critical distance R_c (inside the barrier) is reached and that there is infinite radial or infinite tangential and radial friction at the critical distance. From these assumptions follows

$$P_l = \begin{cases} 1 & l \leq l_{cr} \\ 0 & l > l_{cr}, \end{cases}$$ (11)

where the critical angular momentum l_{cr} is determined in this case by

$$\frac{\hbar^2}{2\ \mu\ R_c^2}\ l_{cr}^2 = E - V(R_c) \qquad \text{only radial friction} \qquad (12)$$

or

$$\frac{\hbar^2}{2\ \Theta\ (R_c)}\ l_{cr}^2 = E - V(R_c), \qquad \text{radial and tangential} \qquad (13)$$
$$\text{friction.}$$

where $\Theta(R_c)$ is the rigid moment of inertia of the fusing configuration at centers-of-mass distance R_c.

For energies well above the barrier the fusion cross section is determined in this model by the P_l- factors in (10) rather than the transmission coefficients T_l. From (10) - (12) follows the sharp-cutoff-type formula

$$\sigma_{fus} = \pi\ R_c^2\ (1 - \frac{V(R_c)}{E}) \qquad \text{for } E \gg V_B.$$

Eq. (12) requires the knowledge of the potential for strongly overlapping nuclei. Soft-core potentials obtained from the "sudden approximation" in the formalism of ref. [59] are used to obtain the high energy branch of fusion excitation functions [9, 31]. The straight line on the high energy side of fig. 3 is obtained from these calculations. However, as will be explained in the next chapter, soft-core potentials seem to be somewhat unrealistic for fusion.

The concentration of friction far inside the interaction barrier in the critical-radius model does not allow a description of deep-inelastic reactions. Therefore radial and tangential friction have been given radial formfactors and l_{cr} (to be used in (1)) is obtained from solving classical equations of motion for the orbital and rotational degrees of freedom

$$\mu\ddot{r} = -\frac{\partial V}{\partial r} + \mu r \dot{\vartheta}^2 - f_r(r) \dot{r}$$

$$\mu r^2 \dot{\vartheta} = l_i - l_1 - l_2, \qquad\qquad l_j = \theta_j \dot{\vartheta}_j, \quad j=1,2 \qquad (14)$$

$$\dot{l}_j = -C_j \left(\frac{r}{C_1 + C_2}\right)^2 f_\vartheta(r) \sum_{k=1}^{2} C_k (\dot{\vartheta}_k - \dot{\vartheta}), \quad j=1,2$$

The rotation angles of the two nuclei and the orbital angle are denoted by ϑ_1, ϑ_2, and ϑ respectively, $f_r(r)$ and $f_\vartheta(r)$ are the radial and tangential friction form-factors respectively l_j, θ_j, C_j are the angular momentum, moment of inertia and half-density radius of nucleus j respectively, and l_i is the initial orbital angular momentum of the system. Orbits which are trapped behind the Coulomb barrier are counted as leading to fusion in this model.

The set of equations (14) is based on the following assumptions: (I) The motion can be described by classical, deterministic equations. This excludes, for example, the possibility that the flux for a given initial energy and angular momentum is split between fusion and a binary exit channel. It also restricts the model to energies somewhat above the barrier. (II) There is instant thermalisation of the dissipated energy, which excludes memory effects. (III) Inertial parameters are in practice kept constant, even for rather close distances between the nuclei.

The numerous calculations performed in this scheme [35 - 40] differ essentially only in the choice of moments of inertia (they are e.g. assumed to be infinite in ref. [36]) and radial formfactors for the conservative potential and the friction terms. Clearly fusion data are insufficient to determine all formfactors on a strictly phenomenological basis. Therefore the potential is usually taken from other sources. But even then are the friction formfactors not unambiguously determined by the fusion data alone so that deep-inelastic cross sections are required to be described within the same scheme [36]. The freedom in the choice of phenomenological form factors seems to be sufficiently large (even after inclusion of deep-inelastic reactions) to reproduce the data reasonably well, despite the somewhat oversimplifying assumptions on which eqs. (14) are based.

Attempts have also been made to use theoretical rather than phenomenological form factors. Using the proximity force [41] together with the proximity friction [42] leads to a fair overall agreement with the existing fusion data [35], though (with very few exceptions) the theory tends to underestimate the data slightly. The proximity friction is on the other hand known to yield somewhat insufficient energy dissipation in deep-inelastic reactions [43].

A lower l-window as schematically shown in fig. 7 can be obtained in classical trajectory calculations only for unrealistically weak friction [35, 36] in conjunction with soft-core potentials. There does not seem to be undisputed experimental evidence for the existence of a lower l-window either [44].

3.3   Coupled Channel Calculations

Some of the surface modes coupled to the orbital motion have typical times of some 10^{-22} sec. Their influence on the orbital motion is therefore inadequately described by friction coefficients. In particular surface multipole vibrations of the quadrupole type [38] or of various multipolarities [45] have been added to the equations (14) and coupled via appropriate potentials to the orbital motion. The surface vibrations are treated as classical, harmonic oscillators (and cannot easily be quantized unless the coupling is - somewhat unrealistically - assumed to be linear). Therefore the same caveats apply to this model as discussed above under (I) and (III) in connection with the classical trajectory model. It has been assumed [45] that the oscillators are damped and

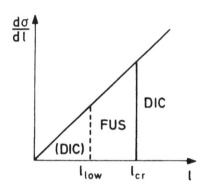

Fig. 7   Lower l-window in a
         classical sharp cutoff-
         model.

the proximity friction is used
to account for the effect of particle exchange on the orbital motion. These various damping terms
in addition to the inertial parameters, spring constants and coupling potentials of all the explicitly treated degrees of freedom
amount to a rather large number
of input parameters. The model also depends on the assumption that
the surface modes are harmonic and
stay orthogonal even for the considerable distortions of the surface encountered in actual calculations.

One would naively expect that the neck formation is the first and most
important intrinsic mode affecting the entrance channel. Only one preliminary calculation has been reported [46] in which the neck degree of
freedom is explicitly coupled to the orbital motion and treated in the
framework of classical mechanics. It should however be kept in mind that
the formation as well as the eventual rupture of the neck probes the
tensil strength of nuclear matter [47] and depends, among other things,
on the nuclear compressibility. It is therefore not simply a surface-
shape mode and cannot be completely described by a superposition of surface multipole vibrations.

Fusion has also been treated as a problem of classical hydromechanics
[48, 49]. The infinitely many degrees of freedom have been reduced in
[48] to three symmetric surface-mode parameters one of which is the
distance between centers of mass. The flow pattern was assumed in [48]
to be that of an incompressible, irrotational flow, though a friction
term resulting from ordinary two-body viscosity was included in the
Lagrange equations for collective motion. Sharp-surface boundary conditions for the velocity field are used in both of these hydrodynamical
treatments. In view of the fact that the fusion barrier lies outside
the touching point for most systems this assumption does not seem to
represent the actual mass flow in the neck region during the decisive
stage of the fusion process.

An additional complication arises if one or both of the colliding nuclei have deformed ground states. Two effects can be distinguished, which arise from non-spherical shapes. (I) there is a static, geometrical effect. The interaction potential and therefore the barrier height and position depend on the orientation of the symmetry axis of the deformed nucleus with respect to the radius vector of the orbital motion. In general the latter does not remain in a plane because of the non-central interaction potential. (II) Before the trajectory reaches the top of the barrier, rotational states can be excited. This reduces the energy of the orbital motion, which leads to an increase of the apparent barrier height. At the same time the transfer of angular momentum reduces the centrifugal potential and therefore lowers the apparent barrier height.

Generally only the first effect is treated to some approximation. For unpolarized target and projectile Wong proposed [24] to calculate the fusion cross section with eqs. (1), (4) - (7) for each orientation of the symmetry axis and finally to average the cross section over all orientations. The multipole moments of the interaction potential as functions of the orientation (needed in (5) and (6)) were calculated from the intrinsic quadrupole moment.

For aligned , deformed nuclei the geometrical effect has also been treated [50] in the barrier-penetration model. However, the interaction potential was calculated with the expectation value of the mass and charge distribution of the deformed nucleus in its ground state, i.e. with the spectroscopic quadrupole moment. The prescription is easily generalized to a situation where the initial state is given by the polarisation tensor.

Both recipes are based on a classical approximation with respect to the rotational degree of freedom, however at somewhat different places of the barrier-penetration calculation. A correct description of fusion, which also accounts for the dynamical effects (II), is in terms of a quantum-mechanical coupled-channel calculation. Such a calculation seems to be feasable only for a system with vanishing ground state spin of the deformed nucleus. Only one, somewhat schematic coupled channel calculation for the system $^{12}C + ^{24}Mg$ has been performed [51]. A conventional optical potential, with $\beta_{Mg} = 0.4$ was used and rotational states up to $I^{\pi} = 10^+$ were included. Fig. 8 shows the contribution of the various angular momenta to fusion and to the inelastic cross section

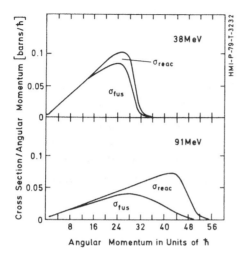

Fig. 8   Total reaction and fusion cross sections for the system $^{12}C + ^{24}Mg$ obtained by a coupled channel calculation. From ref. [51].

for two center-of-mass energies. It is clearly seen that the concept of a critical angular momentum looses even its approximate relevance in this calculation for higher energies.

The example may serve as a warning that classical approximations , which lead to a sharp cutoff in angular momentum, can fail to describe the situation even qualitatively, in particular when fusion is only the smaller part of the total reaction cross section. There are of course also experimental indications that this situation occurs [52].

It is useful to remember under which circumstances a large deviation is to be expected between the solution of stochastic equations for an ensemble or quantum-mechanical equations on the one hand and of corresponding classical, deterministic equations on the other hand. Whenever a mountain ridge or saddle point in the potential landscape splits the classical trajectories of a system into two qualitatively different classes (binary and fusion reactions in our case) they are separated by a sharp cutoff in the space of initial conditions if deterministic equations of motion are used. However the fluctuation within an ensemble smoothens the cutoff, the more the larger the fluctuation is. For this argument it does not matter whether the origin of the fluctuation is quantal or thermal. In the above-mentioned example the increasing excitation of various rotational states with increasing bombarding energy leads to a growing quantum uncertainty of the orbital trajectory near the barrier and hence to a stronger washing-out of the sharp-cutoff angular-momentum in fig. 8.

4. Interaction Potentials

Almost all fusion models require a heavy-ion interaction potential as input. The various methods to derive such potentials can be grouped into five broad classes: (I) the cluster and generator-coordinate model [53 - 57], (II) the folding model [58] (and references given there), (III) the Thomas-Fermi model [41, 59], (IV) the generalized liquid-drop model [8], and (V) the interpolating potential models [25, 60, 61]. The first three methods start in principle with a two-nucleon interaction or G-matrix and attempt to solve the A-body problem in one of the standard approximation schemes. In the last two methods the potential is constructed on a more phenomenological level. Since heavy-ion potentials have been dealt with in several review articles [43, 58, 62] it will be sufficient to restrict this discussion to a few points of particular relevance for fusion.

In methods (IV) and (V) the nuclear potential is required to satisfy some or all of the following four conditions

$$V(0) = Q \qquad\qquad (15)$$

$$V'(C_1+C_2) = \bar{C}\, r_0^{-2} C_s \qquad\qquad (16)$$

$$V''(C_1+C_2) = 0 \qquad\qquad (17)$$

$$V(r_s) = \mathrm{Re}\, U_{opt}(r_s) \qquad\qquad (18)$$

In (15) - (18) Q is the fusion reaction Q-value, C_j are the half-density radii of the two nuclei, $\bar{C}$ is the reduced radius $C_1 C_2 / (C_1+C_2)$, $r_0 = 1.18$ fm is the radius constant related to the saturation density of uncharged nuclear matter, and C_s is the surface-energy constant of the liquid-drop model [41]; r_s is the strong absorption radius and U_{opt} the optical potential fitted to low-energy elastic scattering of light nuclei.

A very early derivation of (16) is given in ref. [63], a more recent one in ref. [41]. The condition (17) is a consequence of the saturation of nuclear forces and has been discussed in ref. [8, 64]. The proximity relations (16) and (17) are only valid in the limit of large nuclear radii (large compared to the range of the nuclear interaction). This introduces some ambiguity in the strength of the attractive force at the contact point for lighter nuclei. But generally the four conditions determine the interaction potential with sufficient accuracy as long as the shapes of the fusing nuclei are and remain spherical. This seems to be

the case for light, spherical nuclei and energies of a few MeV around
the barrier. The inner slope of the barrier is however not sufficiently
determined by the conditions (15) - (18), which leads to ambiguities in
the prediction of S-factors for fusion reactions well below the barrier
[23].

In a cluster-model treatment of elastic heavy-ion scattering the potential is highly non-local in the region of strongly overlapping densities
[54]. The effect of the non-locality is to keep the amplitude of the ela-
stic-channel wave-function small in this region. This is simply an expression of the Pauli principle and the small compressibility of nuclear
matter. For the elastic channel the non-locality may be approximated by
an increased mass or equivalently a soft-core local potential for short
centers-of-mass distances. How useful this substitution may be for the
elastic channel, it does not apply to the fusion reaction. In fusion the
system is followed as it passes through a sequence of inelastic reaction
channels which connect the entrance configuration of two spheres with the
compound-nucleus configuration of one larger sphere through a continuous
sequence of shapes. An "adiabatic" potential in the sense that it corresponds to a preservation of the nuclear volume during fusion is therefore more appropriate for fusion than a soft-core potential.

5. Time-Dependent Hartree-Fock Calculations

The use of the TDHF method [65] seems to be an elegant way to avoid all
the ambiguities of more or less phenomenological parameters in the fusion
models discussed so far. The only input of the method is the effective
two-body interaction potential. No sequence of shapes, friction form factors, heavy-ion potentials or reaction mechanisms have to be prescribed
ad hoc. There are however some serious deficiencies of TDHF, which war-
rant some caution.

As a result of the complete neglect of residual two-body interactions in
TDHF, energy and angular-momentum transfer out of the orbital motion is
due to the excitation of (undamped) surface modes or to one-body "dissipation" in the strict sense of the word. It should however be remembered
that all dissipation models require two-particle collisions either implicitly or explicitly, including the one-body wall-and-window formalism [66]. In the latter one has to assume randomisation of energy
and momentum of each nucleon between consecutive collisions with the
moving wall. If the wall is not "corrugated" two-particle collisions are

a natural reason for the required fast loss of memory. Likewise the coupled-oscillator model [45] introduces two-particle collisions in-directly by attributing a damping width to the oscillators. In general all dissipation theories based on the idea of a Markov chain (which leads to a master equation) require two-particle collisions to estab-lish statistical equilibrium fast enough. It may therefore be expected that TDHF calculations overestimate memory times and lead to surface vi-bations with too large amplitudes.

In the TDHF method the dynamics is restricted to the set of Slater-de-terminantal wave-functions, which do not form a linear space. The dif-ficulties in constructing a scattering matrix under these circumstances have often been discussed [65, 67]. A particular example of this problem is the fact that for given initial conditions the TDHF equations lead either to fusion or to a binary exit channel. They are therefore not able to describe the situation shown in fig. 8, where the flux of a given partial wave splits between fusion and inelastic channels.

We have discussed the conceptual basis of several fusion models. Apart from their success or failure to describe a set of heavy-ion reaction data (after some parameter fitting) neither one incorporates all features which can be expected to be important for the description of a phenome-non as complex as heavy-ion fusion. Limited ranges of mass numbers and bombarding energies can however be specified in which some particular models describe the situation adequately. Entrance-channel break-up pro-cesses have deliberately been excluded from the discussion. They will be dealt with in other contributions to this conference.

References

[1]   G.H.Rawitscher, J.S. McIntosh, and J.A. Polak, Proc. of the
       Third Conf. on Interactions between Complex Nuclei, ed. by
       A. Ghiorso, R.M. Diamond, and H.E. Conzett (University of Cali-
       fornia Press, Berkeley, California, 1963) p. 3
       V.M. Strutinsky, Nucl. Phys. 68 (1965) 221
       S.L. Tabor, B.A. Watson, and S.S. Hanna, Phys.Rev. C14 (1976) 514

[2]   D.L.Hill and J.A. Wheeler, Phys.Rev. 89 (1953) 1102

[3]   J. Dauk, K.P. Lieb, A.M. Kleinfeld, Nucl.Phys. A241 (1975) 170

[4]   R.L. Kozub, N.H.Lu, J.M. Miller, D. Logan, T.W. Debiak, and
       L. Kowalski, Phys.Rev. C11 (1975) 1497

[5]   Y. Eisen, J. Tserruya, Y. Eyal, Z. Fraenkel, and M. Hillman,
        Nucl.Phys. A291 (1977) 459

[6]   B.B. Back, R.R. Betts, C. Gaarde, J.S. Larsen, E. Michelsen,
        and Tai Kuang-Hsi, Nucl.Phys. A285 (1977) 317

[7]   R. Rascher, W.F.J. Müller, and K.P. Lieb, Phys.Rev. C20 (1979)1028

[8]   H.J.Krappe, J.R.Nix, and A.J. Sierk,Phys.Rev.Lett. 42 (1979) 215,
        Phys.Rev. C20 (1979) 992

[9]   U. Mosel, contribution to this symposium and to be published in
        Heavy Ion Collisions, ed. by R.Bock (North Holland Publishing
        Company) Vol. 2

[10]   L.C. Vaz and J.M. Alexander, Phys.Rev. C18 (1978) 2152

[11]   Z.E.Switkowski, R.G.Stockstad, and R.M. Wieland, Nucl.Phys.
        A279 (1977) 502

[12]   H.H.Gutbrod, N.G.Winn, and M. Blann, Nucl.Phys. A213 (1973) 267

[13]   B. Sikora, J. Bisplinghoff, W. Scobel, M.Beckermann, and M. Blann,
        preprint (Hamburg 1979)

[14]   F. Videbaek, R.B. Goldstein, L. Grodzins, S.G. Steadman, T.A.Belote
        and J.D. Garrett, Phys.Rev. C15 (1977) 954

[15]   P. David, J. Bisplinghoff, M. Blann, T. Mayer-Kuckuk, and A. Mig-
        nerey, Nucl.Phys. A287 (1977) 179

[16]   P. Sperr, T.H.Braid, Y. Eisen, D.G.Kovar, F.W.Prosser, J.P. Schif-
        fer, S.L.Tabor, and S. Vigdor, Phys.Rev.Lett. 37 (1976) 321
        P.Sperr, S. Vigdor, Y. Eisen, W. Henning, D.G.Kovar, T.R. Ophel,
        and B. Zeidman, Phys.Rev.Lett. 36 (1976) 405

[17]   B. Fernandez, C. Gaarde, J.S.Larsen, S. Pontoppidan, and F. Vide-
        baek, Nucl.Phys. A306 (1978) 259

[18]   I. Tserruya, Y. Eisen, D. Pelte, A. Gavron, H. Oeschler, D. Berndt,
        and H.L. Harney, Phys.Rev. C18 (1978) 1688

[19]   K.U.Kettner, H. Lorenz-Wirzba, C. Rolfs, and H. Winkler, Phys.
        Rev.Lett. 38 (1977) 337

[20]   R.G. Stokstad, Z.E. Switkowski, R.A. Dayras, and R.M. Wieland,
        Phys.Rev.Lett. 37 (1976) 888
        Z.E. Switkowski, Shiu-Chin Wu, J.C. Overley, and C.A. Barnes,
        Nucl.Phys. A289 (1977) 236

[21]   M.D. High and B. Čujec, Nucl.Phys. A259 (1976) 513,
        Nucl.Phys. A282 (1977) 181

[22]   W.A.Fowler, G.R. Caughlan, and B.A. Zimmermann, Ann.Rev.Astron.
        Astrophys. 13 (1975) 69

[23]   C. Rolfs and H.P. Trautwetter, Ann.Rev.Nucl.Sci. 28 (1978) 115

[24]   C.Y. Wong, Phys.Rev.Lett. 31 (1973) 766

[25]   R. Bass, Phys.Rev.Lett. 39 (1977) 265

[26]   S. Cohen, F. Plasil, W.J. Swiatecki, Ann.Phys.(N.Y.) 82 (1974) 557

[27]   R.L. Hatch and A.J. Sierk, preprint (1978) MAP-3

[28]   S. Åberg, S.E. Larsson, P. Möller, S.G. Nilsson, G. Leander, and
        I. Ragnarsson, Proc.Fourth IAEA Symp. on Physics and Chemistry of
        Fission, Jülich 1979, paper SM/241-C4

[29]   A. Faessler, K.R. Sandhya Devi, F. Grümmer, K.W. Schmidt, R.R.Hil-
        ton, Nucl.Phys. A256 (1976) 106
        M. Ploszajczak, K.R. Sandhya Devi, A. Faessler, Z.Phys. A282
        (1977) 267

[30]   H.J. Krappe, Nucl.Phys. A269 (1976) 493

[31]   D. Glas and U. Mosel, Nucl.Phys. A237 (1975) 429

[32]   R. Bass, Nucl.Phys. A231 (1974) 45

[33]   J. Galin, D. Guerreau, M. Lefort, and X. Tarrago, Phys.Rev.C9
        (1974) 1018

[34]   M. Lefort, Lecture Notes in Phys. 33 (1975) 275;
        Rep.Prog.Phys. 39 (1976) 129

[35]   J.R. Birkelund, L.E. Tubbs, J.R. Huizenga, J.N. De, and D. Sperber,
        Phys.Rep. C56 (1979) 108

[36]   D.H.E. Gross and H. Kalinowski, Phys.Lett. 48B (1974) 302
        Phys.Rep. C45 (1978) 175
        D.H.E. Gross, H. Kalinowski, and J.N. De, Symposium on Classical
        and Quantum Aspects of Heavy Ion Collisions, Heidelberg 1974,
        Lecture Notes in Phys. 33 (1975) 194
        J.N.De, Phys.Lett. 66B (1977) 315

[37]   J.P. Bondorf, M.I. Sobel, and D. Sperber, Phys.Rep. C15 (1974) 83
        J.P. Bondorf, J.R. Huizenga, M.I. Sobel, and D. Sperber, Phys.Rev.
        C11 (1975) 1265

[38]   H.H. Deubler and K. Dietrich, Phys.Lett. B56 (1975) 241

[39]   C.F. Tsang, Phys.Scrip. 10A (1974) 90

[40]   F. Beck, J. Błocki, M. Dworzecka, and G. Wolschin, Phys.Lett. 76B
        (1978) 35

[41]   J. Błocki, J. Randrup, W.J. Swiatecki, and C.F. Tsang, Ann.Phys.
        (N.Y.) 105 (1977) 427

[42]   J. Randrup, Ann.Phys. (N.Y.) 112 (1978) 356

[43]   W.U.Schröder and J.R. Huizenga, Ann.Rev.Nucl.Sci. 27 (1977) 465

[44]   H.C.Britt, B.H. Erkkila, P.D. Goldstone, R.H. Stokes, B.B. Back,
        F.Folkmann, O.Christensen, B. Fernandez, J.D. Garrett, G.B.Hage-
        mann, B. Herskind, D.L. Hillis, F. Plasil, R.L.Ferguson, M.Blann,
        and H.H.Gutbrod, Phys.Rev.Lett. 39 (1977) 1458

J.B. Natowitz, G. Boukellis, B. Kolb, G. Rosner, and Th.Walcher, Preprint (Max-Planck-Institut für Kernphysik, 1978)

[45] R.A. Broglia, C.H.Dasso, G. Pollarolo, and A. Winther, Phys.Rev. Lett. 40 (1978) 707
R.A. Broglia, C.H. Dasso, and A. Winther, Proc.Intern.School of Phys. Enrico Fermi, LXXVII Course, Varenna 1979

[46] W.J. Swiatecki, private communication 1979; cf. also M. Lefort, contribution to this conference

[47] G.F. Bertsch, Nuclear Physics with Heavy Ions and Mesons, Vol. 1, eds. R. Balian, M. Rho, and G. Ripka (Les Houches Lectures 1977) (North Holland Publ.Comp. 1978) p. 177

[48] A.J. Sierk and J.R. Nix, Proc.Third IAEA Symp. on Phys.and Chem. of Fission, Rochester 1973, 2 (1974) 273; Phys.Scrip. A10 (1974) 94; Phys.Rev. C15 (1977) 2072

[49] C.T. Alonso, Proc.Int.Colloquium on Drops and Bubbles, Pasadena, Cal. 1974, 1 (1976) 139

[50] H.J. Krappe and H. Massmann, Proc. XVth. Intern.Winter Meeting on Nucl.Phys., Bormio 1977 (Universita' di Milano 1977) p.227
Z.Phys. A286 (1978) 331
H.J. Krappe, Proc. XVII Intern. Winter Meeting on Nucl.Phys. Bormio 1979 (Universita' di Milano 1979) p. 253

[51] D. Pelte and U. Smilansky, Phys.Rev. C19 (1979) 2196

[52] G.B. Hagemann, R. Broda, B. Herskind, M. Ishihara, S. Ogaza, and H. Ryde, Nucl. Phys. A245 (1975) 166
D.G. Sarantites, J.H. Barker, M.L. Halbert, D.C. Hensley, R.A. Dayras,E. Eichler, R.N. Johnson, and S.A. Gronemeyer, Phys.Rev. C14 (1976) 2138

[53] H.W.Wittern, Nucl.Phys. 62 (1965) 628

[54] K. Wildermuth and Y.G. Tang, A. Unified Theory of the Nucleus (Vieweg, Braunschweig 1977) chap. 11

[55] T. Fliessbach, Nucl.Phys. A194 (1972) 625; Z.Phys. A272 (1975) 39; Z.Phys. A278 (1976) 353

[56] P.G. Zint and U. Mosel, Phys.Rev. C4 (1976) 1488
G.H. Göritz and U. Mosel, Z.Phys. A277 (1976) 243
P.G. Zint, Z.Phys. A281 (1977) 373

[57] S. Saito, S. Okai, R. Tamagaki, and M. Yasuno, Progr.Theor.Phys. 50(1973) 1561
T. Ando, K. Ikeda, and Y. Suzuki, Progr.Theor.Phys. 54 (1975)119;
A. Tohsaki, F. Tanabe, and R. Tamagaki, Progr.Theor.Phys. 53 (1974) 1022
L.F. Canto, Nucl.Phys. A279 (1977) 97
H. Friedrich and L.F. Canto, Nucl.Phys. A291 (1977) 249

[58] G.R. Satchler and W.G. Love, Phys.Rep. C55 (1979) 185

[59] K.A.Brueckner, J.R. Buchler, and M.M. Kelly, Phys.Rev. 173 (1968) 944

C. Ngô, B. Tamain, J. Galin, M. Beiner, and R.J. Lombard, Nucl. Phys. A240 (1975) 353
C. Ngô, B. Tamain, B. Beiner, R.J. Lombard, D. Mas, and H.H. Deubler, Nucl. Phys. A252 (1975) 237
Fl. Stancu and D.M. Brink, Nucl. Phys. A270 (1976) 236; Nucl. Phys. A299 (1978) 321

[60] R. Bass, Phys. Lett. 47B (1973) 139; Proc. Int. Conf. on Reac. between Complex Nuclei (Nashville 1974) 1 (1974) 117

[61] K. Siwek-Wilczyńska and J. Wilsczyński, Phys. Lett. 74B (1978)313

[62] H.J. Krappe, Lecture Notes in Physics 33 (1975) 24
D.M. Brink, Journ. de Phys. 11 (1976) C5 - 47
U. Mosel, Proc. Symp. Macroscopic Features of Heavy-Ion Collisions (Argonne, Ill. 1976) ANL-PHY-76-2, p. 341

[63] R. Bradley, Phil. Mag. 13 (1932) 853

[64] H.J. Krappe, Proc. Intern. Workshop VI on Gross Properties of Nuclei and Nuclear Excitations, Hirschegg, Austria 1978 (TH Darmstadt Report AED-Conf-78-007-014)

[65] S.E. Koonin, Proc. Intern. School of Nuclear Physics, Erice, Italy 1979, preprint MAP-5

[66] J. Błocki, Y. Boneh, J.R. Nix, J. Randrup, M. Robel, A.J. Sierk, and W.J. Swiatecki, Ann. Phys. (N.Y.) 113 (1978) 330

[67] J.J. Griffin, Proc. Topical Conf. on Heavy-Ion Collisions, Fall Creek Falls, Tenn. 1977 (CONF-770602) p. 1

[68] W. Nörenberg and H.A. Weidenmüller, Introduction to the theory of Heavy-Ion Collisions, Lecture Notes in Phys. 51 (1976) chap.3

PREEQUILIBRIUM EMISSION

IN HEAVY-ION INDUCED FUSION REACTIONS

F. Pühlhofer

Fachbereich Physik der Universität Marburg, Germany

1. Introduction

Fusion reactions induced by heavy ions between ^{46}Ti and ^{84}Kr were studied extensively by our group in a collaboration between the University of Marburg (B.Kohlmeyer, F.Busch, M. Canty, W. Pfeffer) and GSI Darmstadt (W.Schneider, H.Freiesleben). In various reactions medium-weight compound nuclei with masses between A=70 and 110 and excitation energies ranging from 50 to 200 MeV were produced, and their decay was studied by measuring mass and Z-distributions of the evaporation residues using the time-of-flight method.

In this report we discuss a single aspect of these investigations, the evidence concerning the change of the reaction mechanism as a function of the energy. This subject seems interesting especially in the context of the extensive discussion of the incomplete-fusion mechanism in reactions with lighter projectiles during this meeting.

2. Experimental data

The reaction considered in the following is the formation and the decay of the compound nucleus ^{70}Se:

5.0 MeV/u	^{58}Ni on ^{12}C	CN = ^{70}Se	E_x = 49.8 MeV
4.7 MeV/u	^{46}Ti on ^{24}Mg	CN = ^{70}Se	E_x = 77.4 MeV
5.8 MeV/u	- " -	CN = ^{70}Se	E_x = 94.5 MeV
8.5 MeV/u	- " -	CN= ^{70}Se	E_x = 136. MeV

The measurements were performed at the UNILAC heavy-ion accelerator at GSI. The method of investigation was the direct identification of the recoiling evaporation residues in a time-of-flight $\Delta E-E$ telescope. In order to obtain a sufficiently high recoil velocity required for a good mass and Z-resolution the heavier reaction partner had to be used as projectile. The higher recoil obtained with a more asymmetric system also was the reason for using a heavier projectile when producing ^{70}Se at the lowest excitation energy. It is assumed that this change of the entrance channel does not have a serious influence on the reaction mechanism.

The main results of these measurements are shown in fig.1, which contains the mass distributions of the heavy products from the reactions mentioned above. From the experience with similar reactions one is sure that at the lower energies these products can be interpreted as residues left over from a highly excited compound nucleus formed in statistical equilibrium after complete fusion of projectile and target. The mass distributions exhibit the typical, strongly structured shape, which one knows from measurements of similar reactions and which is obviously due to the competition of nucleon and α-emission during the deexcitation process. Decay chains can be assigned to the peaks as shown in the figure using the usual rules. One sees, for example, that at 50 MeV the compound nucleus most likely evaporates 3 particles, either 3 nucleons (3N) or 1 α-particle and 2 nucleons (1α 2N), whereas at 77 MeV the average number of emitted particles becomes 4 to 5, and so on.

At the higher energies the mass distributions shift to lighter masses as expected, but their structure is observed to become increasingly washed out. (At 200 MeV excitation we observed a pure Gaussian distribution). Qualitatively this can be understood in the compound nucleus picture. The greater length of the decay chains causes an increasing variation of the number of emitted particles. In addition to that, other decay modes like d-emission become more and more important at higher temperature and their different energy consumption must lead to slightly different lengths of the decay chains. Concerning the left hand tail of the mass distribution at 136 MeV one may think of an interference with products from deep inelastic reactions. Due to their different kinematic behaviour (spectra and angular distributions) those products can be distinguished and they were subtracted. Also, the smoothness of the mass distribution has nothing to do with the experimental mass resolution, which, in fact, becomes much better at higher beam energies.

Fig. 2 contains an attempt to learn something about the reaction mechanism, in particular about its dependence on the energy, without resorting to evaporation calculations. However, it turns out that quantities like the average evaporated mass $\overline{\Delta A}$, the average particle multiplicy $\overline{X}$ and the relative number of α-particles $\overline{X}_\alpha / \overline{X}$, which all can be derived directly from the experimental mass distributions, do not show any dramatic changes as a function of excitation energy. The only surprising fact may be that the slope of $\overline{X}$ increases, which means that the amount of energy needed to evaporate a particle decreases slightly at high temperature. However, because of the changing relative importance of nucleon, α- and γ-emission this fact is barely conclusive without a quantitative model.

Fig. 1 Mass distribution of the evaporation
residues at excitation energies of 50, 77,
95 and 136 MeV

Fig. 2 The average evaporated mass and the
average particle multiplicity as a function
of the excitation energy in the compound nucleus. The data are obtained from the experimental mass distributions.

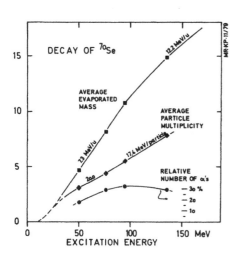

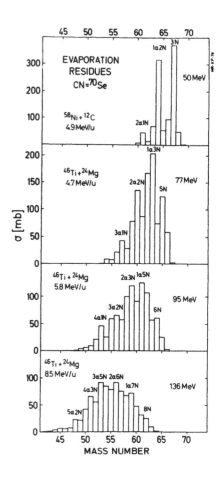

3. Evaporation calculations

In the last section it was suggested that if one looks at the data superficially one
could not discern any change in the reaction mechanism at the higher energies.There-
fore, since there is a good basis for the assumption that at the lower energies (say
below 80 to 90 MeV in this mass range) the reaction can be described quantitatively
assuming formation of a compound nucleus in full statistical equilibrium, one would
expect the same to be true at all energies used here. This impression is proven
wrong by quantitative evaporation calculations.

A set of "standard" calculations performed using the code CASCADE [2]) is contained in
fig. 3. The comparison with the data shows the expected agreement at the low energies; at the two highest energies, however, there are systematic deviations, which
increase with energy, and which consist in an underestimate of the intensity of the
multiple-α decay chains (and a corresponding overestimate of nucleon emission).The
conclusion is: one needs more α-particles and - as details show - of less kinetic

energy in the model. One degree of freedom which is still in the calculations was
left out so far, intentionally, just to make the point. It is the deformation of the
emitting system. It is easy to see that deformation is exactly what is needed to cure
the problem (and it can be shown with some confidence that it is the only means). Deformation lowers the Coulomb barrier, therefore enhances α-emission and lowers the
kinetic energies of the α-particles at the same time. The message from the above
comparison is consequently: the emitting system must be deformed, strongly deformed
as it turns out. The question is only: is this deformation to be interpreted as equilibrium deformation of the rotating compound nucleus or as evidence for precompound
emission.

Concerning the first possibility: From the rotating-liquid-drop theory [1]) one obtains
estimates of the shape changes of a nucleus as a function of angular momentum. Indeed, this theory predicts rather moderate deformations up to about 43 ℏ , which is
the maximum angular momentum leading to fusion at 77 MeV excitation energy in ^{70}Se.
At higher spins, however, the nucleus switches over to a considerably deformed pro-
late shape. At 54 ℏ , the maximum angular momentum obtained at 136 MeV, an axis ratio
of more than 2:1 is predicted.

These deformations enter into the evaporation calculations essentially in two ways.
First, the yrast line and, with it, the level densities at high spins have to be modified. As due to the limitation of fusion by deep inelastic reactions the compound
nucleus population does not come close to the yrast line in the present case, this
does not have a major effect on the deexcitation process. Secondly, however, the
transmission coefficients for the light emitted particles have to be modified, and
this turns out to be of significant influence on the mass distributions. If one includes the predicted deformations using some approximations, which can not be discussed here in detail (but which certainly tend to overestimate the effect), then
one obtains the result displayed in fig.4A. The comparison with the data shows that
the deformation was a step into the right direction. The center of gravity of the
calculated mass distribution shifted somewhat to lighter masses, but it did not move
sufficiently. One would need considerably more deformation than the already large
ones predicted for the equilibrium shape of rotating compound nucleus.

This result suggests to consider the second one of the possible explanations mentioned above. Indeed, there are additional arguments which make one suspicious against
the compound nucleus picture. They are based on the values estimated for the lifetime
of the excited compound nucleus using the statistical theory:

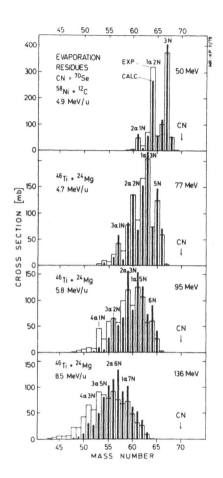

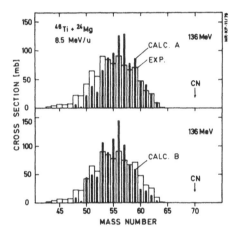

Fig. 3 Evaporation calculations assuming a sherical compound nucleus.

Fig. 4 Attempts to obtain better fits at the highest energy by assuming strong deformation of the emitting system: top: equilibrium shape (deformation 10%/40% above 43 ℏ); bottom: 40% deformation for all angular momenta.

Calculated compound nucleus lifetimes (for ^{70}Se)

E_x	kT(for J=0)	τ (J=0)	τ (J=Lmax)
77 MeV	2.9 MeV	$1.0 \cdot 10^{-21}$ s	$10 \cdot 10^{-21}$ s
96	3.3	$0.8 \cdot 10^{-21}$ s	$5 \cdot 10^{-21}$ s
136	3.9	$0.4 \cdot 10^{-21}$ s	$1.3 \cdot 10^{-21}$ s

These values have to be compared to the time needed to form the compound system. The fusion process itself, being a collective process involving a rearrangement of many nucleons, will have a time constant of the order of $2 \cdot 10^{-21}$ s, which is similar or even longer than the decay times at the highest excitations. Consequently, there must be particle emission already during the fusion process.

This is illustrated in fig. 5 in a somewhat simplifying manner. In the first stage of the collision projectile and target start to interact and to dissipate the energy of the relative motion into internal degrees of freedom. There might in principle be something like an emission of fast particles during this stage, although our data immediately rule out that this is the cause of our concern. In the second stage the reaction partners have attained temperature equilibrium, but there are still two individual nuclei. Finally, there will be a compound nucleus, in equilibrium also with respect to the shape, but deformed due to its rotation.

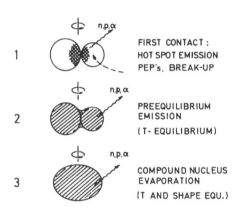

1 FIRST CONTACT:
HOT SPOT EMISSION
PEP's, BREAK-UP

2 PREEQUILIBRIUM
EMISSION
(T-EQUILIBRIUM)

3 COMPOUND NUCLEUS
EVAPORATION
(T AND SHAPE EQU.)

Fig. 5 Emission during the fusion process

The calculated lifetimes show that there must be considerable particle emission already during stage 2 of the fusion process, and the question arises, if this type of emission may have features which distinguish it from compound nucleus emission. The answer is: as long as there is temperature equilibrium already in stage 2, there is barely any basic difference expected. However, the deformation in the intermediate stage might be larger, and its dependence on angular momentum different.

Therefore, we interpret the fact that the data indicate larger deformations of the emitting system than predicted for the shape equilibrated compound nucleus as an experimental evidence for strong contributions of a certain type of preequilibrium emission, namely the one illustrated by fig. 5, stage 2. At present, this statement can not be based upon a quantitative model. Calculations would require additional assumption concerning details like the angular momentum dependence of the deformations and the relative time constants for fusion and decay.

4. Final remarks

In fusion reactions induced by lighter heavy ions (^{12}C, ^{16}O, ^{19}F, ^{20}Ne) evidence for deviations from the standard compound-nucleus mechanism was found [3-6] at beam energies higher than 5 MeV/u above the Coulomb barrier. A direct break—up of the projectile in the first stage of the collision followed by fusion between fragment and target was the process suggested to interfer. One is tempted to compare this incomplete fusion mechanism with the one proposed here for the ^{46}Ti+^{24}Mg reaction, namely preequilibrium emission from a temperature equilibrated di-nuclear system.

In both cases, the main effect in comparison with the compound nucleus decay is a strong change of the branding ratio in the first step of the deexcitation, usually in favour of α-emission. Therefore, one may think that in priciple we are dealing with the same process, the one picture being a more adequate description for lighter projectiles, the other one for heavier ones. As the lighter reaction partner in the Ti+Mg reaction is not drastically heavier than some of projectiles mentioned in connection with incomplete fusion, the true description might lie in between. One should be able to obtain additional information on the mechanism by measuring light particles in coincidence with fusion, as it has been done in refs. [4,5]. At present there are only very limited data for the Ti+Mg reaction, but the few α-spectra we have (taken at 12 MeV/u at angles around 20° LAB) look perfectly as and behave kinematically as evaporation spectra.

As in the case of the reaction ^{12}C+^{160}Gd discussed by Wilczynski et al. [4] one may also speculate about a possible feedback of the preequilibrium decay on the total evaporation residue or fusion cross section in our case. The argument is that a high partial wave, which would not lead to fusion because of a non-attractive interaction potential, could do so after one reaction partner has emitted an α-particle, thereby lowering the fissility of the composite system. Indeed, in the reaction 12 MeV/u ^{48}Ti+^{24}Mg a total evaporation residue cross section exceeding the limits imposed by compound —nucleus fission was found [7].

Acknowledgement

This work was financially supported by the BMFT Bonn.

References

1. S. Cohen, F. Plasil, W.J. Swiatecki, Ann. of Phys. 82(1974)557

2. F. Pühlhofer, Nucl. Phys. A280(1977)267

3. B.Kohlmeyer, W. Pfeffer, F. Pühlhofer,Nucl. Phys. A292(1977)288

4. K. Siwek-Wilczynska, E.H. du Marchie van Voorthuysen, J. van Popta, R.H.Siemssen and J. Wilczynski,Phys. Rev. Lett. 42(1977)1599; and J. Wilczynski, contribution to this conference

5. M.L. Halbert, contribution to this conference

6. H. Lehr, W. von Oertzen, contribution to this conference

7. B. Kohlmeyer et al., to be published

LIGHT PARTICLE CORRELATIONS AND LIFETIME MEASUREMENTS

W. Kühn

Max-Planck-Institut für Kernphysik, Heidelberg, W.-Germany

1. INTRODUCTION

The study of highly excited nuclei is one of the main points of interest in heavy ion physics. In order to understand the reaction mechanisms
involved in producing highly excited compound nuclei (CN) in fusion
reactions or excited fragments in deep-inelastic collisions, a knowledge
of the relevant time scales is essential. At excitation energies above
100 MeV, the statistical model predicts CN lifetimes, which are of the
same order of magnitude or even shorter than the estimated relaxation
times in nuclear matter. Thus the validity of the CN picture becomes
questionable.

At high excitation energies, very little is known experimentaly
on CN lifetimes, as presently available methods are limited to values
above 10^{-18}s. In this talk, I would like to present an experiment, which
is part of a systematic effort[1] at the Max-Planck-Institute to study
nuclear lifetimes below 10^{-20} s by measuring energy and angular correlations of light particles.

2. THE METHOD

2.1 Correlations of identical particles

In 1972, Kopylov and Podgoretzki[2] have proposed to study energy and angular correlations between identical particles emerging from a highly
excited nucleus in order to investigate its space-time history. This
method is based on an experiment by Hanbury Brown and Twiss[3] , who have
employed intensity correlations to measure the size of stars.
The method has also been used in high energy physics to determine the
size of the fireball produced in proton - proton collisions[4].
Fig. 1 illustrates the basic idea of an analogous nuclear physics experiment. Let us consider a nucleus which emits two identical particles at
points A and B and let us detect these two particles in detector 1 and 2,

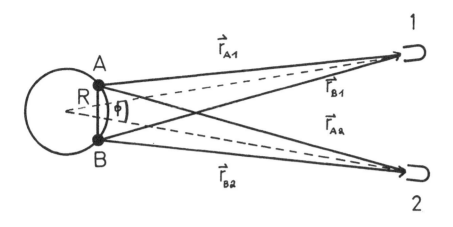

Fig. 1.   Geometry in a Hanbury Brown and Twiss like experiment

in coincidence. Then there are two ways to generate a coincidence :
(i) the particle emerging from A is observed in detector 1 and the particle emerging from B is observed in detector 2; (ii) the particle emerging from A is observed in detector 2 and the particle emerging from B
is observed in detector 1. For identical particles (k_A = k_B) and small
observation angle ϕ the quantum mechanical uncertainty in the position
of the emission points exeeds the nuclear dimensions : the two amplitudes to generate a coincident event cannot be distinguished, and therefore the coincidence probability is the square of the sum of both amplitudes. If we describe the particle wave functions by plane waves, we
obtain the following expression for the coincidence probability :

$$C_{12}(k,R,\phi) \simeq \left| \exp(i\vec{k}_A \vec{r}_{A1})\exp(i\vec{k}_B \vec{r}_{B2}) \pm \exp(i\vec{k}_A \vec{r}_{A2})\exp(i\vec{k}_B \vec{r}_{B1}) \right|^2$$

(+ for bosons, - for fermions)

$$\simeq \begin{array}{ll} 1 + \cos kR\phi & \text{(bosons)} \\ 1 - \dfrac{1}{2} \cos kR\phi & \text{(fermions, spin 1/2)} \end{array}$$

For difference angle $\phi \to 0$, we obtain an enhancement for identical bosons and a suppression for identical fermions. In a typical nuclear experiment (α particles, E_α(surface) $\sim$ 2 MeV, $R \sim$ 5 fm), we get almost

complete coherence within a reasonable angular width of $\sim 10^{\circ}$:

$$\cos kR\phi > 0.9 \quad \text{for} \quad \phi < 8.9^{\circ} \quad .$$

For short-lived states - which we want to study - this description is not appropriate. Instead of plane waves, the particle wave functions have to be described by wave packets of width $\Gamma = \hbar / \tau$, where τ is the nuclear lifetime. Using this description, Kopylov and Podgoretzki[1] have derived the following expression for the coincidence probability as a function of the observation angle difference ϕ, the particle energy difference $\Delta E = E_1 - E_2$ and the lifetime τ :

$$C_{12}(\Delta E, \tau, \phi) \approx 1 + \left\{ \frac{J_1(kR\phi)}{kR\phi} \right\}^2 \left\{ \frac{1}{1+(\Delta E)^2 \tau^2/\hbar^2} \right\} \left\{ \begin{matrix} 1 \\ -1/2 \end{matrix} \right\} \begin{matrix} \text{(bosons)} \\ \text{(fermions)} \end{matrix} \qquad (1)$$

$$\text{"spatial"} \quad \text{"temporal"}$$

Let us first consider the term labeled with "spatial". In contrast to the simplified assumption of two point sources on the surface of a nucleus, the derivation of equation (1) takes into account, that the whole nuclear surface may act as a source of particles. In this case, the spatial distribution of coincidences is given by the Airy function, which is just the Fourier transform of the source distribution connected with an uniformly radiating disk . In the limit $\phi \to 0$, we have maximum spatial coherence, i.e. maximum enhancement for identical bosons and maximum suppression for identical fermions.

Let us now look at the second term in equation (1), labeled with "temporal". At a given lifetime τ, it becomes large in the limit of $\Delta E \to 0$. This is due to the fact that coherence can only be achieved for particles which cannot be distinguished with respect to their energies. In terms of wave packets that means, that the size of the energy overlap between the wave packets describing the particle wave functions measures the degree of coherence. On the other hand, this energy overlap is not only dependent on the difference energy ΔE, but is also a function of the width Γ of the wave packets. For fixed ΔE, we expect the overlap to increase with Γ, that is, to decrease with lifetime $\tau = \hbar/\Gamma$. As a result, the observation of the coincidence yield, as a function of ΔE, contains the desired information on the nuclear lifetime.

In order to maximize the interference effect, we have to measure at small difference angles ϕ and at small difference energies ΔE to simultaneously fulfill the condition of "spatial" and "temporal" coherence.

2.2 Final state interaction

So far, we have completely neglected effects due to final state interaction (FSI) among the outgoing particles. This is of course justified
in the case of photon-photon correlations. In a nuclear experiment, where
we observe particles undergoing strong and electromagnetic interaction,
the FSI may have a large influence on the energy and angular correlation
of light particles. I would like to point out that there is a striking
similarity between the coherence effects of identical particles discussed
so far in section 2.1 and the effects due to FSI : both effects are expected to be strong, if the spatial and temporal separations between the
two light particles are small. Moreover, both effects will only play a
role for sufficiently small difference energies ΔE. As a consequence,
an experiment set up to study coherence effects of identical particles,
will also be sensitive to the presence of FSI. On the other hand, since
the FSI depends also on the temporal separation between the two detected
particles - which is given by the nuclear lifetime - , the experimental
observation of FSI meets our goal to measure lifetimes as well as the
observation of coherence effects.

3. EXPERIMENT

We have investigated the two-particle inclusive reactions

$$^{19}\text{F} + {}^{51}\text{V} \rightarrow {}^{70}\text{Ge}^* \begin{cases} \alpha\alpha \\ p\alpha \\ pp \end{cases}$$

with a 144 MeV ^{19}F beam provided by the postaccelerator at the MPI, Heidelberg. The excitation energy in the compound system ^{70}Ge was 122 MeV.
Fig. 2 shows a schematic view of the experimental setup. It consisted
of three light particle detectors located on a cone of constant scattering angle $\theta = 120^\circ$. The difference angle ϕ between detectors 1 and 2
was 4°, the difference angle between detectors 1 and 3 was 176°. Particle
identification was performed by time of flight measurement with respect
to the beam microstructure. The time resolution achieved was better than
500 ps. We have chosen a symmetric setup consisting of three detectors
for the following reason : for compound nucleus decay, the statistical
model predicts the symmetry $C(\theta,\phi) = C(\theta,\pi-\phi)$. Therefore we expect for

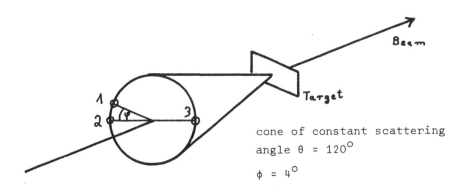

cone of constant scattering
angle θ = 120°

φ = 4°

Fig. 2.   Experimental setup

compound nucleus decay $C_{12}(\Delta E) = C_{13}(\Delta E)$. Coherence effects for identical
particles as well as effects due to FSI are only expected in the narrow
geometry C_{12}. They can be identified by comparing the distribution $C_{12}(\Delta E)$
with the reference distribution $C_{13}(\Delta E)$ (wide geometry). The comparison
will be done by calculating the "surprisal function"

$$S(\Delta E) \; = \; \frac{C_{12}(\Delta E)}{C_{13}(\Delta E)} \; . \tag{2}$$

We note, that the functional behaviour of S does not depend on the
relative normalisation between C_{12} and C_{13}.
      Fig. 3 shows the ΔE - distribution C_{13} for α-α coincidences taken
with the wide geometry. The distribution has a Gaussian like shape centered arround ΔE = 0, i. e. it is most probable, to observe two α particles with equal energies. This is exactly what one would expect for
the decay of a highly excited compound nucleus, where the energy carried
away by a single α particle is only a small fraction of the total excitation energy available. In this case we do not expect the energy conservation law to cause a strong energy correlation between two subsequent emissions. Instead, the correlation is determined by the phase

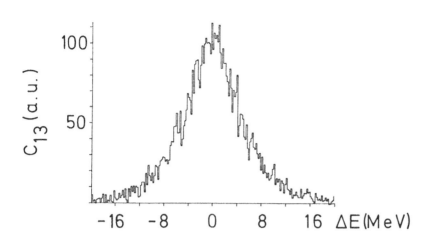

Fig. 3. Distribution of α-α coincidences taken with the wide geometry
as a function of $\Delta E = E_{\alpha 1}-E_{\alpha 3}$. The random coincidences have
been subtracted. All energies refer to the center-of-mass.

space distribution for a single α particle emission process. From CN
studies it is well known, that this distribution peaks near the Coulomb
barrier for α particles. In the limit of vanishing energy correlations
between two subsequent emissions we obtain the distribution of energy
differences ΔE just by folding the singles distribution with itself.
This will produce a symmetric distribution with respect to $\Delta E = 0$ and
with most probable value $\Delta E = 0$. This procedure can be done experimentaly by observing random coincidences. Fig. 4 shows the ΔE - distribution
of random α-α coincidences in the wide geometry. Comparing fig. 4 with
fig. 3, we note that both distributions look in fact very similar.

Fig. 5 compares α-α coincidences in the narrow geometry (fig. 5a)
with α-α coincidences in the wide geometry (fig. 5b). We observe two
prominent peaks in the narrow geometry. They are due to the decay of
$^{8}Be_{g.s.}$. Fig 5c shows the surprisal function defined in equ. (2).We find
a suppression of α-α events in the narrow geometry in comparison to the
reference yield, which can also be seen by direct comparison between
fig. 5a and fig. 5b.

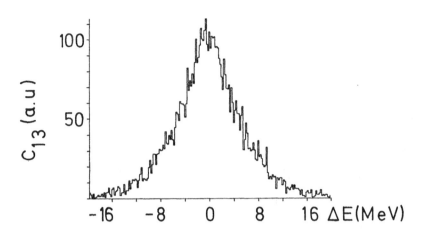

Fig. 4.   Distribution of random α-α coincidences in the wide geometry
            as a function of ΔE.

Fig. 6 shows the results for p-p coincidences. Again, the narrow geometry
(fig 6a) shows a suppression of coincidences in comparison to the reference yield (fig. 6b). The surprisal function (fig. 6c) indicates that
the suppression is maximum at ΔE = 0.

    Fig. 7 shows the results for the p-α coincidences. Here we have
plotted the distribution of events as a function of $\Delta E = 2E_p - E_\alpha$. Then,
ΔE = 0 corresponds to particles with equal velocities, a condition which
is expected to enhance the FSI between the α - particle and the proton.
If we compare the narrow geometry (fig. 7a) with the wide geometry
(fig. 7b), we find again a strong suppression in yield for the narrow
geometry, which is also present in the shape of the surprisal function
(fig. 7c). As in the p-p case, we observe maximum suppression for ΔE = 0.

3.   CONCLUSION

Let me first summarize the experimental facts :

    (a)   α-α, p-p and p-α coincidences show suppression in the
          narrow geometry with respect to the reference yield.
          This suppression is dependent on the energy difference,

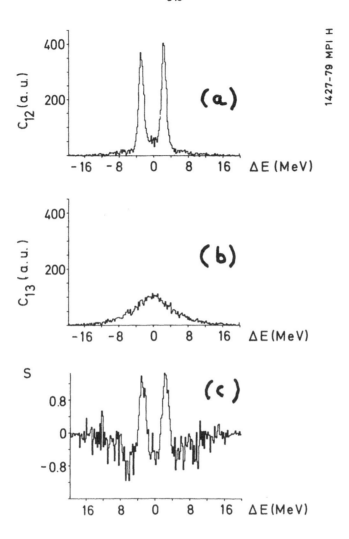

Fig. 5.   α-α coincidences
        (a) ΔE - distribution in the narrow geometry (detectors 1-2)
        (b) ΔE - distribution in the wide geometry (detectors 1-3)
        (c) surprisal function

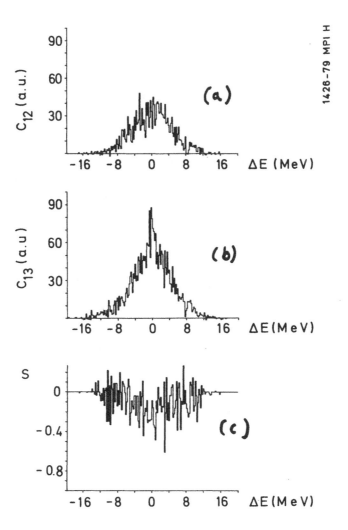

Fig. 6. p-p coincidences

    (a) ΔE - distribution in the narrow geometry (detectors 1-2)

    (b) ΔE - distribution in the wide geometry (detectors 1-3)

    (c) surprisal function

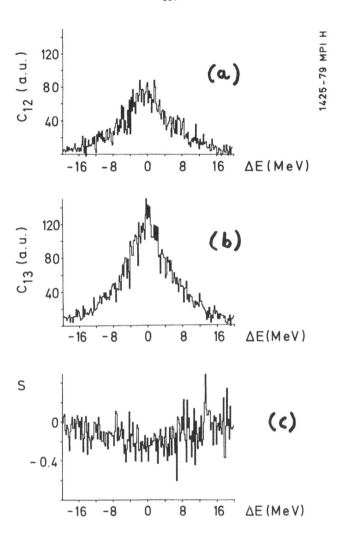

1425-79 MPI H

Fig. 7. p-α coincidences as a function of $\Delta E = 2E_p - E_\alpha$
   (a) ΔE - distribution in the narrow geometry (detectors 1-2)
   (b) ΔE - distribution in the wide geometry (detectors 1-3)
   (c) surprisal function

the strongest suppression is observed near $\Delta E = 0$.

(b) The $\alpha-\alpha$ coincidences show a strong contribution of $^8Be_{g.s.}$ decay in the narrow geometry.

The enhancement which would be expected from coherence effects in the $\alpha-\alpha$ coincidences is not found experimentally. Instead, the signature of FSI is present in all three kinds of coincidences. The FSI leads to a suppression near $\Delta E = 0$ (in the narrow geometry) and to the formation of intermediate systems such as ^8Be. However, we cannot exclude that part of the ^8Be which we observe is due to other processes such as evaporation from the compound nucleus or production in deep-inelastic collisions. The latter process does not seem to contribute strongly, since we are measuring at backward angles ($\theta_{lab}=120^\circ$).

We cannot explain the observed suppression in the narrow geometry without FSI. Since the FSI is dependent on the nuclear lifetime, we can in principle calculate the lifetime from the observed FSI effects. In order to obtain quantitative results, a quantum mechanical calculation would be needed. Such a calculation would also include the coherence effects for identical particles as well as coulomb effects. Up to now we are lacking such calculations. In the case of high energy physics, Koonin[4] has performed similar calculations for p-p coincidences. Since the assumptions which have been made in ref. 4 are not valid in our case, I would like to present two simple estimates of the lifetime. They are based on the assumption that the ^8Be which we have observed, is formed from two α particles in the exit channel as a result of the FSI. In the first estimate, we assume that ^8Be is formed as soon as the two α particles approach each other more than 4 fm, which is about the radius of ^8Be. Assuming that the α particles have mean energies at the nuclear surface, which are equal to the nuclear temperature of 3.8 MeV, we obtain:

$$\tau = \frac{distance}{velocity} = \frac{4 \text{ fm s}}{1.35\times10^{22}\text{fm}} = 3\times10^{-22} \text{ s}$$

The second estimate is based on the fact that the suppression of $\alpha-\alpha$ coincidences is visible till $\Delta E \sim 8$ MeV. Describing the α particle wave function by gaussian wave packets and assuming that $\Gamma > \Delta E/4$ in order to have nuclear FSI, we obtain :

$$\tau = 4\hbar/\Delta E = \hbar/2MeV \sim 3 \times 10^{-22} \text{ s}.$$

These two rough estimates yield the same numbers, which are in fact quite close to the lifetimes predicted by the statistical model in this regime of masses and excitation energies. These results show, that the observed FSI in the correlation of light particles emerging from highly excited nuclei may provide a tool to measure lifetimes in the unexplored region below 10^{-20} s.

The author gratefully acknowledges the collaboration with J.Aichelin, H.Damjantschitsch, H.Ho, J.Slemmer and J.P.Wurm.

REFERENCES

1. J.Aichelin, H.Damjantschitsch, H.Ho, W.Kühn, J.Slemmer, J.P.Wurm
    MPI Heidelberg, Annual Report 1978, Diploma Thesis J.Aichelin,
    Heidelberg, 1979, and to be published.

2. G.I. Kopylov, M.I.Podgoretzki, Sov.J.Nucl.Phys. 15, 219(1972)

3. R. Hanbury Brown, R.Q. Twiss, Nature (London) 178, 1046(1956)

4. C. Ezell, L.J. Gutay, A.T. Laasanen, F.I. Dao, P. Schübenlin
    and F. Turkot, Phys.Rev.Lett. 38, 873(1977)

5. S.E. Koonin, Phys.Lett. 70B, 43(1977)

EVIDENCE FOR INCOMPLETE FUSION IN A LIGHT HEAVY ION REACTION

H. Lehr, W. von Oertzen, W. Bohne, H. Morgenstern, K. Grabisch
Hahn-Meitner-Institut für Kernforschung Berlin GmbH
and F. Pühlhofer, Fachbereich Physik, Universität Marburg

Abstract

Evaporation residue mass distributions and the ^{6}Li yield of the fusion reaction ^{20}Ne on ^{26}Mg are measured and analyzed for ^{20}Ne energies from 4-15 MeV/u. The comparison of the data with statistical model calculations assuming complete fusion shows discrepancies, which are explainable by the assumption of incomplete fusion.

Introduction

Several models have been applied to describe the behaviour of excitation functions for fusion of light heavy ions. Most assume that target and projectile completely fuse populating in the compound nucleus all angular momenta below a limiting value. This quantity is normally extracted from the experimental total fusion cross section. Recently evidence has been found for incomplete fusion (IF) of light projectiles with heavier targets [1-3]. This process seems to be localized at high angular momenta, whereas for even higher L-values a break-up of the projectile into two or more constituents gives the natural extension of IF. Since the α-γ coincidence technique used in refs. [1-3] is not suitable to study IF for lighter systems, one has to use another experimental procedure that is sensitive to the maximum L-values of the compound nucleus (CN).

Extensive studies have been done measuring the total yields of light particles like ^{6,7}Li, ^{7,9}Be emitted by the CN ^{26}Al [4]. For not too high energies good agreement was found between measured and calculated cross sections, which are very sensitive to a cut-off L-value in the entrance channel. However, for higher energies the Hauser-Feshbach calculations underestimate the cross sections for light particles. The shape of the angular distributions as well as the energy spectra suggest a contribution of direct processes to the light-particle yield. It cannot be excluded either, that evaporation residues down to ^{6}Li result from the CN ^{26}Al.

It was the aim of the present study to obtain detailed information on the formation and decay of the CN formed in the fusion of ^{20}Ne + + ^{26}Mg, measuring light-particle yields and mass distributions of evaporation residues (ER). The data were then analyzed using the statistical decay code CASCADE [5], which was modified to include an additional decay channel, which was chosen to be ^{6}Li in this study.

Experimental Procedure

Beams of ^{20}Ne at energies of 85, 120, 150, 200 and 290 MeV from the VICKSI accelerator at the HMI Berlin were used to bombard ^{26}Mg targets of 350 ± 15 μg/cm^2 thickness. The ER were identified with the time-of-flight method in the angular range of 2-30 degrees in lab system . Simultaneously, we measured the yield of light particles with a solid-state ΔE-E telescope. Depending on the beam energy the thickness of the ΔE detector was chosen such as to allow light-particle identification for energies as low as possible. In all cases isotopic resolution was obtained up to mass 12. The error of the absolute cross section for the summed yield of ER is typically 10 % due to uncertainties in target thickness, detector solid-angle and integrated beam current. The error for the ^{6}Li cross sections is of the order of 15 % due to counting statistics, extrapolation of the energy spectra to lower energies at larger angles, and the extrapolation of the angular distributions.

Experimental Data and Complete Fusion Analysis

TABLE 1.   ER cross sections ^{20}Ne + ^{26}Mg

E_{Lab} (MeV)	E_{CM} (MeV)	σ_{ER} (mb)
85	48	1220 ± 150
120	68	1180 ± 100
148	84	1020 ± 100
202	114	920 ± 90
290	164	750 ± 100

The angle-integrated absolute cross sections of the ER are given in table 1. At a bombarding energy of 85 MeV we do not have enough points for a complete angular distribution. We therefore took the shape of the angular distribution from the fusion of ^{20}Ne + ^{27}Al [6], measured at the same energy and normalized it to our data. The angle-integrated absolute cross sections of the masses of the ER for three energies

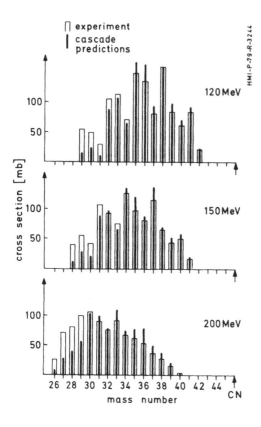

experiment
cascade
predictions

HMI-P-79-R-3244

120 MeV

150 MeV

200 MeV

cross section [mb]

mass number

CN

Fig. 1: Comparison of measured and calculated (CASCADE) mass yields for three energies, assuming complete fusion.

are given in fig. 1. The experimental errors of the mass yields are typically 10 % for large and medium masses and 20 % for the very low masses. Also shown is a comparison with the calculations of CASCADE, for which we used the same parameters as in ref.[5]. The author analyzed the reaction $^{19}F + ^{27}Al$, leading to the same CN ^{46}Ti. The described calculation assumes that all L-values in the entrance channel are populated like $(2L + 1)T_L$, up to a maximum value L_m, which in the sharp cut-off approximation is given by the measured ER cross section.

$$(1) \quad \sigma_{ER} = \pi \lambda^2 (L_m + 1)^2$$

One sees in the diagram, that for all three energies the calculation overestimates the larger masses, whereas it underestimates the smaller ones. This discrepancy increases with energy.

In the new CASCADE version available one can calculate

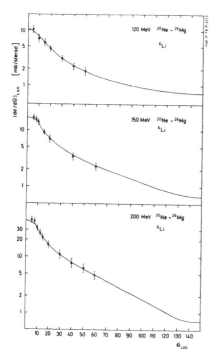

120 MeV $^{20}Ne \cdot ^{26}Mg$
^6Li

150 MeV $^{20}Ne \cdot ^{26}Mg$
^6Li

200 MeV $^{20}Ne \cdot ^{26}Mg$
^6Li

$(d\sigma/d\Omega)_{Lab}$ [mb/sterad]

θ_{Lab}

Fig. 2: Angular distributions for 6Li at three energies. The full lines are calculations, based on the statistical model, normalized to the data points.

the CN decay not only by n,p,α and γ-emission, but also by an optional additional decay channel. We analyzed up to now the ^{6}Li yield, which should be a very sensitive test to the L-population of the CN. In order to compare the calculated ^{6}Li yields from CASCADE with the measured ones, we have to integrate the experimental points over all angles. This is done in the following way: We took the energy distribution for ^{6}Li in the c.m. system calculated by CASCADE and the angular distribution in the c.m. system chosen as 1/sinθ with a cut-off angle [7] as an input for a monte carlo program to give us the shape of $(d\sigma/d\Omega)_{Lab}$. These curves normalized to the data points are shown in fig. 2. It turns out that the cut-off angle stays constant at $\theta_o = 20^o$ for all three energies shown.

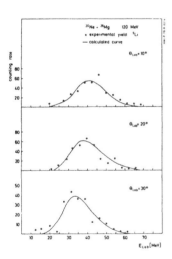

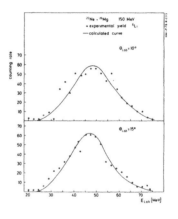

Fig. 3: Energy distributions for ^{6}Li nuclei from ^{20}Ne + ^{26}Mg fusion for several angles and three energies. The points are the experimental values. The curves are calculated as described in the text.

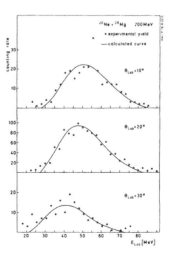

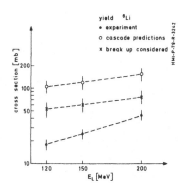

yield ^{6}Li
● experiment
○ cascade predictions
× break up considered

HMI-P-79-R-32/2

cross section [mb]

Fig. 4: Total ^{6}Li yield as a function of E_L. The errors in the calculations are due to uncertainties in the transmission coefficients.

From these calculations we also obtained $(d\sigma/dE)_{Lab}$ as a function of θ_{Lab} (see fig. 3). The good agreement with the data both of the angular distributions and the energy spectra confirms the CN origin of ^{6}Li. A comparison of the angle integrated measured absolute cross sections for ^{6}Li and those calculated by CASCADE with an L_m as described above shows that an overestimation of the ^{6}Li yield for all energies is obtained. Furthermore a wrong energy behaviour is predicted (see fig. 4). It is well-known, that ^{6}Li is likely to break up even by Coulomb excitation. Therefore, we also expect that a considerable part of the ^{6}Li flux in the emission process will decay. In the Hauser-Feshbach expression the time-reversal of the scattering of ^{6}Li on ^{40}K is used to calculate decay probabilities. Doing this, one implicitly assumes, that the CN cross section equals the reaction cross section. The break-up of ^{6}Li has been measured for a number of targets [8]. It is roughly 50 % of the reaction cross section and almost independent of the energy in the interesting interval. In order to include the loss due to break-up we reduce the ^{6}Li yield as calculated from CASCADE by a factor of 2. The result of this procedure is shown in fig. 4. But still the energy dependence and the absolute values of calculation and experiment are in disagreement.

Interpretation of the Data with "Incomplete Fusion"

Up to now, incomplete fusion is expected to exist at higher energies with light projectiles and heavy targets, especially if there are loosely bound constituents. There are also indications for a localization of this process at higher L-values [1-3, 9]. We suggest that in a very early stage of the reaction the projectile ^{20}Ne is excited. Because of the low binding energy of an α-particle (4.8 MeV), ^{20}Ne is likely to decay in ^{16}O + α. This can happen, if there is enough rotational energy in the dinuclear system to allow the escape of an

α-particle, while ^{16}O is able to fuse. The relative velocity in the c.m. system should be nearly the same as between ^{20}Ne and ^{26}Mg. This means that we have for ^{16}O the c.m. energy

$$(2) \quad E_0 = \mu_0/\mu_{Ne} \cdot (E_{Ne} - E_{sep} - E_B)$$

Here μ_0 and μ_{Ne} stand for the reduced masses of the systems ^{16}O + ^{26}Mg and ^{20}Ne + ^{26}Mg, respectively, E_{Ne} denotes the asymptotic c.m. energy for the original system, whereas E_{sep} and E_B refer to the separation energy of the α-particle and the energy to overcome the Coulomb barrier of ^{16}O. The angular momentum at the onset of the incomplete fusion is chosen correspondingly and gives for ^{16}O + ^{26}Mg

$$(3) \quad L_0 = L_{Ne} \cdot \mu_0 \cdot v_0/\mu_{Ne} \cdot v_{Ne}$$

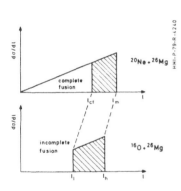

Fig. 5: Sharp cut-off approximation for incomplete fusion. The shaded areas must be equal (see text, relation (3)).

where v_0, v_{Ne} denote the velocities as given by equ. (2). This means that we have an L-population for the two different fusion entrance channels as shown in fig. 5. In the upper half we show the entrance channel population for ^{20}Ne + ^{26}Mg extending up to L_m as it was given by relation (1). We now assume complete fusion only up to L_{cf}. The fusion channel ^{16}O + ^{26}Mg populates an L-window from L_1 to L_h, which is determined by the condition:

$$(4) \quad \sigma_{ER} = \sigma_{CF} + \sigma_{IF}$$

The measured ER cross section must be the sum of the complete fusion σ_{CF} and the incomplete fusion cross section σ_{IF}. The only free parameter in (4) is L_{cf}, respectively L_1. In a first guess we assume, that L_{cf} = const for all energies, although this need not be the case, because of possible entrance channel effects. We therefore tried to determine L_{cf} by fitting the mass distributions and the ^{6}Li yield simultaneously by varying L_{cf}. The best value we obtained in the framework of this model is $L_{cf} = 31 \pm 2$ ℏ for all three energies. The mass distributions added for complete and incomplete fusion are given in fig. 6. They now show a satisfying agreement with the data. One should emphasize that the CASCADE parameters are kept constant for all three energies during the fitting process. The ^{6}Li yields calculated with this procedure are compared to the experiment in fig. 7. Due to the reduction of the high L-value population (cf. fig. 5) the calculated ^{6}Li yield is strongly reduced, and an excellent agreement both in slope and absolute values is obtained taking the break-up of ^{6}Li as mentioned above, into account.

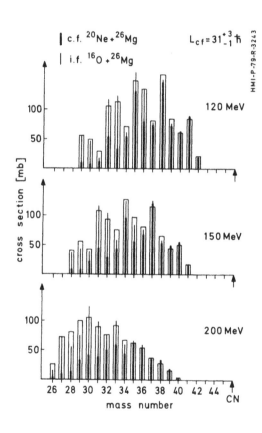

Fig. 6: Angle integrated mass yields for complete fusion (thick lines) and incomplete fusion (thin lines) as calculated by CASCADE compared to the experimental values.

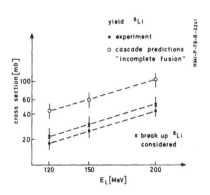

Fig. 7: ^{6}Li yield for "incomplete fusion". The inclusion of ^{6}Li break-up gives the right values in the limits of experimental and theoretical errors.

References

[1] T. Inamura, M. Ishihara, T. Fukuda and T. Shimoda, Phys. Lett. 68B (1977) 51

[2] D.R. Zolnowski, H. Yamada, S.E. Cala, A.C. Kahler and T.T. Sugihara, Phys. Rev. Lett. 41 (1978) 92

[3] K. Siwek-Wilczynska, E.H. du Marchie van Voorthuysen, J. van Popta, R.H. Siemssen and J. Wilczynsky, Phys. Rev. Lett. 42 (1979) 1599

[4] R.G. Stokstad, M.N. Namboodiri, E.T. Chulick and J.B. Natowitz, D.L. Hanson, Phys. Rev. C16 (1977) 2249

[5] F. Pühlhofer, Nucl. Phys. A280 (1977) 267

[6] H. Morgenstern, W. Bohne, K. Grabisch, HMI Berlin, private communication

[7] T. Ericson, V. Strutinsky, Nucl. Phys. 8 (1958) 284

[8] K.O. Pfeiffer, E. Speth and K. Bethge, Nucl. Phys. A206 (1973) 545

[9] K.A. Geoffrey, D.A. Sarantites, M.L. Halbert, D.C. Hensley, R.A. Dayras and H.J. Barker, Phys. Rev. Lett. 43 (1979) 1303

FUSION EXCITATION FUNCTIONS FROM NEUTRON YIELD
MEASUREMENTS INSIDE THE CYCLOTRON

U. Jahnke, S. Kachholz, and H.H. Rossner

Hahn-Meitner-Institut für Kernforschung Berlin GmbH
D 1000 Berlin 39, Gliendcker Straße 100

Our split pole cyclotron in Berlin is computer controlled and in addition
it has a good beam diagnostic system and reliable magnetic field map
data. Because of these qualities we are able to use the internal beam
for excitation function measurements. We have chosen the system ^{40}Ar +
^{110}Pd to demonstrate the method of getting fusion excitation functions
from neutron yield measurements inside the cyclotron.

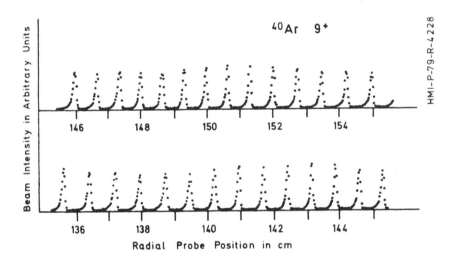

Fig. 1   Typical orbit pattern of the internal beam

Part of a standard turn pattern for ^{40}Ar projectiles is shown in fig. 1.
This orbit pattern was taken with the radial differential probe by recording the beam intensity versus radial probe position. In case of a
well centered beam we can calculate the beam energy at turn i using
the relations

$$E_i = \frac{E_o}{\sqrt{1 - \beta_i^2}} - E_o \; ; \qquad \beta_i = \frac{2\pi}{c} R_i \frac{f_{rf}}{n}$$

with E_o being the rest mass of the accelerated particle, R_i the average radius of turn i and $f_{rf}/n = f_p$ the particle revolution frequency. To check the calculated energy values, we measured the beam energy of some turns by the recoil proton technique, having the corresponding standard experimental set-up [1] mounted on the radial probe head. It turned out that for radii $R_i > 80$ cm the measured and calculated energies coincide within 0.5 %.

To measure excitation functions we have to mount a target onto the radial probe head and then move the target from one turn to the next. Now, being close to a strong magnetic field and rather limited in space for putting target and detection system onto the probe head, we decided to measure neutron yields with two long counters that are fixed inside the cyclotron vacuum chamber.

The calibration of the neutron detectors was achieved by means of calibrated neutron sources (Pu-Be, Cf). When we replace the target by a neutron source and then move the radial probe parallel to the sector magnet, we will get a normalization curve for neutron sources with isotropic angular distribution. In case of a moving compound nucleus we have to trace back the measured neutron yield to the yield of a neutron source at rest, that is to a source with isotropic angular distribution [2].
To get an estimate about the neutrons being reflected from the surroundings, we fixed our calibration source to the long counter and then moved this set perpendicularly to the sector magnet. The simple model of a mirror detector finally provides us with a first order correction to account for the reflected neutrons.

When we stop the ^{40}Ar beam on a sheet of tungsten, then we get the neutron yield excitation function which is represented in fig. 2 by open circles. The neutron counts are normalized to the beam current and to the solid angle, and the turn position is transformed into beam energy. In this energy range we measure two neutron barriers: ^{40}Ar + C and/or O at $E_{lab} \approx 80$ MeV and ^{40}Ar + W at $E_{lab} \approx 160$ MeV. The barrier for the reaction ^{40}Ar + ^{16}O is very close to the barrier of ^{40}Ar + ^{12}C because in case of a projectile that is heavy compared to the target, the cou-

Fig. 2: Neutron yield excitation functions for ^{40}Ar projectiles being stopped in a sheet of tungsten (open circles) and on a ^{110}Pd target with tungsten backing (filled circles). The curves represent linear least square fits through the background data points.

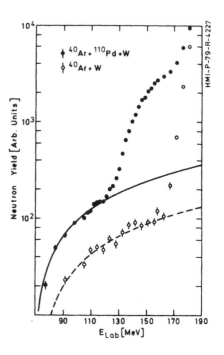

lomb barrier depends mostly on the neutron to proton ratio of the target. The filled circles in fig. 2 represent the neutron yield excitation function we get when we evaporate 60 μg/cm² of ^{110}Pd onto our sheet of tungsten. It becomes quite obvious that we have problems in preparing carbon- and oxygen-free targets. Thus, we lose sensitivity in the ^{110}Pd data for subcoulomb energies. At higher energies the experimental error increases because of the contribution of the tungsten backing.

E_{lab} (MeV)	$\sigma_{fus} = \sigma_{fus}^{MB\,2}$		$\sigma_{fus} = 2\,\sigma_{fus}^{MB\,2}$		Exp. ref. 4
	a=A/9	a=A/12	a=A/9	a=A/12	
140	4.20	3.96	4.01	3.80	3.62
150	4.35	4.10	3.77	3.56	4.08
160	4.44	4.17	3.81	3.60	4.31
170	4.52	4.24	4.30	4.05	4.39

Table 1: Neutron multiplicity

We want to go one step further now, and analize the data in terms of fusion cross sections. In this case, however, we have to know the average neutron multiplicity. Table 1 is supposed to demonstrate that the average neutron multiplicity, calculated with the computer code MB2 [3], does not very much depend on the level density and the fusion cross section. Changing the level density parameter by 30 % and the initial spin distribution of the compound nucleus by 100 % will cause a change of the neutron multiplicity by 10 - 20 % for energies between 140 and 170 MeV. Furthermore, a comparison with experimental values, measured by Della Negra et al. [4], indicates that the evaporation code with standard input parameters will reproduce the neutron multiplicity within 15 % at least in this mass and energy region.

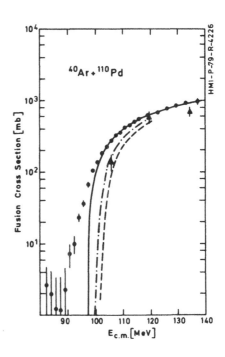

Fig. 3: Measured and calculated fusion cross sections for ^{40}Ar + ^{110}Pd. Circles: present work, triangles: ref. 4, line: linear fit on 1/E-scale, dash-dot curve: prediction based on Krappe-Nix-Sierk potential, dashed curve: prediction based on proximity potential.

After background subtraction and normalization we get our fusion cross sections. These are shown in fig. 3 in the c.m.-energy range between 80 and 140 MeV. When we compare our data with those of Della Negra et al. [4], we see that both methods, the neutron yield measurement and the measurement of the residual nuclei cross sections by specific decay modes, do give similar results for the fusion cross section.

We should mention that for $E_{c.m.}$ > 130 MeV we expect a considerable contribution of the neutron yield coming from the fusion-fission process. Yet, for the sake of simplicity in the calculations of the kinematic factor and the neutron multiplicity we neglected this process. Therefore, we have to regard our data for c.m.-energies beyond 130 MeV with restrictions. For $E_{c.m.}$ < 130 MeV we consider the systematic error to be in the order of 23 %.

In fact we have to accept a rather big uncertainty on our fusion cross section scale, whereas the error on the energy scale is half a percent. So, our method is especially suitable and convenient to measure the height of the fusion barrier. From the data in fig. 3 we determine the height of the barrier to be at about 90 MeV. However, when we extract the barrier height in the classical way by extrapolating the data on the $1/E_{c.m.}$-scale, then we get a value of 97 MeV. The corresponding linear fit is represented by the continuous line in fig. 3. This energy difference is due to the quantum-mechanical effect of the barrier penetration, which, in turn, depends on the shape of the barrier [5].

A comparison with calculated fusion cross sections based on the proximity potential [6] (dashed curve) or the Krappe-Nix-Sierk potential [7] (dash-dot curve) shows that our data might indicate a deviation of the spherical shape of the nuclei before they fuse. We believe that our neutron yield measurements are appropriate to study these kinds of processes like the formation of a neck between the particles before they end up in the compound nucleus. Of course, we know, that we have to investigate these effects in a systematic way before we come to a final conclusion.

References

[1] D.K.Olsen et al., Nucl. Instr.Meth. 114 (1974) 615
     R. Bimbot and D. Gardès, Z.Phys. A286 (1978) 327

[2] S.K. Allison, Nucl.Phys. 77 (1966) 541

[3] M. Beckerman and M. Blann, UR-NSRL-135 (1977)

[4] S. Della Negra et al., Z.Phys. A282 (1977) 65
     S. Della Negra et al., Z.Phys. A282 (1977) 75

[5] H. Gaeggeler et al., Z.Phys. A289 (1979) 415

[6] J. Blocki et al., Ann. Phys. 105 (1977) 427

[7] H.J. Krappe et al., Phys.Rev.Lett. 42 (1979) 215

ENTRANCE CHANNEL VERSUS COMPOUND NUCLEUS LIMITATIONS IN THE FUSION
OF 1p AND 2s-1d SHELL NUCLEI

S. Harar

DPh-N/BE, CEN Saclay, BP 2, 91190 Gif-sur-Yvette, France

At the Caen Conference [1], the Saclay group presented some results concerning the fusion cross-sections (σ_F) of 1p and 2s-1d shell nuclei. At that time, two interesting features were observed as illustrated in Fig. 1.

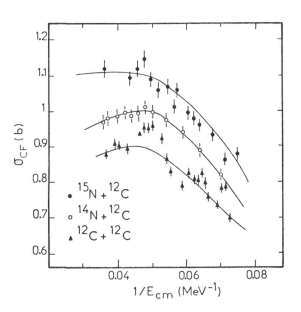

. Oscillatory structure for the $^{12}C + ^{12}C$ system as the one observed previously by the Argonne group [2] for the $^{16}O + ^{12}C$ system.

. Systems differing by only one valence nucleon presented significantly different fusion excitation functions.

Fig. 1 - Fusion cross-sections versus $1/E_{cm}$ for different entrance channels. Lines are from the Glas and Mosel model as discussed in ref. [3].

These results suggested that the detailed structure of colliding nuclei should play an important role in the fusion process. To investigate these aspects systematically, σ_F have been measured for a number of systems as shown in table 1. Most of the experiments have been achieved using the Saclay Tandem Van de Graaff over an energy, ranging from 1.5 to 3 times the Coulomb barrier. These systems were choosen on one part of investigate structures in fusion excitation functions and on other part to discriminate between entrance channel and compound nucleus effects in limiting the fusion process by comparing different systems leading to the same compound nucleus.

So far oscillations have been observed for the $^{12}C + ^{12}C$, $^{16}O + ^{12}C$ and $^{16}O + ^{16}O$ systems. When adding or substracting an extra nucleon to these nuclei, oscillations are smeared out as shown in Fig. 1 and 2 as examples.

The next symmetrical system which can be studied is $^{20}Ne + ^{20}Ne$ which is difficult experimentally for obvious reasons. So we measured the $^{24}Mg + ^{24}Mg$ and as shown in Fig. 3, no structure was observed. As we mentionned already elsewhere [3] :

. Oscillations are predominantly observed in the α decay channels which are fed by high values of the angular momentum distribution of the compound nuclei.

Table 1

List of systems whose fusion cross sections have been measured by the Saclay group

Systems	Energy range (c.m. MeV)	Compound nucleus	References
$^{12}C + ^{12}C$	13 - 26	^{24}Mg	[3]
$^{14}N + ^{12}C$	14 - 27	^{26}Al	[3]
$^{15}N + ^{12}C$	14 - 27	^{27}Al	[3]
$^{14}N + ^{13}C$	15 - 30	^{27}Al	[13,14]
$^{17}O + ^{10}B$	12 - 22	^{27}Al	[14]
$^{17}O + ^{12}C$	13 - 30	^{29}Si	[13,14]
$^{17}O + ^{13}C$	13 - 30	^{30}Si	[13,14]
$^{14}N + ^{16}O$	16 - 32	^{30}P	[13,15]
$^{15}N + ^{16}O$	16 - 32	^{31}P	[13,15]
$^{16}O + ^{16}O$	60 - 70	^{32}S	[8]
$^{20}Ne + ^{12}C$	25 - 60	^{32}S	[8]
$^{17}O + ^{16}O$	15 - 35	^{33}S	[14]
$^{24}Mg + ^{12}C$	25 - 42	^{36}Ar	[16]
$^{24}Mg + ^{24}Mg$	23 - 42	^{48}Cr	[16,17]
$^{24}Mg + ^{26}Mg$	23 - 39	^{50}Cr	[17]

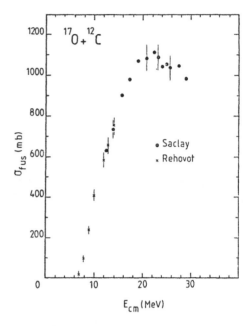

Fig. 2 - Fusion excitation function of the $^{17}O + ^{12}C$ system. Rehovot data are from ref. [19].

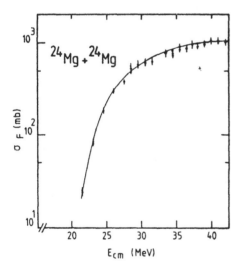

Fig. 3 - Fusion excitation function of $^{24}Mg + ^{24}Mg$ system.

. Gross structures are observed for there systemes whose elastic scattering studies revealed absorbing potentials and so can be associated to shape resonances. The structure are damped for heavier systems or for nuclei with extra valence nucleon on ^{12}C and ^{16}O cores.

My second point concerns the suggestion made by J.P. Schiffer [4] connecting the maximum of the fusion cross-sections (σ_F^{max}) to the shell location of valence nucleons of the colliding ions. This analysis based on the data shown by full circles in Fig. 4, pointed out that σ_F, for 1p shell saturate at less than 1000 mb while for 2s-1d nuclei σ_F^{max} are around 1200 mb ; at that time the only $^{15}N + ^{12}C$ exception was measured at Saclay. The new data reported here and recently elsewhere [5,6] allow to complete the data systematics as shown in Fig. 5 (open circles). It is clear now that the mentionned shell effect is not an important factor in inducing the variations of σ_F^{max}. Nevertheless some σ_F^{max} are be different

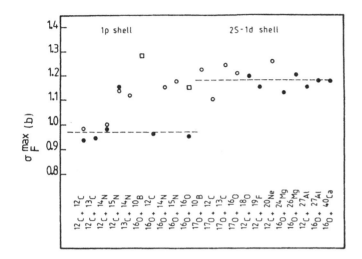

Fig. 4 - Plot of σ_F^{max} measured for various projectiles and targets of 1p and 2s-1d shell. Full points are data discussed by J.P. Schiffer in ref. [4]. Open circles are Saclay measurements. Open squares correspond to $^{16}O + {}^{10}B$ system [5] and $^{16}O + {}^{16}O$ system [6].

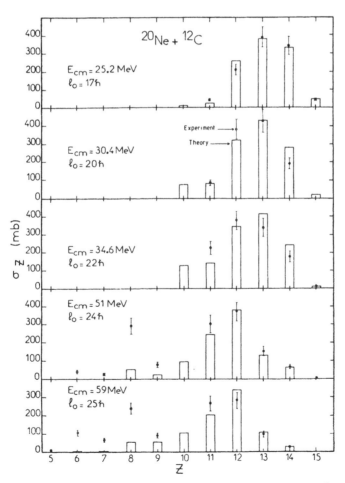

Fig. 5 - Integrated cross-sections of fusion-like products observed in the $^{20}Ne + {}^{12}C$ system at different energies. Histograms are the predictions form the Cascade code with ℓ_o shown in the figure.

for neighbour systems and these effects could be related to the excitation energies at which the compound nucleus properties become the limiting factor for the fusion process.

The formation of the same compound nucleus via different entrance channels is a good way to discriminate between compound nucleus properties and structure effects of colliding ions in fixing the fusion process. So, we decide to study the ^{20}Ne + ^{12}C and ^{16}O + ^{16}O both forming the ^{32}S compound nucleus. The structure of these ions are quite different and the extensive elastic studies revealed drastic differences [7] attributed to surface transparency of interacting potential which can also affect the fusion process. The integrated cross-section of fusion like products measured for the ^{20}Ne + ^{12}C systems at different incident energies are presented in Fig. 5. At incident energies lower than 35 MeV (c.m.) the agreement with the Cascade predictions is rather good ; but at higher energies the low Z experimental cross-sections are strongly underestimated. Reaction products cross-sections measured for ^{20}Ne + ^{12}C and ^{16}O + ^{16}O at 30 MeV and 60 MeV are compared in Fig. 6. At both energies the experimental results look similar. Using the sharp cut off model, one can extract critical angular momenta (ℓ_c) from experimental cross-sections following the expression $\sigma_F = \Pi \lambda^2 (\ell_c + 1)^2$; these values are plotted in Fig. 7 as a function of the compound nucleus excitation energies (E^*). Above 50 MeV there are two series of ℓ_c values (crosses or circles) depending of the definition of σ_F ; indeed σ_F can be defined as the sum of σ_Z for Z > 8 with adding predicted σ_Z for

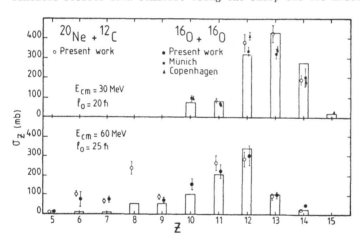

Fig. 6 - Comparaison of integrated cross-sections of fusion-like products measured for the ^{20}Ne + ^{12}C and ^{16}O + ^{16}O systems at 30 MeV and 60 MeV center of mass energies. Histograms are predictions from Cascade.

Z ⩽ 8 by evaporation calculations (circles in Fig. 7) ; σ_F can be also defined as the sum of all fusion-like products (crosses in Fig. 7) with σ_Z from Cascade for Z = $Z_{proj.}$. In both assumption, results showed that the fusion limits don't depend either of the grazing angular momenta (ℓ_g) nor of the ^{32}S yrast line performed assuming a spherical rigid body moment of inertia as presented in Fig. 7. Nevertheless the limits are the same for both systems and this suggest that some compound nucleus properties play a role.

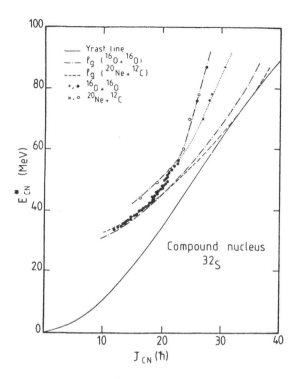

Fig. 7 - Critical angular momenta derided from experimental fusion cross-sections. Circles and crosses as defined in the text. The grazing angular momenta are also shown a well as the ^{32}S yrast line calculated with a moment of inertia given by $J = J_{rig}(1+\delta J^2)$ with $J_{rig} = 2/5$ m R^2 and $r_0 = 1.27$ fm.

Recently the Oak Ridge group studying the ^{16}O + ^{10}B and ^{14}N + ^{12}C systems [5] both leading to the ^{26}Al compound nucleus emphasized on the fact that σ_F values differing by 300 mb around 50 MeV c.m. (Fig. 8) prove the importance of the microscopic aspects of the entrance channels in limiting the fusion process in contradiction with our ^{20}Ne + ^{12}C and ^{16}O + ^{16}O results.

Birkelund et al. [9] developped a dynamical model with a non conservative potential to fit the fusion excitation functions in a wide mass region. As shown in Fig. 8 predictions from this model, fit the ^{14}N + ^{12}C results but not at all the ^{16}O + ^{10}B ones ; in the framework of the dependent friction model the question is : why the dissipative forces have to be much stronger for the ^{16}O + ^{10}B compared to the ^{14}N + ^{12}C system ? [5]. In studying the ^{17}O + ^{13}C system, we obtained also quite different fusion cross-sections compared to the ones derived from the ^{18}O + ^{12}C [2,10] as presented in Fig. 9. So one is attempted to conclude that the structure of colliding ions are important in limiting the fusion mechanism.

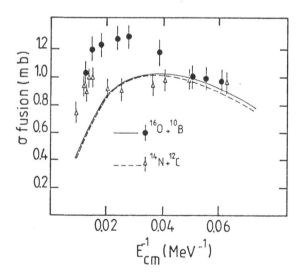

Fig. 8 - Fusion cross-sections data from ref. [5]. The lines are predictions from ref. [9].

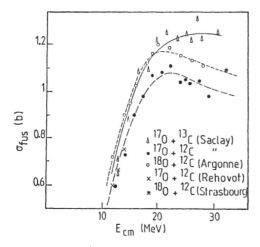

Fig. 9 - Comparison of fusion excitation function for different systems. Data of $^{18}O + ^{12}C$ are from ref. [2] and [10]. Data of $^{17}O + ^{12}C$ as defined in Fig. 2.

Nevertheless other factors might be also important, mainly the compound nucleus properties. If one reduced the data (σ_F, E_{cm}) in the new reference (ℓ_c, E^*), one obtains the Fig. 10 for systems leading to the same compound nucleus. The spectacular differences are made smaller if not disappearing. Than the important feature revealed by this representation is that above the energy corresponding to σ_F^{max}, the ℓ_c values are closed for different entrance channels leading to the same compound nucleus. So, the σ_F values would be determined in fact by the excitation energies reached in the compound system. Indeed for a given center of mass energy, the corresponding compound nucleus excitation energy and the channel spins are higher for $^{16}O + ^{10}B$ than for $^{14}N + ^{12}C$.

As we mentionned already the compound nucleus yrast line is to far from the experimental ℓ_c values to be a limiting factor. But the statistical yrast line performed assuming a spherical rigid body is located near the experimental ℓ_c values. In

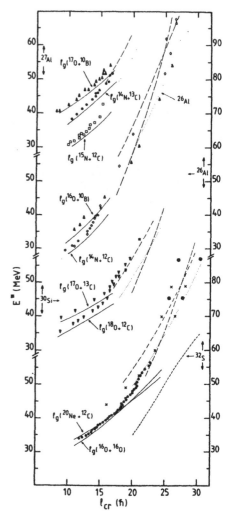

Fig. 10 - Experimental values for the critical angular momenta versus excitation energies of the compound nucleus. Full lines are the grazing angular momenta of entrance channels. The dotted lines correspond to 400 levels per MeV and the dotted-dashed lines to 4000 levels per MeV. Long dashed lines are the yrast lines assuming spherical rigid body moment of inertia ($r_o = 1.25$ fm). Short dashed lines is the ^{32}S yrast line from ref. [18].

order to give an idea of the level densities, in the compound nucleus at excitation energies where the experimental ℓ_c are localized, calculations were performed using the Lang formula [11] and standard parameters for this mass region [12]. The dotted line is for 400 levels/MeV and the dotted-dashed line for 4000 levels/MeV. Whatever the absolute values of these level densities, the important fact is that the values are similar for the studied systems. This suggest that a kind of a statistical limit in the compound nucleus play an important role in the fusion process at energies above the saturation region.

It is a pleasure to thank M. Conjeaud, S. Gary, F. Saint-Laurent, C. Volant and J.P. Wieleczko for making available their most recent data.

REFERENCES

[1] M. Conjeaud, S. Gary, S. Harar, J.P. Wieleczko, Conference on nuclear physics with heavy ions, Caen (1976) communication p. 116.

[2] P. Sperr, T.H. Braid, Y. Eisen, D.G. Kovar, F.W. Prosser, J.P. Schiffer, S.L. Taylor and S. Vigdor, Phys. Rev. Lett. 37 (1976) 321.

[3] M. Conjeaud, S. Gary, S. Harar, J.P. Wieleczko, Nucl. Phys. A309 (1978) 515.
S. Harar, Proceedings of the International Conference on resonance in heavy ion reactions (Hvar, 1977).

[4] J.P. Schiffer, Colloque Franco-Japonais de Spectroscopie Nucléaire et Réactions Nucléaires, Dogashima, Japan (ed. by Y. Shida, Inst. Nucl. Study, Univ. Tokyo, 1976) p. 176.
J.P. Schiffer, Proceedings Intern. Conf. on Nucl. Structure, Tokyo, Sept. 5-10, 1977 (ed. by T. Marumori) p. 66.

[5] J. Gomez del Campo, R.A. Dayras, J.A. Biggerstaff, D. Shapira, A.H. Snell, P.H. Stelson and R.G. Stokstad, Phys. Rev. Lett. 43 (1979) 26.

[6] B. Fernandez, C. Gaarde, J.S. Larsen, S. Pontoppidan and F. Wideback, Nucl. Phys. A306 (1978) 259.

[7] R. Vandenbosch, M.P. Webb and M.S. Zirman, Phys. Rev. Lett. 33 (1974) 842.

[8] F. Saint-Laurent, M. Conjeaud, S. Harar, J.M. Loiseaux, J. Menet and J.B. Viano, Nucl. Phys. A327 (1979) 517.

[9] J.R. Birkelund, Proceedings of the Conference.
J.R. Birkelund, L.E. Tubbs, J.R. Huizenga, J.N. De and D. Sperber, Preprint University of Rochester NSRL 193, submitted to Physics Reports.

[10] J.P. Coffin, P. Engelstein, A. Gallmann, B. Heusch, P. Wagner and H.E. Wegner, Phys. Rev. 17 (1978) 1607.

[11] D.W. Lang, Nucl. Phys. 77 (1966) 545.

[12] F. Pülhofer, Nucl. Phys. A280 (1977) 267.

[13] C. Volant and J.P. Wieleczko, Contribution to the XVII International Winter meeting on nuclear physics, Bormio 1979, p. 308 (ed. by I. Iori).

[14] J.P. Wieleczko, S. Harar, M. Conjeaud, F. Saint-Laurent, submitted to Physics Letters.

[15] M. Conjeaud, S. Gary, S. Harar, S. Janouin, F. Saint-Laurent, C. Volant, J.P. Wielczko, Compte rendu d'activité du Département de Physique Nucléaire 1977-1978, Note CEA-N-2070, p. 36.

[16] S. Gary, C. Volant, Compte rendu d'activité du Département de Physique Nucléaire 1977-1978, Note CEA-N-2070, p. 41.

[17] S. Gary, H. Oeschler, C. Volant, Compte rendu d'activité du Département de Physique Nucléaire 1978-1979, to be published.

[18] D. Glas and U. Mosel, Phys. Lett. 78B (1978) 9.
M. Diebel, D. Glas, U. Mosel and H. Chandra, preprint 1979.

[19] Y. Eyal, M. Bekerman, R. Chechick, Z. Fraenkel and H. Stocker, Phys. Rev. C13 (1976) 1517.

THE ROLE OF THE YRAST LINE IN HEAVY ION FUSION[*]

U. Mosel and M. Diebel

Institut für Theoretische Physik, Universität Giessen
6300 Giessen, West Germany

ABSTRACT

Yrast lines for light nuclei have been calculated with the Strutinsky procedure for rotating nuclei. It is found that in all cases analyzed the yrast line lies below the empirical fusion band. However, level-density limitations may be effective in certain cases to limit fusion at high energies.

[*]Work supported by BMFT and GSI Darmstadt

1.) Introduction

Experimental developments have led to an increased interest in theoretical calculations of yrast lines in the mass range of sd-shell nuclei beyond the highest spins known so far (I $\approx$ 8). Interest in the
spin range 10 $\leq$ I $\leq$ 20 is also strongly triggered by the possible
existence of yrast line limitations seen in heavy-ion fusion cross sections[1,2]. In this paper we discuss calculations of yrast lines in this
mass-range and their reliability in connection with results from fusion
experiments.

2.) The model

We have extended the standard method of calculating yrast lines by means
of the Strutinsky method to the region of light nuclei. For those nuclei an adequate treatment of the surface diffuseness was important.
Furthermore we have employed for the first time a rotating basis that
diagonalizes the cranked anisotropic harmonic oscillator

$$ H = \frac{p^2}{2m} + \frac{m}{2} \sum_{i=1}^{3} \omega_i^2 x_i^2 - \omega \ell_x $$

exactly and therefore makes a restriction to good principal quantum
numbers unnecessary. Details of the calculation as well as an assess-
ment of the reliability of the calculations can be found in ref. 3.

Calculations along these lines do not include pairing and have to be
identified with an unpaired band. In the high spin range (I $\geq$ 10) the
true excitation energy is, therefore, higher than the calculated one
just by the pairing-correlation energy. In order to get a feeling for
the magnitude of this energy we have performed cranked HFB calculations
for ^{24}Mg (ref. 4). With particle number projection taken into account
we obtain roughly 6 MeV total pairing correlation energy in this case.
Shifting our calculated yrast line up in excitation energy by this
amount makes the calculated 8$^+$ state coincide with the experimental
one[5]. The assumption then is that pairing in this nucleus breaks down
at I $\approx$ 8 and that for higher spins the shifted yrast line is correct.
That this nucleus indeed goes through a major structural rearrangement
at I = 8 is also indicated in large-scale shell model calculations[6]
that show that the groundstate band terminates at I = 8 although the
maximum angular momentum that could be formed by the eight nucleons in
the sd-shell is I = 12 . In our calculations we identify this re-

arrangement with the break down of pairing.

That this interpretation is indeed correct is indicated by the experimental observations of Gomez del Campo and J.L. Ford and collaborators[7]. These authors have determined moments of inertia at excitation energies $E^* \sim 30$ MeV that are about two times as large as those of the groundstate band and agree well with the ones calculated by us in this energy range. This agreement strengthens our arguments for a transition to an unpaired state around 20 MeV of excitation so that the known gs-bands in this mass region cannot simply be extrapolated in spin but have instead to assume a significantly (factor ~ 2) smaller slope at $I \gtrsim 10$.

3.) Comparison with experiment

A comparison between yrast line calculations and results of fusion experiments is stimulated by the observation that cross sections for heavy ion fusion reactions show some indications for Q-value effects[1]. These could be explained by assuming that the yrast line is responsible for the observed limitation of complete fusion at high bombarding energies[2].

While for heavy nuclei such a limitation of σ_{fus} due to properties of the compound nucleus was conclusively ruled out in experiments[8] the situation is not so clear in light nuclei.

Figure 1 shows as an example the calculated results for ^{24}Mg together with the experimental fusion bands. These are obtained from the published fusion cross sections by means of the sharp cut off approximation. In this approximation that assumes that all partial waves up to a maximum angular momentum l_{max} fuse the fusion cross section reads:

$$\sigma_{fus} = \pi \lambdabar^2 (l_{max} + 1)^2$$

Measuring σ_{fus} as a function of energy thus allows one to translate the $\sigma_{fus}(E)$-dependence into an $E^*(l_{max})$-dependence where E^* is the excitation energy in the compound nucleus ($E^* = E_{cm} + Q$). This experimentally determined dependence $E^*(l_{max} \triangleq I)$ is plotted in the figure.

Fig. 1 also contains - indicated by vertical bars - the observed molecular resonances, known up to $I = 18$. Below $I \sim 12$ these resonances fall on the fusion-band. Since fusion up to $l = 12$ is dominated

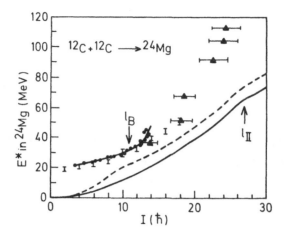

<u>Fig. 1</u> Yrast line for ^{24}Mg. The solid line shows the re-
sults of a calculation without residual interaction, the
dashed line indicates the effects of the inclusion of pairing (see ref. 4). The full triangles and solid points give
the limiting angular momenta for fusion of ^{12}C + ^{12}C from
refs. 9 and 10. The vertical bars give position and width
of molecular resonances in ^{12}C + ^{12}C . ℓ_B marks the
position of the bend in a plot of σ_{fus} vs. $1/E_{cm}$. The curve
through the fusion data gives an entrance model fit.

by effects of the (outer) interaction barrier this agreement just reflects the fact that both phenomena are determined by the grazing partial waves. For higher angular momenta (between I = 12 and 14) the
fusion band seems to bend upwards away from the molecular band. This
bend that is not indicated in the data of ref. 9 is contained in the
fusion cross sections of the Argonne group[10]. The bend in the fusion
band just reflects the standard bend in a plot of σ_{fus} vs. 1/E and,
therefore, the limitation of fusion at high energies. On the basis of
this bend in the fusion band that has now been confirmed by the Notre
Dame-Strasbourg[11] collaboration one has to conclude that fusion at
higher energies does not proceed through the molecular states as doorway states.

The calculated solid line in fig. 1, that does not contain any pairing
correlations, lies well below the fusion band ^{12}C + ^{12}C . The same
situation appears for all other systems analyzed (e.g. ^{26}Al, ^{28}Si, ^{30}Si,
^{32}S). If the pairing correlations are estimated as indicated above and

explained in ref. 3 this is still true for all systems. However, in the
case of $^{12}C + ^{12}C \rightarrow ^{24}Mg$ the experimental fusion band comes quite
close to the pair-correlated yrast-line, in particular, if the general
uncertainty of the method ($\Delta I \sim 2$) is kept in mind. This can be seen
by looking at the dashed curve in fig. 1 that contains in an ad hoc
matter the effects of the pairing correlations. The closest point lies
at $I \sim 12 - 14$, i.e. just at the point where the experimental fusion
band bends upward.

A possible explanation for the observed bend may, therefore, be the
low number of states available at the relatively low energies above
the yrast line that are reached in this experiment. Thus fusion in this
case could be hindered by phase-space arguments. This argument has recently also been invoked for an explanation of the resonance structure
in $^{12}C + ^{12}C$ (ref. 16).

4.) Summary

The question of a limitation to fusion due to the yrast lines of the
compound nuclei formed is still not conclusively answered. For all nuclei calculated by us the empirical fusion bands lie above the (pair-
corrected) yrast lines. That indeed for example for the reaction
$^{14}N + ^{12}C$ fusion is not limited by an yrast line limitation is indicated by the agreement of critical angular momenta obtained on one hand
from evaporation-residue measurements and on the other hand from Hauser-Feshbach analyses[15]. In the specific case of $^{12}C + ^{12}C \rightarrow ^{24}Mg$,
however, the yrast line comes quite close to the fusion band just at
a point where this latter shows a clear irregularity. Thus for this
system fusion reactions may be sensitive to the location of the yrast
line and could provide a valuable tool to explore the high spin yrast
states in this nucleus. In other systems, however, even the pair-correlated yrast lines are well below the fusion bands. Therefore, based
on this result the question of a possible yrast line limitation to
fusion may not have an universally valid answer.

Although two recent studies[12,13] both indicate that the yrast line
does not limit fusion, they disagree on the importance of entrance
channel effects. In view of our result for ^{24}Mg it would be extremely
interesting to populate this nucleus in two different entrance channels to see whether here the limiting angular momenta for both channels
agree and lie close to the yrast line. At the same time this particular

case stresses the need for a reliable determination of the yrast line of ^{24}Mg for spins $I \gtrsim 12$, both experimentally and theoretically. The latter is only possible if pairing correlations - including those between protons and neutrons - are taken into account. A method to do so within the Strutinsky approach was recently developed by us[4].

References:

1.) S. Harar, in: Molecular Phenomena, Proceedings of the International Conference on Resonances in Heavy Ion Reactions, Hvar 1977 (North Holland, Amsterdam, 1978).
2.) C. Volant, M. Conjeaud, S. Harar, S.M. Lee, A. Lépine and E.F. Da Silveira, Nucl. Phys. A238 (1975) 120.
3.) M. Diebel, D. Glas, U. Mosel and H. Chandra, Nucl. Phys. A (1979), in press
4.) M. Diebel and U. Mosel, Z. Physik A (1979), in press
5.) A. Szanto de Toledo, M. Schrader, E.M. Szanto and H.V. Klapdor, Phys. Rev. C19 (1979) 555.
6.) A. Watt, D. Kelvin and R.R. Whitehead, Phys. Lett. 63B (1976) 385; R.R. Whitehead, A. Watt, B.J. Cole and I. Morrison, Advances in Nucl. Phys. 9 (1977) 123.
7.) K.R. Cordell, S.T. Thornton, L.C. Dennis, P.G. Lookadoo, J.L.C. Ford, Jr., J. Gomez del Campo and D. Shapira, University of Virginia and Oak Ridge Nat. Lab., Preprint, 1978.
8.) A.M. Zebelman and J.M. Miller, Phys. Rev. Lett. 30 (1973) 27.
9.) M.N. Namboodiri, E.T. Chulick, J.B. Natowitz, Nucl. Phys. A263 (1976) 491.
10.) P. Sperr, T.H. Braid, Y. Eisen, D.G. Kovar, F.W. Prosser, Jr., J.P. Schiffer, S.L. Tabor and S. Vigdor, Phys. Rev. Lett. 37 (1976) 321.
11.) J.J. Kolata, R.M. Freemann, F. Haas, B. Heusch, A. Grallmann, Centre de recherches nucléaires de Strasburg, preprint CRN/PN 79-13 (1979)
12.) F. Saint-Laurent, M. Conjeaud, S. Harar, J.M. Loiseaux, J. Menet, J.B. Viano, Saclay preprint, May 1979.
13.) J. Gomez Del Campo, R.A. Dayras, J.A. Biggerstaff, D. Shapira, A.H. Snell, P.H. Stelson and R.G. Stokstad, Oak Ridge preprint, March 1979.
14.) D. Glas and U. Mosel, Phys. Rev. C10 (1974) 2620; Nucl. Phys. A237 (1975) 429.
15.) U. Mosel, "Fusion of 'Light' Heavy Ions", in: Proc. Int. Conf. Nuclear Interactions, Canberra, 1978, Springer Lecture Notes Vol. 92, p. 185.
16.) R. Vandenbosch, "On the origin of oscillations in the fusion cross section of ^{12}C + ^{12}C", Seattle preprint, 1979, to be published.

Time Dependent Hartree Fock Theory for Heavy Ions

J. A. Maruhn[*]
Department of Physics & Astronomy
Vanderbilt University
Nashville, TN 37235

and

Physics Division
Oak Ridge National Laboratory[+]
Oak Ridge, TN 37830

1. Introduction

Although the time-dependent Hartree-Fock (TDHF) approximation has been known for a long time [1], it has only been applied to the calculation of the behaviour of nuclei in a heavy-ion collision in the last few years. After the initial proof of feasibility and the first one-dimensional calculations [2], there was a surprisingly rapid progress in the technology of the calculations that led to a realistic two- and three-dimensional calculations, more realistic interactions and heavier systems [3-17].

At the same time, understanding about the consequences of the approximations made has deepened, and it was found that only a very limited set of physical quantities calculated in TDHF can reasonably be compared with experiment.

In this paper I shall discuss the main consequences of the TDHF approximation and the present status of comparison with experimental data. I hope that this will help to answer the question of whether the results obtained from the method are in reasonable proportion to the effort invested.

2. Derivation of TDHF

The simplest derivation of the TDHF equations involves the truncation of the equation of motion for the one-particle density matrix,

$$i\hbar \frac{\partial}{\partial t} \rho(\vec{r},\vec{r}') = -\frac{\hbar^2}{2m} (\nabla^2 - \nabla'^2)\rho(\vec{r},\vec{r}')$$
$$+ \int d^3 r'' (V(\vec{r}-\vec{r}'') - V(\vec{r}'-\vec{r}''))\rho^{(2)}(\vec{r},\vec{r}'';\vec{r}',\vec{r}''). \tag{2.1}$$

This equation still contains a general two-body interaction $V(\vec{r}-\vec{r}')$ that may, of course, be spin- and isospin-dependent, and also the two-particle density matrix $\rho^{(2)}$, in whose equation of motion in turn the three-particle density matrix appears, etc.

[*]Permanent address: Institut für Theoretische Physik, der Universität Frankfurt, Frankfurt am Main, West Germany.

[+]Research sponsored by the Division of Basic Energy Sciences, U. S. Department of Energy, under contract W-7405-eng-26 with the Union Carbide Corporation.

The TDHF approximation may now be obtained simply by assuming the absence of two-body correlations in $\rho^{(2)}$, which in this case can be expressed in terms of ρ only

$$\rho^{(2)}(\vec{r}_1,\vec{r}_2;\vec{r}_3,\vec{r}_4) = \rho(\vec{r}_1,\vec{r}_3)\rho(\vec{r}_2,\vec{r}_4) - \rho(\vec{r}_1,\vec{r}_4)\rho(\vec{r}_2,\vec{r}_3). \tag{2.2}$$

In this case the equation (2.1) becomes self-contained and determines the time-dependence of ρ. We shall see, however, that the approximation (2.2) has far more serious consequences than is apparent at this stage.

Equation (2.1) can now be rewritten with this approximation

$$i\hbar \frac{\partial}{\partial t} \rho(\vec{r},\vec{r}') = - \frac{\hbar^2}{2m} (\nabla^2 - \nabla'^2)\rho(\vec{r},\vec{r}')$$

$$+ \overline{V}(\vec{r})\rho(\vec{r},\vec{r}') - \overline{V}(\vec{r}')\rho(\vec{r},\vec{r}')$$

$$- \int d^3 r'' \ [V(\vec{r}-\vec{r}'') - V(\vec{r}'-\vec{r}'')]$$

$$\times \rho(\vec{r},\vec{r}'')\rho(\vec{r}'',\vec{r}'). \tag{2.3}$$

In the terms involving the two-particle interaction, we now have a direct term containing the average potential

$$\overline{V}(\vec{r}) = \int d^3 r' \ V(\vec{r},\vec{r}')\rho(\vec{r}',\vec{r}') \tag{2.4}$$

and an exchange term that is usually much too complicated to handle in a calculation. For this reason, all TDHF calculations up to now utilized some form of zero-range interaction, usually Skyrme forces, in which case the exchange term becomes similar to the direct one. The average potential in that case can be written as a functional of such quantities as the density, spin density, and so on.

Most TDHF calculations also involved additional, non-zero range, potentials like a Yukawa and a Coulomb interaction. In all of these cases the corresponding exchange contribution was neglected.

It is advantageous to express the one-particle density matrix in terms of single-particle wave functions,

$$\rho(\vec{r},\vec{r}') = \sum_{\text{occupied}} n_i \psi_i(\vec{r})\psi_i^*(\vec{r}'). \tag{2.5}$$

n_i is unity for standard TDHF; however, in some cases, it is useful to have fractionally occupied orbits, e.g. to produce spherical ground states for non-magic nuclei ("filling approximation"). Then the system is no longer in a pure state.

Inserting Eq. (2.5) into Eq. (2.3), we get the TDHF equations in terms of the single-particle wave functions

$$i\hbar \frac{\partial}{\partial t} \psi_k(\vec{r}) = - \frac{\hbar^2}{2m} \nabla^2 \psi_k(\vec{r}) + \overline{V}(\vec{r})\psi_k(\vec{r})$$

$$- \sum_m \psi_m(\vec{r}) \int d^3 r' \; V(\vec{r}-\vec{r}')\psi_m^*(\vec{r}')\psi_k(\vec{r}')$$

(2.6)

where the indices k and m run over all occupied states.

At this point, one may already discuss some of the limitations of TDHF apparent from the derivation.

One trivial observation, but one that should be stressed nevertheless, is that we have a time-dependent description that involves an approximation to the change of the system at each point in time. This implies that as we let time go on our approximation will deviate arbitrarily much from the true solution no matter how good the description was during the initial stage.

Let us now discuss the approximation introduced explicitly: the omission of two-body correlations implies the complete neglect of two-body collisions during the reaction. This should be valid at low ion energies, small compared to the Fermi energy, where the Pauli principle restricts the final states available decisively and the nucleons have extremely long mean-free paths. For higher energies in the several tens of MeV per nucleon range, however, that restriction is lifted and two-body collisions may not be negligible any more. We thus have an upper limit in energy, as well as in time for the validity of TDHF.

The TDHF equations are being solved numerically by two quite different methods [7,9]. Although a comparison has shown differences between the solutions, these are very small in view of the complexity of the problem.

3. Dissipation and Thermalization

An interesting question to be asked about TDHF is to what extent it allows for a thermalization of the incoming kinetic energy.

A qualitative idea of what is happening may be obtained by examining the behaviour of the single-particle wave functions during a collision. Initially, all wave functions translate with the same uniform velocity given by the ion kinetic energy. As the collision proceeds, their translational motion becomes randomized and finally approaches something quite similar to a random thermal distribution.

The problem with this argument is, of course, that we deal with wave functions and probability distributions translating in space and not with the motion of real particles. The velocity of translation is not even observable. Still, it shows convincingly that some thermalization is going on, although we cannot determine it quantitatively as yet. The determination of a thermal energy is very difficult because quantum-mechanical uncertainties and collective motion should not be included in the thermal energy.

One further problem is that thermalization proceeds only within the space of

Slater determinants which is a very small subspace of all states accessible to the system in principle. Thus, thermalization at best corresponds to partial equilibration that will be followed by complete equilibration once all the degrees of freedom neglected in TDHF come into play.

The mechanism responsible for this equilibration is the "single-particle dissipation" proposed by Swiatecki [18]. It is the dissipation mechanism operating in a gas with mean-free path comparable to the dimensions of the system.

For the case of a heavy-ion collision, there are two idealized variants of single-particle dissipation. The "window" type describes dissipation of relative momentum of the two ions through the exchange of nucleons through the neck (or "window") joining the ions. The "wall" variant considers dissipation of kinetic energy from a moving wall that reflects the nucleons producing a net increase in their thermal energy.

There are several problems about applying these ideas to a realistic heavy-ion collision. First, it has to be assumed that there is no correlation between subsequent collisions of a nucleon with the wall or between the nucleon momenta and the wall velocity. These conditions are certainly violated e.g. for collective vibrations, and in any case the "wall" in nuclei is the average potential produced by the nucleons themselves, so that there is a correlation a priori from self-consistency. The problem of self-consistency has been investigated heuristically by Sierk, Koonin, and Nix [19] with some success, whereas Randrup and Koonin [20] tried to develop a formalism for the correlation between subsequent reflections off the wall.

An unfortunate feature of single-particle dissipation is that it is not a local effect. Because of the long mean-free path, it cannot be said where in space the corresponding thermal energy is deposited. This precludes the use of single-particle dissipation in hydrodynamical models of the microscopic type.

One may conclude from the foregoing discussion that TDHF is still the only practical method for computing single-particle dissipation in a non-idealized situation, i.e. for real heavy-ion collisions.

4. Final-State Distributions and Spurious Cross-Channel Correlations

It is in the description of the final state of a heavy-ion reaction that the restriction to a single Slater determinant is felt most strongly.

The real final state should contain all the exit channels corresponding to different angular momenta, fragment masses, fragment excited states, and so on. All of these should propagate freely towards their asymptotic limits.

In the TDHF approximation, almost all of these requirements are not fulfilled. Although TDHF contains many different breakup channels in its final states, none of these are described properly and the widths of the pertinent distributions are always found to be far too small.

Let us examine these problems in some more detail.

The initial state in TDHF is made up from two Slater determinants, one for each fragment, combined to form a larger Slater determinant for the total system. In the language of density matrices, we can write

$$\rho = \rho_1 + \rho_2 \;,\; \rho_1^2 = \rho_1 \;,\; \rho_2^2 = \rho_2 \;,\; \rho^2 = \rho \qquad (4.1)$$

where all density matrices for the combined system, as well as for each individual nucleus, are idempotent. This implies that both fragments have definite mass number.

Now if we propagate ρ in time, it will remain idempotent by virtue of the TDHF equations, but if it is dissected into a ρ_1 and a ρ_2 for the final state fragments by just cutting up configuration space, neither ρ_1 nor ρ_2 will be idempotent, so that there is a spread in fragment masses. It is found in the calculations, however, that this spread is much smaller, usually about an order of magnitude, than the experimental spreads, even if subsequent evaporation is allowed [7].

If the mass spread came out in the right order of magnitude, there would be another problem destroying confidence in the results: all of these final channels interact with each other through the average potential, an effect named "spurious cross-channel correlation" by Griffin [21]. This would certainly lead to incorrect kinetic energies and binding properties for the fragments.

It is thus advisable to accept, for the present, the narrow mass spreads in TDHF, hoping that the theory will describe the average behaviour of the reaction.

The other principal limitation to a realistic scattering theory based on TDHF is its failure to describe isolated nuclei as free particles. Nuclei remain localized indefinitely and do not spread out like wave packets in scattering theory should do. This is because although the TDHF equations are translation invariant, the non-linearity in the Hamiltonian makes all the results for the usual free particle solutions applicable to the center of mass of a TDHF nucleus.

A serious consequence of this is that the scattering angle for a given initial impact parameter and energy is precisely defined. Thus, we get essentially classical scattering behaviour from a fully quantum-mechanical theory. The classical cross sections turn out to be quite unrealistic [8].

One final problem concerns fusion especially. Although, as we saw, TDHF can incorporate a mixture of different channels in the final state, albeit unrealistically, there is never enough spread to get totally different channels -- like fusion and deep inelastic -- mixed in one collision event. Fusion will thus always be described in a sharp cut-off approximation; for each impact parameter the system fuses or does not fuse, tertium non datur.

It is clear from the above discussion that all that can reasonably be expected from TDHF is a description of the average features of the reaction, and in practice, this means fusion cross sections and the gross features of Wilczynski plots.

All of these problems and the seemingly meager area of contact with experiment should not obscure the fact, however, that TDHF has many advantages compared to other theoretical descriptions of heavy reactions and is certainly a very worthwhile pursuit. I shall come back to a discussion of this point in the final chapter. First, though, let us examine some recent results and get an impression of the quality of results in TDHF.

5. Fusion Cross Sections and Wilczynski Plots

To give an impression of the type of agreement with experiment that can be achieved in TDHF, I here discuss some recent results of Davies et al. [22] on the ^{86}Kr + ^{139}La reaction.

Figures 5.1 - 5.3 show the experimental Wilczynski plots and the TDHF curve at

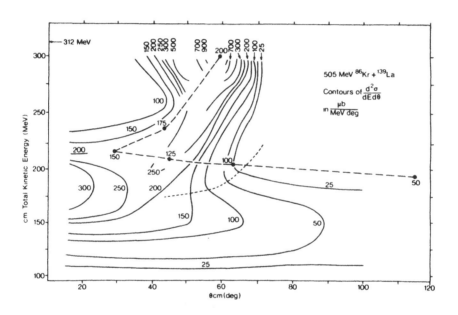

Fig. 5.1. Experimental Wilczynski plot [23] and TDHF scattering result (long dashes) for ^{86}Kr + ^{139}La at 505 MeV. Figure taken from Ref. [22].

three different laboratory energies. In all three cases the TDHF result reproduces the main structure of the plots quite well qualitatively, but gives insufficient energy dissipation in the deep inelastic branch. Since the calculations employed a restriction to axially symmetric shapes, there may not be sufficiently many states available, so that dissipation is reduced because of it. Still, the agreement is quite impressive considering the fact that the nucleon-nucleon interaction, in this case a full Skyrme force, is the only input to the calculations.

The spread in the final fragment masses in this case was found to be between 5 and 7 mass units, certainly quite small compared to experiment, but ensuring, on the

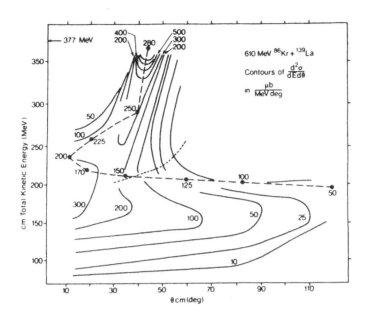

Fig. 5.2. Experimental Wilczynski plot [23] and TDHF scattering result (long dashes) for ^{86}Kr + ^{139}La at 610 MeV. Figure taken from Ref. [22].

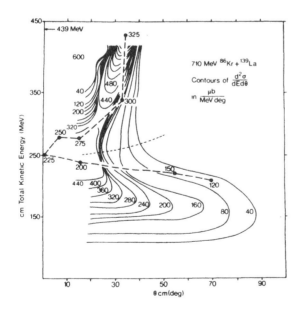

Fig. 5.3. Experimental Wilczynski plot [23] and TDHF scattering result (long dashes) for ^{86}Kr + ^{139}La at 710 MeV. Figure taken from Ref. [22].

other hand, that cross-channel correlations do not distort the results too much.

A very interesting phenomenon occurred in the fusion behaviour for this system. No fusion was observed for the two lower energies, but at the highest energy the system fused with a cross section of 118 mb. This would place the threshold for fusion between 120 and 200 MeV above the Coulomb barrier and should provide an interesting test for the TDHF method. This prediction cannot yet be asserted very strongly, though, because the insufficient dissipation seen in the Wilczynski plots may also have reduced the fusion cross section.

A more detailed study was made recently of the fusion cross sections of ^{16}O + ^{40}Ca and ^{28}Si + ^{28}Si [24]. For these, a comparison of the different types of approximations and nucleon-nucleon interactions used in various calculations was also performed. I have to refer to the original paper for a discussion of these, but I hope the different results will give an impression of the accuracy achievable at present.

Figure 5.4 shows the fusion cross section as a function of energy for ^{16}O + ^{40}Ca.

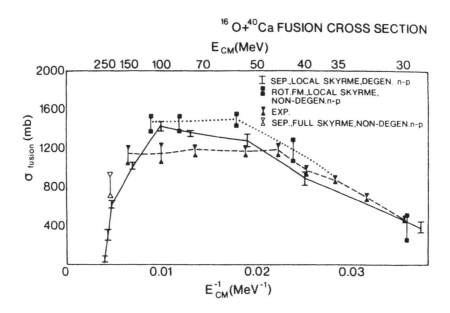

Fig. 5.4. Fusion cross section for ^{16}O + ^{40}Ca as a function of center-of-mass energy. The experimental curve [25] is compared with theoretical results obtained using various TDHF approximations. Figure taken from Ref. [24].

Apparently, all methods produce the gross behaviour quite well, but there are noticeable differences in detail, and all of them overestimate the fusion cross section at the higher energies. The difference between experimental and theoretical curves is of the same order of magnitude as that between the different theoretical ones, so that with these calculations one should not really expect higher accuracy. It still remains to be seen whether a calculation incorporating none of the symmetry restrictions present in all of the results cited here will give more accuracy.

Figure 5.5 shows the upper and lower angular momentum limits to fusion for the

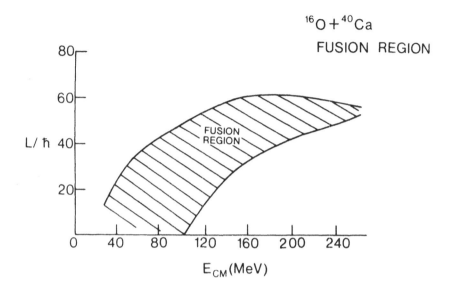

Fig. 5.5. Upper and lower angular momentum limits for fusion of $^{16}O + ^{40}Ca$, as a function of center-of-mass energy. Figure taken from Ref. [24].

same system. Above 100 MeV in the center of mass there is a fusion window, and the more rapid rise in the lower limiting angular momentum seems to be an important factor in reducing the cross section at higher energies. If that lower limit were not present, the cross section should go approximately as $1/E_{cm}$ at high energies, but in its presence the decrease is more rapid.

6. TDHF Results for a Very Heavy System

It is quite interesting to study the TDHF predictions for very heavy systems, where a closer approximation to liquid-drop behaviour is expected. Recently, calculations were carried out for the $^{238}U + ^{238}U$ system [26], and I shall here discuss some of the features appearing in those results.

Obviously, such calculations are still just barely feasible with even the fastest computers available today. Therefore, in this particular case many symmetries could not be removed. The calculation employed quartet symmetry (isospin and spin degeneracy) and a simplified Skyrme interaction, but with Yukawa and Coulomb fields included. Pairing was included in a BCS-type treatment.

Because of these restrictions, the calculations should not be regarded as representing ^{238}U nuclei realistically. They rather describe large nuclei with shell
effects and pairing present. On the other hand, it is quite possible that both of
these have little influence on the reaction so that the results could be applicable
also to real ^{238}U nuclei.

Figure 6.1 shows a sequence of shapes for the collision at a laboratory energy
of 7.5 MeV per nucleon and an angular momentum of 300ℏ. We here see how after a
relatively uneventful neck formation and rotation in shapes similar to the first one
in this series the two nuclei separate again. There are complicated fluctuations in
density inside the fragments, but these seem to correspond more to thermal fluctuations, and the really important degrees of freedom apparently are the surface shapes.
But the most fascinating feature is the actual breakup of the neck. The neck becomes
extremely elongated, but in the third frame the density at its center begins decreasing and this leads to a rapid snapping in the next two frames. Note that the
snapping appears to start at the central region of the neck in the third frame, the
lower density isolines show no constriction as yet. This is due to the equation of
state of nuclear matter implied by the Skyrme force: below a certain density nuclear
matter becomes unstable with respect to a further decrease in density.

The time evolution of several characteristic quantities is plotted in Fig. 6.2
for the same collision. I shall only discuss the most interesting features here.

The behaviour of the pairing gap is quite surprising. Although it decreases
monotonously throughout the collision, there is still a sizable pairing gap of 0.2 MeV
in the final state with about 300 MeV of excitation energy in the fragments. Whether
this indicates an unexpected stability of the pairing correlations or is an artifact
of the symmetries in the calculation or the BCS approximation will have to be checked
in future calculations.

The orbital quantities, the separation distance and the radial kinetic energy,
show an extremely smooth behaviour. Essentially this collision is dominated by
Coulomb repulsion, and the neck formation and rupture are just perturbations.

The density at the center of the neck shows the characteristics discussed above;
at a certain point it decreases much more rapidly than the neck radius. There is
also very little compression during the collision.

Summarizing these results, one is tempted to say that a liquid-drop description
using surface deformation parameters should provide a quite reasonable approximation
to this reaction once the dissipative mechanisms can be inserted reliably, but the
process of neck rupture certainly requires the introduction of other than mere shape
coordinates.

7. Conclusions

The preceding discussion has shown that TDHF, although fraught with many intrinsic problems restricting its applicability, still provides some insight into

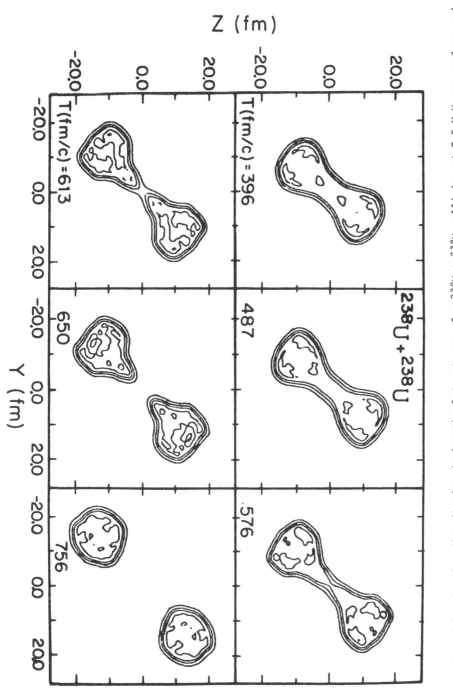

Fig. 6.1. Density distribution during the late stages of a ^{238}U + ^{238}U collision at 7.5 MeV per nucleon and L = 300ℏ.

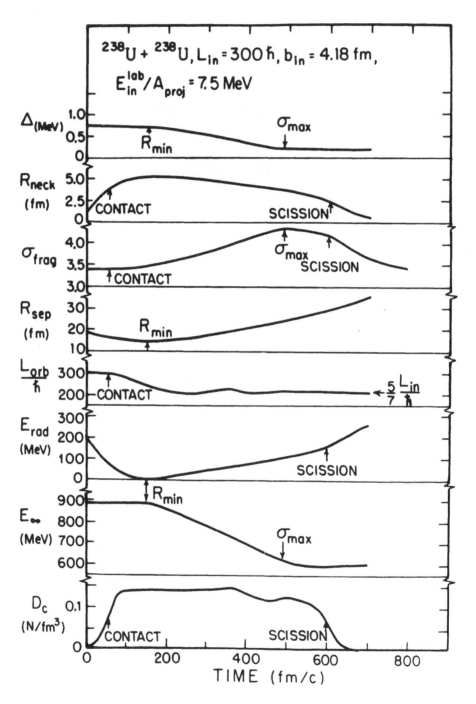

Fig. 6.2. Time dependence of various quantities during a ^{238}U + ^{238}U collision at 7.5 MeV per nucleon and L = 300$\hbar$. Plotted from top to bottom are the pairing gap Δ, the neck radius, the fragment deformation σ, the separation distance R_{sep}, the orbital angular momentum, the radial kinetic energy (in the c.m.), the asymptotic c.m. kinetic energy, and the density at the center of the neck.

heavy-ion collisions on a microscopic basis. Its most important advantages compared to other theoretical descriptions are the almost unlimited freedom given to the many-body system to evolve as it likes, without questionable restrictions such as frozen densities or frozen shapes, and using only the nucleon-nucleon interaction as an input. This is felt most strongly in the behaviour of the neck which, as it involves both density and surface degrees of freedom, is extremely hard to treat in more macroscopic formulations.

The fact that contact to experiment appears to be restricted to fusion cross sections and gross behaviour of Wilczynski plots is certainly a shortcoming of the method, but if these could be reproduced systematically throughout the periodic table and for a wide range of bombarding energies with just a single nucleon-nucleon interaction as input, that would certainly constitute a major achievement in understanding heavy-ion reactions.

It may be argued that the huge numerical effort is not in proportion to the results obtainable. I would counter that by stating that we should not judge the complexity of a theory by the amount of work done by the computer, but by the human effort going into it. And in this sense, once the numerical methods were understood, TDHF has become one of the simplest theories in nuclear physics, both conceptually and in practice.

TDHF can be used additionally as a starting point for considering more advanced approximations. Since these are not yet at the stage of practical application, I just refer the interested reader to the proceedings of the Paris workshop [27] for an overview.

I would like to express my gratitude to Vanderbilt University and Oak Ridge National Laboratory for their kind hospitality during my stay. Also, I am grateful to K. T. R. Davies for permission to reproduce figures from Refs. [22] and [24].

References

[1]   P. A. M. Dirac, Proc. Cam. Phil. Soc. 26, 376 (1930).

[2]   P. Bonche, S. Koonin, and J. W. Negele, Phys. Rev. C13, 213 (1976).

[3]   R. Y. Cusson and J. Maruhn, Phys. Lett. 62B, 134 (1976).

[4]   S. Koonin, Phys. Lett. 61B, 227 (1976).

[5]   J. A. Maruhn and R. Y. Cusson, Nucl. Phys. A270, 437 (1967).

[6]   R. Y. Cusson, R. K. Smith, and J. Maruhn, Phys. Rev. Lett. 36, 1166 (1976).

[7]   S. E. Koonin, K. T. R. Davies, V. Maruhn-Rezwani, H. Feldmeier, S. J. Krieger, and J. W. Negele, Phys. Rev. C15, 1359 (1977).

[8]   V. Maruhn-Rezwani, K. T. R. Davies, and S. E. Koonin, Phys. Lett. 67B, 134 (1977).

[9]   R. Y. Cusson, J. A. Maruhn, and H. W. Meldner, Phys. Rev. C18, 2589 (1978).

[10] J. W. Negele, S. E. Koonin, P. Möller, J. R. Nix, and A. J. Sierk, Phys. Rev. C17, 1098 (1978).

[11] H. Flocard, S. E. Koonin, and M. S. Weiss, Phys. Rev. C17, 1682 (1978).

[12] P. Bonche, B. Grammaticos, and S. E. Koonin, Phys. Rev. C17, 1700 (1978).

[13] K. T. R. Davies, V. Maruhn-Rezwani, S. E. Koonin, and J. W. Negele, Phys. Rev. Lett. 41, 632 (1978).

[14] S. J. Krieger and K. T. R. Davies, Phys. Rev. C18, 2567 (1978).

[15] K. R. Sandhya Devi and M. R. Strayer, J. Phys. G4, L97 (1978).

[16] K. T. R. Davies, H. T. Feldmeier, H. Flocard, and M. S. Weiss, Phys. Rev. C18, 2631 (1978).

[17] S. E. Koonin, B. Flanders, H. Flocard, and M. S. Weiss, Phys. Lett. 77B, 13 (1978).

[18] W. J. Swiatecki, Proc. Int. School-Seminar on Reactions of Heavy Ions with Nuclei and Synthesis of New Elements, Dubna, 1975 (JINR-D7-9734).

[19] A. J. Sierk, S. E. Koonin, and J. R. Nix, Phys. Rev. C17, 646 (1978).

[20] S. E. Koonin and J. Randrup, Nucl. Phys. A289, 475 (1977).

[21] J. J. Griffin, Proc. of the Topical Conference on Heavy-Ion Collisions, Fall Creek Falls State Park, Tennessee, 1977 (CONF-770602) 1977.

[22] K. T. R. Davies, K. R. Sandhya Devi, and M. R. Strayer, ORNL preprint, 1979.

[23] R. Vandenbosch, M. P. Webb, P. Dyer, R. J. Pugh, R. Weisfield, T. D. Thomas, and M. S. Zisman, Phys. Rev. C17, 1672 (1978).

[24] P. Bonche, K. T. R. Davies, B. Flanders, H. Flocard, B. Grammaticos, S. E. Koonin, S. J. Krieger, and M. S. Weiss, ORNL preprint, 1979.

[25] S. E. Vigdor, D. G. Kovar, P. Sperr, J. Mahoney, A. Menchaca-Rocha, C. Olmer, and M. S. Zisman, unpublished.

[26] R. Y. Cusson, J. A. Maruhn, H. Stöcker, and A. Gobbi, to be published.

[27] P. Bonche, B. Giraud, and Ph. Quentin, editors: "Time-Dependent Hartree-Fock Method", Saclay, 1979.

R. Bass

Nuclear Reactions with Heavy Ions

1980. 177 figures, 31 tables. Approx. 370 pages
(Texts and Monographs in Physics)
ISBN 3-540-09611-6

Contents: Introduction. – Light Scattering Systems. – Quasi-Elastic Scattering from Heavier Target Nuclei. – General Aspects of Nucleon Transfer. – Quasi-Elastic Transfer Reactions. – Deep-Inelastic Scattering and Transfer. – Complete Fusion. – Compound-Nucleus Decay. – Appendices.

The last decade has witnessed an astounding increase in heavy ion research. This book presents – from an experimentalist's point of view – a critical and coherent outline of the results of large scale heavy ion research in the area of low energy nuclear reactions in the 5–10 MeV per nucleon range. Using phenomenological models, the author explains these experimental results, achieving a good balance between a critically selected review and a textbook. This makes it attractive for the advanced student and the specialist alike.

P. Ring, P. Schuck

The Nuclear Many-Body Problem

1980. 171 figures. Approx. 800 pages
(Texts and Monographs in Physics)
ISBN 3-540-09820-8

Contents: The Liquid Drop Model. – The Shell Model. – Rotation and Single-Particle Motion. – Nuclear Forces. – The Hartree-Fock Method. – Pairing Correlations and Suprafluid Nuclei. – The Generalized Single-Particle Model (HFB-Theory). – Harmonic Vibrations. – Boson Expansion Methods. – The Generator Coordinate Method. – Restoration of Broken Symmetries. – The Time Dependent Hartree-Fock Method (TDHF). – Semiclassical Methods in Nuclear Physics. – Addendices. – Bibliography. – Author Index. – Subject Index.

This book, while covering a fair amount of physical observations, stresses the methodology and technical aspects of the different theories presently used in the description of the nucleus. The authors present the more modern theories such as Boson expansions, generator coordinates, time-dependent Hartree-Fock method, and semiclassical models which so far have found only limited mention in textbooks. The book also covers subjects like the liquid drop and the shell model, both presented in a updated version in, for example, rotations and random phase approximation. The full presentation of mathematical details, illustrated by observational data, will help the student fully understand the present views on the nuclear many-body problem.

Springer-Verlag
Berlin
Heidelberg
New York

Lecture Notes in Physics

Selected Issues from
Lecture Notes in Mathematics

Lecture Notes in Physics